A SURVE

DATE

A ...SERVATION OF WILDLIFE IN

BRITAIN 1950–2001

*Dedicated to Derek Ratcliffe
for reminding us that means should have ends,
and that behind the posturing it is wildlife that matters*

'The landscape is like a historic library of 50,000 books. Many were written
in remote antiquity in languages which have only lately been deciphered;
some of the languages are still unknown. Every year fifty volumes are
unavoidably eaten by bookworms. Every year a thousand volumes are taken at
random by people who cannot read them, and sold for the value of the
parchment. A thousand more are restored by amateur bookbinders who
discard the ancient bindings, trim off the margins, and throw away leaves
that they consider damaged or indecent. The gaps in the shelves are filled
either with bad paperback novels or with handsomely-printed pamphlets
containing meaningless jumbles of letters. The library trustees, reproached
with neglecting their heritage, reply that Conservation doesn't mean
Preservation, that they wrote the books in the first place, and that none
of them are older than the eighteenth century; concluding with a plea for
more funds to buy two thousand novels next year.'

Oliver Rackham (1986), *The History of the Countryside*

'Writers and politicians may come out with all sorts of edifying sentiments,
but they are what is known as declarations of intent. If I had to say which
was telling the truth about society, a speech by a Minister of Housing or the
actual buildings put up in his time, I should believe the buildings!'

Kenneth Clark (1969), *Civilisation*

The aim of this series is to interest the general reader
in the wildlife of Britain by recapturing the enquiring
spirit of the old naturalists. The editors believe that
the natural pride of the British public in the native
flora and fauna, to which must be added concern for
their conservation, is best fostered by maintaining a
high standard of accuracy combined with clarity of
exposition in presenting the results of modern
scientific research.

The New Naturalist

NATURE CONSERVATION
A REVIEW OF THE CONSERVATION OF
WILDLIFE IN BRITAIN 1950–2001

Peter Marren

With 16 colour plates and over 130 black
and white photographs and line drawings

HarperCollins*Publishers*

HarperCollins*Publishers*
77–85 Fulham Palace Road
Hammersmith
London W6 8JB

The HarperCollins website address is:
www.**fire**and**water**.com

Collins is a registered trademark of HarperCollins*Publishers* Ltd.

First published 2002

100320806 7

ISBN 000 711305 6 (Hardback)
ISBN 000 711306 4 (Paperback)

Printed and bound in Great Britain by the Bath Press
Colour reproduction by Colourscan, Singapore.

)

Contents

Editors' Preface

It is 32 years since Dudley Stamp's New Naturalist on Nature Conservation in Britain appeared. Published posthumously, it summarised events and progress up to 1965, concentrating on the dominant role of the official Nature Conservancy. Since then, nature conservation has become much more fraught and politicised. It is in continual conflict with the forces of modern 'progress', in the increasing demands made on the natural environment by agriculture, forestry, mineral extraction, water use, energy supply, transport, urban development, recreation and defence. The voluntary bodies for nature conservation have also grown to rival the official side in importance. These massive shifts require a completely new book that takes over the story where Dudley Stamp left off.

Peter Marren is the acclaimed author of the 50th anniversary volume, *The New Naturalists* (1995), whose earlier career was in the Nature Conservancy Council. As a regional staff member in Scotland and later England, he saw things from the front line of nature conservation. His later role, in compiling its Annual Report, gave him an overview of the whole wide field of activities and issues in which the NCC became involved, and an insight into the frequently political nature of skirmishes with opposing vested interests. Then the NCC was devolved, and Peter found himself in English Nature, with a much reduced geographical remit and a new management-cum-public relations style. Unable to stand the combination of internal regimentation and external timidity, he resigned, to start a new career as a freelance writer on wildlife and its conservation. To this he brought a distinctive style, of fluent prose allied to sardonic wit, winning many fans who relish the appearance of his next work.

Peter's perceptive eye has ranged over an astonishingly broad field in this new book, and he has told its complicated story with his usual flair. This is an honest appraisal of the net results of all the years of striving on behalf of wild nature in Britain – an assessment of the balance between success and failure. He puts all aspects of the business under the microscope: the organisations concerned, the threats to nature, the measures for dealing with these, the politics involved, and the outcomes on the ground. The book is analytical as well as factual, but enlivened by its author's characteristic flashes of humour. This re-evaluation shows that, despite many successes in saving important wild places, plants and animals, losses have also continued on an unacceptable scale. Nature reserves do, at least, give us some tangible reward for all the effort, but the results from the persuasion approach are often more difficult to measure. In giving us this invaluable reference work, Peter Marren has also conveyed the richness and splendour of our national capital of wild nature, and its importance to our cultural heritage. Its defenders also have sombre lessons to learn from this synthesis if they are to improve their performance during this new millennium.

Author's Foreword

This is the first New Naturalist book on nature conservation since 1969. Its predecessor was written by Dudley Stamp, a geography professor and an influential voice in rural planning matters in the 1940s and 1950s. He was well qualified to report the early development of nature conservation in Britain, having been a long-time Board member of the Nature Conservancy, the original wildlife agency which he helped to set up in 1949. Stamp was also a well-practised writer of textbooks on geography and land use, and had contributed no fewer than four volumes to the New Naturalist library, including the outstandingly successful *Britain's Structure and Scenery*. *Nature Conservation in Britain* was his last book, published after Stamp's untimely death in 1966. It was a mellow look back at the past, broadly successful, quarter-century, in which conservation had developed from the hopes of a few naturalists to a broad-based going concern with its own mini-department in government. It was a relatively short book, concentrating on the work of the Nature Conservancy and the county wildlife trusts, and listing nearly all the nature reserves existing at that time.

Thirty years on, that world has changed utterly. Conservation has become a much more pluralist activity involving many sectors of society, both official and voluntary. It has broadened out from its original base in natural history to address fundamental issues like the future of the countryside and the claims of the urban majority on how land is used, and whether we should be allowed open access to it. It has become ever more

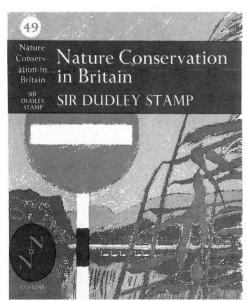

Nature Conservation in Britain (1969). The dust jacket designed by Clifford and Rosemary Ellis was inspired by the Abbotsbury swannery in Dorset, Britain's oldest bird sanctuary.

difficult to say where nature conservation ends and concern about our own futures begins. Indeed, the very phrase nature conservation means something different today. It is now shot through with fashionable concerns like sustainability, animal rights and habitat creation, some of which might have baffled the conservationists of the 1960s, whose perceptions were rooted more in natural sciences and the development of institutions.

Even so, many of the things Dudley Stamp was writing about are still with us. We still conserve wildlife by setting up nature reserves, or designating private land as Sites of Special Scientific Interest. What has changed more is the scale of the resources brought to bear on protecting nature. Bodies like the RSPB and National Trust have more members than any of the political parties or trades unions. Money has flowed into the business (as it has become) from the Heritage Lottery Fund, tax credits and the EU LIFE fund. We stand on the verge of a major shift of agricultural subsidies from food production to sustainable land use. If I was writing this book from the perspective of conservation bodies, that is, in terms of wealth and influence, it would be a story of unalloyed success. Instead, I have chosen to write it, as far as possible, from the more awkward perspective of wildlife. I try to assess to what extent all this power and money has benefited British wildlife, and how far the undoubtedly frenetic activity of the recent past has translated into policies that improve the lot of rare and declining species and wild places. Is the countryside in better shape today than in 1969? Do we have more wildlife now than we had then? Perhaps we should have. The population has hardly grown in that time, and the polluting industries of the 1960s have either cleaned up their acts or disappeared. Persistent pesticides are no longer widely used. And yet the statistics show a steady loss of habitats and species. Although in some ways land use has become more environment-friendly, a short walk with open eyes almost anywhere in Britain is a good antidote to the wilder claims of the prophets of the 'Things are Much Better Now' school.

Writing about the conservation product rather than the process poses an interesting problem. The documents of the conservation industry tend to be aspirational: they tell us, sometimes in impenetrable gobbledegook, sometimes in talk-down, creepy-vicar homilies, what ought to happen and what they would like to happen, but all too often do not tell us what actually happened. The emphasis tends to be on the means – plans, strategies, partnerships – rather than the ends. The Nature Conservancy and the NCC did sometimes review events from the standpoint of ecology and wildlife, but their successors seem much more interested in talking up their 'achievements'. It is rare nowadays that a body tries to take an objective view and balance success and failure. A certain caution about the claims of conservationists is therefore healthy, and wise.

The conservation story would not make sense without a description of the main players, their basic beliefs and actions, and the historical framework against which decisions are taken. I therefore planned this book in three sections. The first concerns the main players, the official agencies, the successors of Dudley Stamp's Nature Conservancy, and the voluntary bodies, and is also about how our wild places are protected. The second is about the playing field itself – the environment in which our wild species live, and what we have done with it over the past half century. The third is about species, especially those we know about and have polices for: the big ones, rare ones and new arrivals. The main text is topped by an introduction summing up the state of our wildlife at the start of the new millennium, and tailed by a look at where current trends might be taking us. I have not felt it necessary to review every wild habitat in detail; for example, I have not devoted much space to river engineering and floodland, partly because that has been done so well already by Jeremy Purseglove in his prize-winning book *Taming the Flood* (1989). Because this book is about the natural world, I have little to say about the human environment – food, renewable resources, radiation, ozone, carbon dioxide – nor 'wider countryside' issues such as access and recreation, except where they affect wild species and living communities. On the other hand, there is a lot about trees, flowers and birds, and, I hope, not too much conservation jargon.If it has an emergent theme, it might be the gap between aspiration and achievement. The political and physical conditions of a crowded island make the conservation of nature extraordinarily difficult, and wildlife survives largely despite us.

I am lucky to have friends in the business who agreed to cast their eyes over what I had written and put me straight on things, especially Desmond Thompson, James Robertson, Gary Mantle and Graham Bathe. I am grateful also to the three national conservation agencies, English Nature, SNH and CCW for the loan of photographs and other material, and for their patience in answering my queries. Among other bodies that helped in some way are the JNCC, Heritage Lottery Fund, the Countryside Agency and the Wildlife Trusts partnership, especially my own trust, the Wiltshire Wildlife Trust, as well as the RSPB, BTO, WWF-UK and the Marine Conservation Society. My old friend Bob Gibbons came to my rescue in the last-minute scramble for pictures with great generosity. My faithful Maureen Symons processed the script with her usual speed and cheerfulness, and Isobel Smales turned it into illustrated pages with sensitivity and intelligence. Thanks also to Myles Archibald at Collins for making it all possible. My old colleague and friend Derek Ratcliffe read the whole draft, and his comments, on the main themes as well as the detail, were most valuable. He would not necessarily agree with everything in it, but writing this book put me constantly in mind of Derek's unique contribution to the protection of wildlife in Britain, and I am proud to dedicate this book to him.

Peter Marren
Ramsbury, January–June 2001

1

Introduction: Where We Are Now

Wildlife in a crowded island

In his classic book, *Nature Conservation in Britain* (1969), Sir Dudley Stamp began with the world population. By the mid-1960s, wildlife shared the planet with 3,400 million human beings. Thirty years later that number had grown to 5,292 million. Today Asia alone holds most of the world population of the mid-1960s. In 2000, we broke the six billion mark. The United Nations forecast for 2050 is 9,833 million people; if so, in a single century the human population will have nearly trebled.

In Great Britain, by contrast, the population growth is slow. In 1921, nearly 43 million people lived on our island. In 1965, despite the postwar 'baby boom', the population had increased by only 19 per cent to 53 million. Today it stands at 57 million and is virtually static. *Our* population explosion happened early, in Victorian times. In the developing world, most people are young. In Britain we have an ageing population. Mr and Mrs Average will have 2.1 children and will live well into their seventies. If the size of the human population were all that mattered, the countryside and its wildlife would have been under remarkably little pressure in the twentieth century, except around a few cities, mainly in south-east England.

Even so, 57 million people is plenty on an island of only 230,000 square kilometres (88,780 square miles). It gives us an average population densi-

The New Naturalist Board in 1966. From left to right: James Fisher, John Gilmour, Sir Julian Huxley, Sir Dudley Stamp and Eric Hosking. (Eric Hosking)

ty of 3.6 persons per hectare in England, 1.36 in Wales and 0.65 in Scotland. We outnumber every wild mammal found in Britain, with the possible exception of the field vole. We outnumber the commonest wild bird by about five to one. If we all had a decent-sized garden, there would be no countryside.

It is not the size of the human population so much as our changing ways of life that have created the pressure on our wildlife. Despite a near static population, another 4.4 million 'homes' are to be built over the next 20 years, at least half of them in the countryside. Apparently, this is because many people nowadays prefer to live on their own. Our devotion to the car has produced a bonanza of road building so that it can be hard to find a place where the traffic cannot be heard, and the stars shine in a dark sky. It has produced an American-style, road-centred landscape that was only starting to appear in the 1960s: flyovers, filling stations, shopping malls and multistorey car parks. Although roads are thin, they flatten a lot of wildlife sites. A clover-leaf motorway junction was built smack in the middle of Hook Common SSSI in Hampshire. Another motorway cut Aston Rowant National Nature Reserve in half (though zealous geologists promptly designated the road cutting). You get a wonderful view of bisected Kentish SSSIs from the M2.

Even so, if the rest of the countryside was still rich in wildlife, we could lose some of it to further our material convenience without losing any species. However, agricultural improvements, well under way by the 1960s, have transformed the old prewar mixed farms, with their acres of permanent pasture and miles of hedges, into prosperous modern arable units, or rye-grass-based milk factories. The reasons, which might seem inadequate now, made sense then. As a wise giant in Swift's *Gulliver's Travels* had observed, 'whoever could make two ears of corn or two blades of grass to grow upon a spot of ground where only one grew before, would deserve

The New Naturalist Board in 2001. From left to right: Richard West, David Streeter, Derek Ratcliffe and Sarah Corbet. Max Walters is absent from the picture. (Debra Sellman)

better of mankind than the whole race of politicians put together'. We vast-ly exceeded that modest goal, but, essentially, the higher the unit produc-tion, the lower the value for wildlife. By the 1990s we had achieved what would once have seemed impossible: wheat fields with nothing left over for the wild birds to eat, or fields of grass with scarcely a single wild flower. Another crop that was starting to take over much of the poorer land in the 1960s was Sitka spruce. Forestry was once considered by many to be a friend to nature conservation. Unfortunately the industry went the same way as farming – trees were treated purely as a crop, like wheat – and so it lost the favourable reputation it is now struggling hard to regain. The nation could have fed itself without destroying important wildlife habitats, and the heav-ily subsidised home-grown trees scarcely dented Britain's import bill for tim-ber. The destruction of our wild places, it seems now, was unnecessary. If it had a single cause it was an airtight Ministry of Agriculture, the client min-istry of the big farming unions and the agro-chemicals industry, but no one else. When the Ministry was finally put to sleep in the government reor-ganisation of June 2001, no one, least of all the farmers, had a good word to say for it. The officials of MAFF, it is said, were the most blinkered and obstinate in the entire civil service (ask anyone in a conservation agency – or for that matter, on a farm). They made a mess of everything, from BSE to ESAs. The chaos that engulfed the farming industry in the 1990s was the endgame of decades of preventable idiocy.

A high-density human population makes every acre precious. As Dudley Stamp noted, 'nature conservation must work out its own salvation in cramped conditions'. Good wildlife habitat, such as chalk grassland and heath, has been reduced to scraps within a predominantly arable or urban environment. Many of our nature reserves seem ridiculously small to visi-tors from large, continental countries like the United States. Yet the same visitors also marvel at how attractive much of our countryside still is, despite all the pressures upon it. The island scale of the British landscape means that small is often beautiful. Britain has an amazingly complex geol-ogy, producing variations in landforms and building stone within just a few miles. A traveller from London to Brighton passes over at least seven dif-ferent landscapes on gravel, clay, greensand and chalk, past downs, heaths, woods and secluded lakes. The lowland agricultural landscape is a patch-work, and small nature reserves may be big enough to enclose some of the best examples of natural habitats. Even so, in the long run they may still be too small. Even quite large nature reserves cannot do much for wide-rang-ing, low-density species such as wildcats or golden eagles. Small reserves tend to lose more species than large ones. A famous example is the inabil-ity of Woodwalton Fen, a small, brick-shaped nature reserve in Cambridgeshire, to sustain a population of the large copper butterfly. The butterfly survived through the 1930s and 1940s, when the weather was kinder, but the reserve was just not big enough when the going got tough later on. Because they are small, British nature reserves tend to be highly planned, with various zones and prescriptions designed in effect to squeeze the maximum of wildlife out of the minimum of space. Like farms, every acre has to count, which is why there are few places where wildlife is simply left to look after itself. A special site (a steep ravine) had to be found to study the natural development of woodland in Britain. So ingrained is the concept of management that in Britain we do not seem very interested in how the natural world actually works.

This would seem odd in a big, wild country such as Canada or Brazil, but, as we ecologists never tire of pointing out, every inch of Britain is used. Most of the land has a history. Most of it is privately owned, and managed to provide income. British wildlife has long been used to living with the British, and exploiting such opportunities as we offer it. Presumably any species unable to make that crucial accommodation, such as the wolf perhaps, or *Cerambyx cerdo*, a huge beetle of giant trees in virgin forests, died out. But it is quite wrong to assert that because he has planted a few trees and hedges, and farms most of it, that the landscape is therefore man-made. This is a sinister reasoning, because if man has created a landscape there is an implication that he is entitled to do with it as he likes, and, if necessary, destroy it. In fact, many of our wilder places are almost unfathomably ancient, and were never planned or created. Most of our woods and commons, and even some hedges, evolved naturally in ways that are still mysterious. Whoever coined the term 'natural history' must have realised this instinctively. In Britain, nature does indeed have a history that runs parallel with that of humankind, often in harmony with it (man and nature in harmony is the subject of much of British art). Our relationship with the natural world has a history of its own (see, for example, Keith Thomas' masterpiece, *Man and the Natural World*). Nature conservation is only the most recent phase in a long-running love affair.

Britain's oldest farmed habitats can in fact be more ancient than natural landforms. For example, the fields of West Penwith, in Cornwall, with their strange polygonal patterns, are much older than the shingles at Dungeness or the wet levels at Pevensey, which were still under the sea when William the Conqueror invaded in 1066. A few woods have been managed in much the same way for as long as historic records permit us to see (and to carry on managing them in exactly the same way is absolutely the right thing to do: indeed, it is almost a duty to history). Past land uses were not necessarily ideal for wildlife, but they tended to leave plenty of opportunities. Butterflies could lay their eggs in sunny glades created by woodmen, before the canopy closed over again. Meadows were often full of wild flowers because farmers lacked the means to drain them dry or improve the soil (flower fields also made better hay). The highly regulated management of medieval commons could almost have been invented by a latter-day nature reserve manager. Modern conservationists are to some extent stepping into the vacated shoes of farm labourers, shepherds, woodmen and peasants, who would not have been able to read a conservation manual but knew more about conservation in practice than most of us. The challenge today is to obtain similar results by different means. Recent advances include the creative use of bulldozers, JCB diggers and suction dredgers.

Some millennial stocktaking

About 30 years ago, when I was just setting out into the nature conservation world, we used to be slightly apologetic about British wildlife. Almost anything we had, other European countries had better, we thought. Britain has few endemic species worth mentioning, apart from *Primula scotica* (the Scottish primrose) and the good old red grouse (which, ornithologists insisted, was the same species as the bigger, better-looking continental willow grouse), nor much of world importance, apart from the grey seal (which we were then culling by the hundred) and maybe the gannet. Since then, Britain's stock has risen considerably. We have places such as the New Forest, St Kilda and the Cairngorms, which are important after all, and more old trees than anywhere north of the Alps. Our marine life, estuaries and Atlantic oak woods are pretty special, and Britain is right at the top of the European liverwort league. I was slightly staggered to hear we have 40 per cent of Europe's wrens. If Britain sank beneath the waves, quite a lot of species would be sorry.

Britain's wildlife is important in another way. Natural history is very popular – we have 4,000 RSPB members for every species of breeding bird – and we are very good at it. (Who wrote *The Origin of Species*? Who lived at Selborne?) Britain probably has the best documented wildlife in the world – all right, we might tie with the Dutch and the Japanese. Our 60-odd butterflies are nothing on the world stage, but the expertise acquired in studying them has been exported worldwide. British bittern experts are in demand internationally, though we have only a handful of bitterns. We might not be much good at protecting wildlife, but we certainly know a lot about it. How well, in fact, are our wild animals and plants faring at the start of the new millennium?

There are some grounds for optimism. The loss of wild habitats with which we are sadly so familiar can now be seen in a historical context – as a feature of the domination of intensive agriculture (including forestry)

The red grouse, Britain's most celebrated endemic animal, adapted for life on heather moors. (Derek Ratcliffe)

between 1940 and 1990. The statistics of habitat loss put together by the NCC in 1980 have been much repeated, and certainly paint a grim picture. I will not march through them all again, since statistics tend to have a numbing effect on our jaded twenty-first century minds, and they are meaningless without context. For example, what does the loss of 95 per cent of 'neutral grassland' mean? It does not tell us whether all the lost ground was of importance for wildlife, nor whether what remains is of the same value as it was. It does, however, imply that something pretty far-reaching has happened to the landscape, and that perhaps only one unim-proved meadow in 20 has survived. Meadows are the biggest losers of the habitat loss stakes, but we have lost almost as much wet peatland and 'acid grassland', mainly on former commons and village greens, and about half the chalk downs and natural woods that existed before 1939.

The official nature conservation agencies publish annual estimates of habitat loss and damage in their annual reports. The loss is ongoing, and although it has fallen since the peak in the agricultural high noon of the 1970s, there is now less habitat to damage. The problem with the figures is that, however they are defined, damage assessment is relative. Some types of ecological damage, such as eutrophication, need a trained eye, while others, such as surface disturbance, may look like damage, but may be harmless or even beneficial. Moreover, the agencies have a bad habit of changing their method of calculation from year to year, and since devolu-

The decline of lowland heath in six parts of England over two centuries. (From *Nature Conservation in Great Britain*, NCC 1984)

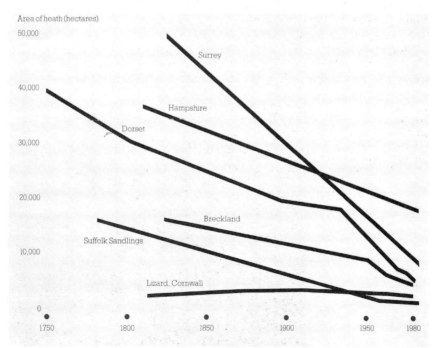

tion it has become maddeningly difficult to compare what is happening in England with Scotland or Wales or vice versa. All the same, the losses indicate that statutory protection of wild habitats has not been as effective as many had hoped.

The concensus at the turn of the millennium is that loss and damage has slowed considerably and for some habitats may have halted. SSSIs have acquired a greater land value because of the grants they now attract, and SSSI protection is also more effective than formerly. Also more of them are owned or managed by conservation bodies or benign private owners. For at least one well-documented habitat, hedges, removal is now offset in terms of length by newly planted or repaired hedges (quality is not assessed). Hedgerow removal peaked between 1955 and 1965, when many of the open barley prairies of eastern England were created. The process went on in lower gear, with about 2,000 miles (3,200 kilometres) of hedges removed per year, mainly by arable farmers, but also for urban development, roadworks and reservoirs. The Government's Countryside Survey found virtually no overall decline in hedges between 1990 and 1998, except in neglected 'remnant hedges': about 10,000 kilometres (2 per cent) of hedges were removed, and a similar amount planted. But the survey found evidence for some loss of plant diversity, especially tall grasses and 'herbs'. The reason why hedges have stabilised is because government no longer pays the farmer to pull them up, but pays him to plant them. Stewardship and other agri-environmental schemes are the main reason why hedges are still planted.

We have lost a lot of wildlife habitat, but conservation bodies saved many of the best places, and the story is by no means all doom and gloom. Today's problem, which I will turn to in the last chapter especially, is the reduction in habitat quality and diversity, which is all the more insidious because it is hard to measure precisely. When we turn to species, the overall picture is the same. Many species – possibly *most* species – have declined over the past half-century, but relatively few have become nationally extinct. Some once believed extinct have since reappeared, or have been reintroduced in biodiversity schemes. The best-known victim of the past half-century, the large blue butterfly, has been reintroduced from Swedish stock, which closely resembles the lost British race. Thanks to thorough research and ground preparation by Jeremy Thomas and David Simcox, the project has been a resounding success, and several introduction sites are now open to the public. But there is a very large disparity between the small number of recorded extinctions and the much larger number that seem to be approaching extinction. For example, the JNCC considers that about 20 native wild flowers are extinct (though my own tally makes it nearer ten), but some 200 face extinction in the short or medium term. Similarly, some 29 lichens are considered extinct, but 148 are 'endangered' or 'vulnerable'. The implication is that conservationists are good at saving species at the last ditch, but bad at preventing them from getting that far.

The status of plants and fungi in Britain (based on details from the JNCC's website)

	Number of native species	Number extinct	Number critically endangered	Number endangered	Number vulnerable
Seed plants	2,230	21	25	43	132
Ferns	70	0	0	1	3
Bryophytes	c. 1,000	15	20	40	65
Lichens	c. 1,700	29	27	30	91
Stoneworts	30	2	1	7	1
Fungi	> 20,000	21	8	44	93
Slime-moulds	c. 350	19	4	24	16

The' threat categories' are defined by the IUCN as follows:
Critically endangered: 'Facing an extremely high risk of extinction in the immediate future'
Endangered:'Facing a very high risk of extinction in the near future'
Vulnerable:'Facing a high risk of extinction in the medium-term future'

Invertebrates are often 'finely tuned' to their environment and are more vulnerable to change than birds or many wild plants. Their fortunes have varied from group to group. Ladybirds and dragonflies are not doing too badly – ladybirds like gardens, dragonflies like flooded gravel pits – but things are currently looking grim for wood ants, and some bumble-bees and butterflies. In a recent review I learned that some 200 species of flies (Diptera) are considered endangered and another 200 or so vulner-able (Stubbs 2001). Granted that flies are mysterious things with rather limited appeal, one wonders what the implications of 400 nose-diving flies could be. If 400 subtle, specialised ways of living are under threat, the environment must be quietly losing variety, losing tiny facets of meaning, like a little-read but irreplaceable book being nibbled away bit by bit by bookworms. Interestingly the author of this review blamed some of the losses on conservationists – 'inappropriate decisions by amenity and con-servation organisations' – as much as farmers and developers. Tidying up is bad for flies.

Not all species are fated to decline. In a detailed assessment of Britain's breeding birds by Chris Mead (Mead 2000), the accounts are surprisingly well balanced: some 118 species are doing well and 86 are doing badly. However, the fortunes of different species fluctuate, reflecting the lack of stability in the modern countryside. Some once rare birds, such as Dartford warbler, red kite and hobby, are among those that Mead awards a smiling face, meaning that they are doing well – stupendously well in the case of the kite. Among other smiling faces are gulls, geese, many water birds and seabirds and most raptors. Several species have colonised Britain naturally, notably little egret, Cetti's warbler and Mediterranean gull, and others, such as spoonbill and black woodpecker, may be on the brink of doing so. The commonest bird in 2000 was the wren, which has benefited from the recent run of mild winters and thrives in suburban gardens. Conservation schemes have probably saved corncrake, cirl bunting, stone curlew, and perhaps woodlark, for now. Perhaps the best news of all is the recovery of the pere-

About 14 per cent of our moss and liverwort flora is considered vulnerable, endangered or extinct. This is JNCC's projection of their respective 'threat status'. (JNCC)

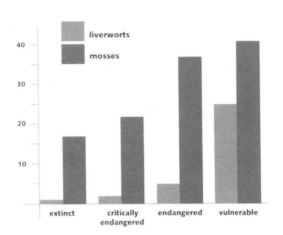

Threat status of British Bryophytes

grine falcon from a dangerous low in 1963, following the ban on organochlorine pesticides, though it still faces persecution in parts of Britain. We are now quite an important world stronghold for peregrines.

The losers – Mead's unhappy faces – include familiar farmland birds such as skylark, song thrush, linnet, grey partridge, lapwing and snipe. Even starling and house sparrow have declined markedly. The problem they all face is lack of food on today's intensively managed, autumn-sown cereal fields. The red-backed shrike has ceased to breed, and the red light is showing for black grouse and capercaillie, which are finding life difficult in the overgrazed moors and upland woods (and no good just putting up deer fences: the capercaillie crashes straight into them). Interestingly, small birds are faring worse, on balance, than big ones. A birder of the 1960s would be shocked at what has happened to lapwings or golden plovers, but pleased and probably surprised at how well many comparatively rare species have adapted to a changing environment. Stranger things lie just ahead. Try imagining green parakeets stealing the food you left out for the disappearing starlings.

Like birds, some of the smaller mammals have declined more than the big ones. Some carnivores, such as polecat and pine marten, are more widespread today than they were in 1966. The grey seal is much more numerous, thanks entirely to the cessation of regular culling. Deer are also more numerous, though this is a mixed blessing. Though increasing, the red deer is threatened genetically by hybridisation with the increasing, introduced sika deer, and may soon be lost as a purebred species. Bats, as a class, have declined. The best counted species, the greater horseshoe bat, is believed to have declined by 90 per cent during the twentieth century. The present population is estimated to be only 4,000 adult individuals. Only 12 colonies produce over ten young per year (Harris 1993). Rabbits have made a slow recovery from myxomatosis, and are back to about 40 per cent of their original abundance, but occur more patchily than before. The otter has staged a slow recovery, aided by reintroductions, but may take another century to recover its former range across eastern Britain. The real losers are red squirrel and water vole, both victims of introduced mammals. Dormouse and harvest mouse are also declining, apparently because of changes in woodland and agricultural land that reduce food availability. Our rare 'herpetiles' (i.e. reptiles and amphibians), sand lizard, smooth snake and natterjack toad, have benefited from site-based conservation and a zealous British Herpetological Society. Most of our freshwater fish seem fairly resilient, but the burbot has been lost and our two migratory shads reduced to rarity status because of pollution and tidal barrages. The char, which likes clear, cold water, has disappeared from some former sites. The powan faces an uncertain future in Loch Lomond following the accidental introduction of a competitor, the ruffe, which eats its eggs. Its relative, the pollan of Lough Neagh, is now threatened by carp, casually introduced nearby to please a few anglers. In the sea, we, with the help of our European friends, have overfished herring, cod and 22 other species, and almost wiped out the skate.

We should not, however, judge the success of nature conservation measures solely by changes in the numbers of well-known animals. Birds are important, because everyone likes them, and because losses and gains among such well-recorded species are important clues to what is happening to their environment. In nature conservation, every bird is a miner's canary. But birds are almost too popular. In the 1960s, many field naturalists specialised in relatively obscure orders, pond and shore life, and difficult insects, such as beetles or bugs. Today, an oft-heard complaint is that taxonomists are an ageing and diminishing band, and that the few professional ones are nowadays tied up in administrative tasks. The number of people who can identify protozoa, or diatoms, or worms is probably fewer now than a century ago. As a result, we have no idea what is happening to them. All too often, biodiversity has been lost from ignorance, even on nature reserves. Britain's nature reserves are run by people who would know a hawk from a handsaw at a thousand paces, but to whom invertebrates are just wriggly things that live in bushes.

Discovering where the wildlife is

In the 1960s, the study of British natural history was in a reasonably healthy state, better in some respects than it is today. Entomology and microscopy were less popular than in their late Victorian heyday, but with the advent of cheap, lightweight binoculars, birdwatching was growing in popularity, and ecology was being taught at schools and universities. Serious naturalists were making connections between a species and its environment, which led, by extension, to conserving and managing natural habitats. Naturalists were well catered for by a wide range of books in print, not least by 40-odd volumes in the New Naturalist library. The now universal field guide had made an appearance, but there were also handbooks on beetles, spiders, bugs, grasshoppers and even centipedes and rotifers, at affordable prices. Naturalists were not infrequently equipped with a hand lens, and specimen collecting was not yet considered a crime. Television natural history had begun, with programmes such as *Look* and *Survival* and, though still in black-and-white, had less manic, less dumbed-down presenters and they were more often about wildlife near at home.

Much less was known about wildlife habitats and sites. Although some places had been thoroughly explored by naturalists, with long typed lists of species bound in massive ledgers, there had been few systematic surveys of habitats or species. The first attempts to census and record the distribution of species had been made in the 1920s and 1930s for certain colonial birds, such as heron and rook. However, the most important mapping scheme to date had been for wild flowers. In 1962, the Botanical Society of the British Isles published the *Atlas of British Flora*, which mapped the nationwide distribution of some 1,400 native or naturalised wild flowers and ferns using dot-maps based on a grid of 10 x 10 kilometre squares. What made the atlas possible was the invention of punch-card computers. It was all done without sizeable grant-assistance or central organisation, although the production of such atlases was later facilitated by a Biological Records Centre, established at Monks Wood under Franklyn Perring and, later, John Heath.

Similar atlases have since been produced and published for other plants and animals, including lichens (from 1982), butterflies (1984), bryophytes (from 1991), dragonflies (1996), grasshoppers (1997) and molluscs (1999). Pre-eminent among them are the bird atlases, *The Atlas of Breeding Birds* (1976) and *The Atlas of Wintering Birds* (1986). The production of a second breeding bird atlas recording breeding birds in 1993 was all the more valuable because it enabled an analysis of change over a 20-year period, providing a temporal dimension to the maps. The same has been done for only one other group, the butterflies, which received perhaps the most lavish and detailed atlas of all, *The Millennium Atlas*, in 2001. Distribution maps of many British species are now published on the Internet, via the National Biodiversity Network.

Of course, the relatively crude scale of 10 x 10 kilometres does not record actual sites (and so can make a species appear more frequent than it really is). Maps of vascular plants and butterflies have been published on finer scales, and a 2 x 2 kilometre 'tetrad' is now standard for county-scale maps (and represents a stupendous recording effort by local naturalists). These reveal some of the detail of actual distribution, fine enough to show the course of rivers and different strata and soil types. Some rare species have been mapped by actual sites; visually, the problem of actual-scale maps is that the 'dots' all but disappear – fly-specks on a blank canvas. Some insects, including butterflies, seem to occupy tiny patches of land, measured in metres rather than kilometres.

The venerable Victorian tradition for recording the wild plants of a county has been given a new lease of life by computer-based mapping. Grid-mapping was used in most of the county floras published since 1960. Their compilation by dedicated amateur naturalists, sometimes working as a team, sometimes, especially in the more remote regions, representing a titanic solo effort, is one of the wonders of British natural history. The

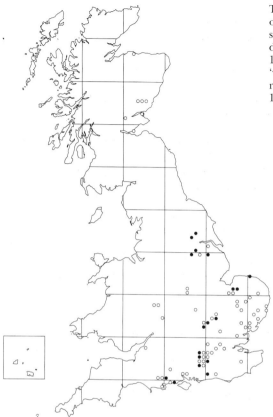

Ten-kilometre square records of *Arnoseris minima* (lamb's succory) plot the gradual decline to extinction in about 1970 of this once widespread 'weed'. Hollow circles are records pre-1930, solid circles 1930–1970. (BSBI/CHE)

majority of English counties have a flora published during the past 20 years, generally mapped on a fine 2 x 2 kilometre scale. By tradition they also include an overview of the county's natural vegetation, progress in conservation and potted biographies of local botanists. The recent *Flora of Cornwall* has an accompanying CD-Rom holding the entire database of records. Some, such as *The Flora of Dorset*, record mosses, lichens and even fungi, as well as vascular plants. Others, such as *A Flora of Norfolk* and *A Flora of Cumbria* use elaborate coloured maps to compare plant distribution with physical features, a valuable advance made possible by modern printing technology.

There are a growing number of 'county faunas', mainly of birds or butterflies, but on occasion extending to other vertebrates and at least the more popular invertebrates. Surrey Wildlife Trust has published a wonderful series of local insect 'faunas', including hoverflies, dragonflies, ladybirds and grasshoppers. Local bird reports have reached a high state of elaboration. They are often produced annually, enabling the present year to be compared with the last and revealing population trends. By contrast, many fine wildlife journals have fallen by the wayside, victims of rising costs and falling subscriptions, such as the long-running *Scottish Naturalist* and

Tetrad records
of Danish
scurvy-grass
from *A Flora of
Norfolk* reveal
the plant's
fidelity to a
natural habitat,
coastal salt
marshes, and
its mimic, the
moist, salty
verges of main
roads. (Alec
Bull & Gillian
Becket)

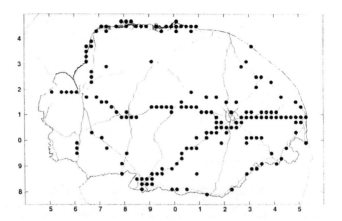

Nature in Wales. Fortunately the magazine *British Wildlife* came in the nick of time to rescue serious naturalists, and includes valuable news reports on all the main groups of flora and fauna. There are welcome signs of a revival in regional or specialised publications, such as *Natur Cymru*, the new 'review of wildlife in Wales', and the lively magazine *Atropos*, which is devoted to moths, butterflies and dragonflies.

A series of Red Data Books has been published on rare and vulnerable species, each one containing a great deal of ecological information on the eternally fascinating question of why some species are rare. The more recent ones include distribution maps. So far there are Red Data Books for *Vascular Plants* (first published 1977, 3rd edition 1999), *Insects* (1987), *Birds* (1990), *Other Invertebrates* (1991), *Stoneworts* (1992), *Mammals* (1993), *Lichens* (in part – 1996) and *Mosses and Liverworts* (2001). There is also *Scarce Plants in Britain* (1993), a kind of 'Pink Data Book' for wild flowers that live on the edge of conservation activity. Together they cover nearly 4,000 species, and undoubtedly many more would qualify, especially if we knew more about fungi, or soil and marine life. It is a daunting thought that perhaps a quarter of our wild plants and animals are rare enough to warrant conservation action – and, if our climate is truly changing, that this proportion will rise.

Site registers and monitoring schemes

To conserve species in a site-based system of nature conservation you need to link species to places. During the 1980s, the NCC embarked on a series of grand, nationwide habitat surveys, a kind of resource account of Britain's wildlife. Cumulatively they produced a detailed documentation of Britain's wild places as a factual basis for conservation decisions. The NCC drew on existing, scattered information, but the surveys also amounted to possibly the most ambitious programme of field survey there has ever been. First there were the habitat surveys: ancient woodland ('the Ancient Woodland Inventories'), the coast ('Sand-dune Survey', 'Estuaries Review'), peatlands ('National Peatland Resource Inventory') and rivers. Beyond the shoreline there was the Marine Conservation Review (q.v.), analogous to the Nature Conservation Review on land, which classifies maritime communities and documents the best sites. The Geological Conservation Review did the same for rock strata and special geological sites over 41 volumes. Then there were the species reviews. The Invertebrate Site Register documented the best sites of insects and other invertebrates. A Rare Plant Survey did the same for rare vascular plants. The long-running Seabirds at Sea project discovered where seabirds went and what they did once they were beyond the reach of land-based binoculars. These surveys formed a mass of raw material that could be tapped for SSSI notification, species protection, drawing up oil-spill contingency plans etc. The Geological and Marine Conservation Reviews were also major contributions to pure science. Most of this work remains unpublished, perhaps destined to remain forever in sagging, spiral-bound folders thronging the shelves of conservation offices or lodged somewhere in the labyrinth of cyberspace. It is almost unsung, this last, vast outpouring of British natural history. The NCC's successor agencies continued some of it, and implemented smaller-scale surveys of their own, though generally for a particular, short-term purpose, but none of them had any interest in praising the work of their predecessor. All the same, it

Recording rare flowers: Barbara Jones, a trained climber, surveying the Snowdon lily, *Lloydia serotina*, (below) on Snowdon.

was a tremendous voyage of discovery with able captains. Those of its mariners that are not retired are now mostly in administrative conservation jobs, or working as consultants.

There were also surveys that established a system of baseline monitoring to record how the natural world was changing. A basic 'phase one' survey that purported to map all Britain's natural habitats started in the early 1980s, with the help of the county wildlife trusts. The NCC also started a National Countryside Monitoring Scheme, which compared present and past land use in selected parts of the countryside. A much finer tool for measuring change was handed to conservation workers with the publication, during the 1990s, of the five-volume *British Plant Communities*, compiled by John Rodwell of Lancaster University, the fruits of a 15-year project to classify all kinds of natural and semi-natural vegetation found in Britain. The vegetation of many SSSIs has been mapped using this National Vegetation Classification, thereby providing a baseline for monitoring change as well as for assessing how much of the different types of vegetation are on protected sites. The species that best lend themselves to detailed census work are birds and butterflies. Birds have long been monitored using the Common Bird Census (now the Breeding Bird Survey)

'Abundance map' of the kestrel from *The New Atlas of Breeding Birds 1988–1991* showing relative numbers as well as distribution. (By permission, BTO)

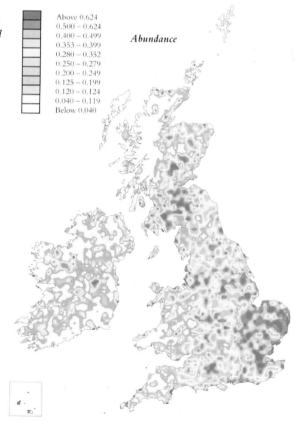

Above 0.624
0.500 – 0.624
0.400 – 0.499
0.353 – 0.399
0.280 – 0.352
0.250 – 0.279
0.200 – 0.249
0.125 – 0.199
0.120 – 0.124
0.040 – 0.119
Below 0.040

Abundance

organised by the British Trust for Ornithology (BTO), along with the regular counts of seabirds by the Seabird Group. The performance of rare birds is the responsibility of the Rare Breeding Birds Panel. For monitoring butterflies, a method was worked out in the 1970s by Ernest Pollard, which involved counting all the species seen along a regularly walked transect. The success of the scheme has meant that fluctuations in butterfly numbers can be recorded, as well as their distribution. Computer techniques have also made possible some novel 'phenograms' (see p. 28) combining numbers with dates of emergence. Further evidence of population changes in insects, especially moths, is provided by the 'Rothamsted traps' operated under standardised conditions across the country. There has also been a revival of phenology – timing the appearance of flowers or frogspawn or migrant birds – in relation to climate change.

From data to action

Surveys and data provision are meat and drink to conservation bodies. It is what they do best. But interesting and often valuable though they are, surveys and monitoring only tell you what is happening. They give a sense of the overall state of health of the patient, but are not in themselves a cure. In practice, looking after wildlife is not based on scientific rationalisation

Red Admiral 1995-9

'Phenogram' of the red admiral butterfly from *The Millennium Atlas of Butterflies*, plotting latitude at 100-kilometre intervals against months, forms a ghostly outline of Britain. (From Butterflies for the New Millennium survey organised by Butterfly Conservation and Biological Records Centre)

alone, but on negotiation and politics. It is rare that a conservation body has full control over a given situation, even on a freehold nature reserve. Decisions are often made in a cloud of ignorance, or in a spirit of compromise with more powerful interests. Indeed, conservation in practice is to a large extent to do with quarrelling. You make the best case you can, you cite your legal and moral rights, you appeal to the more important party's better nature. Then, often with the mediation of a third party, you reach the best deal you can, with or without bitter words and recrimination.

Yet 'quarrel' is a remarkably rare word in conservation literature. I think the first time this particular spade was called a spade was in Professor Smout's book, *Nature Contested*, published in 2000. More often, like politicians, the parties prefer to sweep disagreements under the carpet, using euphemisms like 'discussion', 'debate' or even – a popular choice in the 1990s – 'partnership'. Nature conservation is a quasi-political matter in which the arguing is done as far as possible behind closed doors, and the outcome reported to a supine press with a bland statement. Many conservation bodies have become so used to the self-censorship of uncomfortable facts that they seem to operate on a different plane of reality from the farmhouse or the estate office. Their publications reflect the power of image and presentation in the modern world. It is necessary for conservation bodies to appear slick, dynamic, successful and, above all, relevant. Where the facts of disappearing wildlife appear to contradict this image they can be distorted by the same black arts ('let's focus on the positive') or used to justify an appeal for more money. Ignorance of natural history can be a distinct advantage in this world. Hence, if in the later chapters of this book, I may sometimes seem rather sceptical about the claims of the conservation industry, and its official agencies in particular, it is because I have seen something of this world from the inside. Conservation bodies rarely stoop to deliberate distortion, but their version of events can be coloured by the views of their 'clients' and partners, by the attitude of their political masters or by that of a mass-membership. You do not, for example, hear the RSPB talking much about cats, or the Wildlife Trusts about fox hunting. Nature conservation in a crowded island in which all land is property is bound to be difficult, even when everyone agrees that wildlife is a good thing. Conservation can be seen in different ways, depending on how you are affected by it: as a moral absolute, as a cumbersome, bureaucratic restriction, as an unjust imposition by ignorant outsiders, as a potential source of income. I think wildlife is a good thing. Indeed, in my own life I think it is probably the most important thing. I would like my country to preserve as much wildlife and countryside as possible, but without enmeshing rural life in petty restrictions. The standpoint of this book is a love of wildlife but not necessarily conservationists. In these pages you may therefore find a lot of 'buts'. I hope it will not sound unduly negative. I feel it may be necessary. An account of nature conservation in Britain devoid of individual opinion would be a dull read, indeed not worth reading. I hope this book is worth reading.

PART I

Dramatis Personae

2

The Official Conservation Agencies

The British government has always delegated its responsibilities for nature conservation to a semiautonomous agency. The governments of other European countries tend to keep theirs within agricultural departments or National Park bodies. The reason why Britain behaves differently probably lies in our early start and the influence of science in the 1940s (for a good account of this postwar science boom, see Sheail 1998). The founders of the Nature Conservancy, the first official conservation agency in Britain, saw it as a biological service, comparable with a research council or scientific institution, like the Soil Association. They hoped it would develop as a science-based body, using its own research programme to advise government on land-use policies affecting wildlife. As Professor Smout has pointed out, 'the rule of the bureaucrat guided by the scientific expert has been highly prized in government for most of the twentieth century' (Smout 2000). They were anxious that nature conservation should not be swallowed up in the departments for agriculture and forestry, where, as a newcomer, and so starting at the bottom of the civil service peck order, its influence would be stifled. Max Nicholson, who directed the Nature Conservancy between 1951 and 1965, had influence in high places and ensured that, as a semi-specialised body, it secured a semide-tached status as a research council under the wing of Herbert Morrison, then Lord President of Council. As such, it could not be bossed about by predatory departments of state. There are advantages to ministers in such arrangements. Expertise is 'on tap, not on top', and if anything goes wrong it is the agency's fault, not the minister's. Dispensable Board chairmen can be sacked, but the minister need not resign. Much of the history of the Nature Conservancy and its successor bodies hovers around the tension between the zeal of semiautonomous agency officials and the

Max Nicholson, Director of the Nature Conservancy 1951–1965. (NCC)

brake of government (the appointed Councils of these bodies have tended
to be part of the braking mechanism rather than the zeal). It is there
between the lines of their annual reports and, now that the papers are at
last available under the ridiculous 30-year-rule, you can read about the formative years of that thorny relationship in a fine, detailed book by John
Sheail (1998). But here I need to skip over those, to many, golden, well-
remembered early years with unseemly haste.

The Nature Conservancy is said to have been the first official, science-
based conservation body in the world, and the only one with a large
research arm. Although money was always tight, the Nature Conservancy
under Nicholson tended to box above its weight. It achieved a good deal,
acquiring a nationwide network of National Nature Reserves and research
stations, and gaining an international reputation for sound, science-based
advice on the management of wild species and natural habitats. I begin the
story where Dudley Stamp left off, in 1965, shortly after the Nature
Conservancy lost its independence after becoming a mere committee within the newly formed Natural Environment Research Council (NERC).
'One chapter is concluded,' wrote Stamp, 'but there is every sign of a new
one opening auspiciously.'

If so, it did not stay auspicious for long. The 1960s should have been a
good decade for the Nature Conservancy. The general public had become
more 'environmentally aware' through events like the pesticides scare
(Rachel Carson's book about it, *Silent Spring*, became an international best-
seller) and the Torrey Canyon disaster in 1967. The threat to our wild
places had been underlined by the construction of a nuclear power station
at Dungeness (Plate 11) and a reservoir at Cow Green in Upper Teesdale
(Plate 13). The Conservancy was closely involved in these issues, and its
growing fame was exemplified by the traffic jams on Open Days at Monks
Wood Field Station (not to mention visits on different days by Prince
Charles and the Prime Minister). However, its status within NERC drew
attention to the essential ambiguity of the Conservancy's role: could a body
be scientific, and therefore impartial, and yet advocate a partial view – that
conserving nature is a good idea? The Conservancy itself dealt with this
duality by dividing its administrative responsibilities under one subdirector
(Bob Boote) and its research under another (Martin Holdgate).
Unfortunately the Conservancy no longer had full control over its affairs.
For example, its budget for nature reserves had to compete for funds with
NERC's broader research, including geophysics, oceanography and the
Antarctic. Internal censorship prevented the Nature Conservancy from
speaking out on pesticides and other pollutants. Tensions grew in the
boardroom, where some members thought it was worth making sacrifices
to preserve the link between conservation and fundamental science while
others decided that nothing had been achieved by joining NERC, and that
the Conservancy would be better off going it alone. The Conservancy's new
director, Duncan Poore, was of the latter view.

Unfortunately there was to be no return to the pre-1965 days: the choice
lay between the frying pan of NERC and the fire of a government depart-

ment. The Conservancy's committee split, with an influential group voting to leave NERC. A way out of the impasse was offered by the Government's Central Policy Review under Lord Rothschild – the famous 'Think Tank' – which advocated the separation of customer and contractor. As a 'customer' of the natural sciences, the logic was that the Nature Conservancy should become independent of NERC, but the same logic prevented it from carrying out in-house research. Rothschild proposed that only half of NERC's budget should be paid by its parent Department of Education and Science, with the balance found by commissioning research from other government departments. Most of the Nature Conservancy's own little budget would now come from the new Department of Environment, or, in Scotland, from the Scottish Development Department (SDD). Funds were also transferred from NERC to pay for contract research. By one of life's little coincidences, the Education Secretary who helped to set up this new Nature Conservancy Council (NCC) was the same person who presided over its demise, 16 years later – Mrs Thatcher.

Rothschild's report gave the Conservancy the excuse it needed to make public its wish to leave NERC. Government agreed that the Conservancy's dual role had 'caused stresses difficult to resolve within the present framework' (Sheail 1998). Unfortunately the solution, as Government saw it, was to separate science from administration. The Nature Conservancy would become a quasi-autonomous council of the Department of Environment, but its scientific stations would remain behind in NERC. This divorce, representing the exact moment when field-based natural history began to turn into administrative nature conservation, became known as 'The Split'. The Nature Conservancy Council, usually referred to as the NCC, was established by Act of Parliament in 1973. Its first chairman was a Whitehall mandarin, Sir David Serpell, lately Permanent Secretary at the DoE. He promised to run the new agency on 'a loose rein' (which fooled nobody). As a sop to anguished pleas that the NCC must retain some scientific capacity to function properly, it was allowed to keep a small in-house team of scientists under a 'Chief Scientist', a term coined by Rothschild. But their job would be limited to commissioning and keeping abreast of research, rather than doing it themselves. In the meantime, its erstwhile Research Branch was reconstituted within NERC as the Institute of Terrestrial Ecology (ITE – now renamed the Centre for Ecology and Hydrology).

The Nature Conservancy Council (NCC)

'No one was entirely happy with the outcome of the "Split"' (Sheail 1998). Some saw it as a further demotion that threatened the special relationship between science and land management so carefully fostered by the Nature Conservancy. However, that relationship was already falling apart. While the

White Paper 'Cmd 7122' had talked up the potential of nature reserves as 'outdoor laboratories' and the importance of its advice to land managers, the hard truth was that by the 1970s only a handful of nature reserves were used for fundamental research, and farmers and foresters were not queuing up for the Conservancy's advice (they had their own scientists). Moreover, the crisis in the countryside was growing and it was no longer a matter of experimenting over the best way to manage a wood or a heath but of saving such places from complete destruction. Inevitably this required a shift in emphasis away from scientific research towards site safeguard, which, unless you happen to manage the land yourself, is an administrative task. Most of the Conservancy's research budget now went on cheap, low-key surveys that helped to identify or characterise the places that most deserved safeguarding. Consequently, the split between the NCC and its former science branch broadened into a chasm. ITE gradually ceased to be a significant part of the nature conservation world – to the deep regret of many of its staff, which included New Naturalist authors like Ian Newton, R.K. Murton and Max Hooper.

The 1970s were a bad decade for the natural environment. In Britain, Dutch elm disease and the removal of hedges created stark, arable landscapes, while in the uplands blanket afforestation transformed many square kilometres of open country into sepulchral timber crops of introduced spruce, pine and larch. Limestone pavements were smashed to bits to adorn suburban gardens and corporate offices. The Norfolk Broads, still crystal clear in the early 1950s, became clouded with silt. The heaths went up in flames during the drought years 1975 and 1976, and, apart from the mountain tops, there seemed to be hardly any wild land that agricultural grants could not convert into profitable farmland. Hence, the NCC was overstretched, using what small authority it had to oppose harmful developments, reach agreements and establish nature reserves. On occasion, it stepped back from events to appraise the situation. In 1977, for example, it published a 'policy paper', *Nature Conservation and Agriculture*, containing the NCC's thoughts on how to reconcile increasing food production with the maintenance of 'Britain's rich heritage' of wildlife. Essentially the message was that, while vast amounts of public money were helping farmers plough and drain the land, the incentives to preserve wildlife were negligible. You did not have to travel far to see the consequences. A second policy paper, on forestry, was shelved after reported disagreements on the NCC's Council, which contained members with vested interests in forestry.

In 1977, the NCC at last published *A Nature Conservation Review*, edited by its Chief Scientist, Derek Ratcliffe, describing the range of wildlife and natural vegetation in Britain, and singling out the 735 best examples of coastlands, woodlands, lowland grasslands and heaths, open waters, peatlands and upland habitats, all graded according to their international, national or second-string importance. The Review was, and remains, an astounding *tour de force*, combining a rationale for site selection with a kind of Domesday Book of Britain's wild places (though, as Jon Tinker pointed out in *New Scientist*, it had taken eight times as long to produce as the original

Domesday Book!). The original purpose of the Review had been to provide a reasoned 'shopping list' for nature reserve acquisition. Because of
the obvious sensitivities involved – for by no means every landowner would
have been delighted to find his property on the list – this aspect was played
down, and the Review was presented to the public as a reference book of
important biological sites. In commending it in these terms, the Ministers
for Environment and for Education and Science were careful to avoid com-
mitting themselves to any particular action. The Review sparked no change
in environmental policy, but it did form a necessary reference point for site
protection. Without some means of assessing the relative importance of
wildlife sites, the NCC would be blundering in the dark.

The second key NCC document was its long-term strategic review, published in 1984 and entitled *Nature Conservation in Great Britain* ('NCGB'). It
was in part an assessment of the successes and failures of the nature conservation movement, and in part a set of ground rules for the future. The
failures outnumbered the successes by 21 pages to five, and any impression
given by the glossy pictures of a healthy, vital natural environment was con-
tradicted by the lowering bar graphs that showed 'with stark clarity' how far
wildlife habitats had diminished during the past half-century – a loss of 40
per cent of lowland heaths, for example, and an incredible 95 per cent of
'lowland neutral grasslands'. Behind the statistics lay a detailed analysis of
habitat loss undertaken by Norman Moore – but, as it happened, 'NCGB'
proved to be the only opportunity to publish any of it. Perhaps Council
thought it might depress the minister. The real significance of the review
lay not so much in the detail but in its heightened sense of conviction. For
the first time the NCC explicitly recognised nature conservation as a cultural activity, and not merely as pure or applied science. 'Simple enjoyment
and inspiration from contact with nature' was not a partisan activity: it concerned us all. It followed that we should conserve nature in the same way
that we take care of other essentials like air and water. 'Nature conservation has in the past sometimes conducted its business on too apologetic
and timid a note', declared the NCC, looking back at its own history.
Timidity had too often meant surrender. 'We need to play a hard but clean
game for our side,' said the NCC's new chairman, William Wilkinson. So
there were now 'sides', us against them. The strategy was heartily supported by most of the voluntary bodies, who rightly saw it as a challenge,
heralding a significant change of policy, and expected NCC to honour its
brave words to the letter. But in the freewheeling climate of the 1980s, having the courage of your convictions meant having to fight for them. The
five years of corporate life left to the NCC were hard ones, and led straight
to its destruction.

There were really two NCCs, separated by the watershed year 1981 in
which the Wildlife and Countryside Act reached the statute book. The pre-
1981 NCC was a fairly low-key organisation with a staff of about 500 dispersed thinly about Britain, struggling along on an annual budget of about
£6 million (the NCC had scarcely any income or assets). It advised government on issues affecting wildlife, commented on local plans and devel-

Sir William Wilkinson,
chairman of the NCC
1984–1991. (English
Nature)

opments and grant-aided worthy projects, but was rarely in the headlines. The man in the street had never heard of it, which is not to deny that the NCC achieved a great deal on very little.

The post-Act NCC took a little while to get going, but it became another organisation entirely, more powerful, more centralised, and often in the headlines, especially in Scotland. By 1988, the NCC had 780 permanent staff with a sixfold budget increase to £39 million. An enforced move in 1984 from the old Nature Conservancy's stately quarters at Belgrave Square to a modern office block in Peterborough gave the organisation a chance to centralise its dispersed branches – an England headquarters at Banbury, scientists at Huntingdon, geologists at Newbury, publicists and cartographers at Shrewsbury were all sucked into Peterborough. The organisation also became computerised and corporatised. Corporate planning was introduced in 1985, requiring staff to complete monthly time records, recording (in theory at least) every half-hour of activity. The Act made nature conservation much more expensive. By 1988, nearly a quarter of the NCC's budget was spent on management agreements on SSSIs.

The Wildlife and Countryside Act distorted the NCC's activities for nearly a decade, as its regional staff struggled to notify SSSIs and negotiate agreements over their safeguard. Land not notified as SSSI became known as 'wider countryside', and there was little enough time to devote to it (with the honourable exception of urban conservation, largely a one-man crusade by George Barker). Unfortunately, the wording of the Act forced the NCC to adopt a heavy handed approach on SSSIs, in which an owner or tenant would be presented with a formidable list of 'Potentially Damaging Operations'. Permissions to carry on farming in ways that did not damage the site's special interest were called 'consents'. This sort of language understandably put people's backs up, as did the fact that there was no appeals system and the conviction that notification would lower the land value. Suspicion and potential hostility could be mollified by the farm-to-farm visits of the NCC's regional staff, who were generally speaking more charming and persuasive than the documents they had to deliver. To the extent that the Act was a success it was theirs, not that of the politicians who created a botched system, nor the civil servants and lawyers who insisted on its rigid application. The local NCC staff rapidly learnt that the only way to make the Act work was by goodwill – hardline interpretations of the law and threats of prosecution simply alienated people, and got nowhere.

In some areas, especially in island communities, SSSIs were seen as an alien imposition. In Islay, teeming with wildlife, a severe dose of SSSIs looked like a punishment for farming in harmony with nature. Things came to a head in the small matter of finding a source of peat for an Islay distillery as a substitute for Duich Moss, one of the best raised bogs in the Hebrides. A team of environmentalists led by David Bellamy, intending to plead the cause of peatland conservation, was howled down by angry islanders normally renowned for their hospitality and gentleness of manners. It was not that the community was against nature conservation, only that they did not enjoy being told what to do by a Peterborough-based quango (a place not particularly noted for its teeming wildlife).

Notifying SSSIs was a much more complicated business than anyone had foreseen. To begin with staff had to find out who owned the land, and even that could be a hornet's nest with, in Morton Boyd's words, 'many untested claims to holdings, grazings and sporting rights' (Boyd 1999). SSSIs were an absolute, bureaucratic system imposed on a system of tenure that

J. Morton Boyd, Scottish director of the NCC, at Creag Meaghaidh, which he helped save from afforestation. (Des Thompson)

was often the opposite: communal, fluid, and based on non-Westminster concepts of custom, neighbourliness and unwritten rules. Outsiders blundering into these matters could unwittingly set neighbour against neighbour. They could also make themselves very unpopular. Moreover some SSSIs were already under dedicated schemes for forestry or peat extraction, or an agricultural grant scheme. In the Outer Hebrides an EU-funded Integrated Development Scheme was in progress, while in Orkney an Agriculture Department-funded scheme was encouraging farmers to reclaim moorland. The NCC often found itself outnumbered. When, in 1981, it opposed the extension of ski development into the environmentally sensitive Lurchers Gully in the Cairngorms (see Chapter 10), the NCC found itself ranged against the Highlands and Islands Development Board, Highland Regional Council, and sports and tourism lobbies, as well as local entrepreneurs. It won that particular battle but, in the Highlands and Islands at least, the NCC eventually lost the war.

The break-up of the NCC

Replying to an arranged Parliamentary question on 11 July 1989, Nicholas Ridley told a near-empty House of Commons that he had decided to break up the Nature Conservancy Council. In its place he would introduce legislation for separate nature conservation agencies in England, Scotland and Wales. I well remember the shock. Just the previous week we had attended a ceremony to mark the retirement of Derek Ratcliffe, Chief Scientist of the NCC since its establishment 16 years before. 'Things will never be quite the same again,' we thought, little suspecting just how different they would be. I was in the canteen at the NCC's headquarters at Northminster House as a rumour spread over the cause of the emergency Council meeting upstairs. The hurried patter of feet on the third floor, doors banging, chairs scraping, voices raised, all signalled unusual excitement. NCC's chairman, Sir William Wilkinson, had, it seemed, been given a week's notice of the announcement, but had not been allowed to tell anyone. He used the time to appeal to the Prime Minister, but she backed her minister. Council had had only one day's official notice, although some of them did not seem very surprised, and a few welcomed it. We, the staff, were caught completely unawares. As *Forestry and British Timber* magazine gloated, 'it was, no doubt, to spare the NCC the horrors of anticipation that the Ridley guillotine crashed down upon it last week. There was no warning, no crowds, no tumbrils, no (or very little) mourning. The end of the Peterborough empire came silently and swiftly'.

No mourning from foresters may be, but it sent a seismic shudder, shortly to be followed by an outpouring of rage, through the nature conservation world. 'At no time was NCC given notice of such extreme dissatisfaction with its performance as to register a threat to its corporate existence', wrote Donald Mackay, a former undersecretary at the Scottish Office (Mackay 1995). The only clue in Ridley's statement was that there were apparently 'great differences between the circumstances and needs of England, Scotland and Wales ... There are increasing feelings that [the

present] arrangements are inefficient, insensitive and mean that conservation issues in both Scotland and Wales are determined with too little
regard for the particular requirements in these countries'. Evidently, then,
events in Scotland and Wales had propelled the announcement.

The sentence had been done in haste. Ridley was about to move from
Environment to Energy, where he was sacked a year later for making offen-
sive remarks about the Germans. Nothing had been thought through. The
implication was that, as far as nature conservation was concerned,
England, Scotland and Wales would now go their separate ways, but left
hanging was the not unimportant matter of who would represent Britain
internationally and who would referee common standards within the new
agencies. Moreover, far from being more efficient, a devolved system
implied endless duplication (actually, triplication) and waste. 'What would
you rather have?' asked Wilkinson, 'a peatland expert for Great Britain, or
three under-resourced experts in England, Scotland and Wales? It's obvious isn't it?' Behind Wilkinson's disappointment and frustration was the
knowledge that his Council had been about to introduce a 'federal' system
of administration that, he thought, would largely have answered the genuine problems being experienced in Scotland and Wales.

Some of the smoke from Ridley's 1989 bombshell has since cleared. At
issue was the NCC's unpopularity in Scotland, and in particular its opposition to afforestation. Things came to the crunch in 1987 when, alarmed at
the rate of afforestation in the hitherto untouched blanket bogs of far away
Sutherland and Caithness (see Chapter 7), the NCC called for a moratorium on further planting in the area. Fatally, the NCC decided to hold its
press conference in London, not in Edinburgh or Inverness, lending substance to the accusation that the NCC was an English body, with no right
to ban development in Scotland, especially when jobs were at stake. It is
alleged that there was a reluctance on the part of the NCC's Scottish headquarters to host the press conference; its Scottish director, John Francis,
had taken diplomatic leave. The Scottish media took more interest in a
spoiling statement by the Highlands and Islands Development Board,
whose chief took the opportunity to call for a separate Scottish NCC. The
Scottish press took up the cry, and from that day on another 'split' was
probably inevitable. The MP Tam Dalyell was in no doubt that this was why
the NCC was broken up: 'It originated out of a need that had nothing
whatsoever to do with the best interests of the environment. It was about
another need entirely, that is, the need for politicians to give the impression that they were doing something about devolving power to the Scots as
a sop to keep us happy' (Dalyell 1989).

Just as the Scots resented 'interference' from Peterborough, so the
Secretary of State for the Environment resented having to pay for things
outside his direct control (for DoE's writ ran only in England and Wales).
According to Mackay, Ridley, growing alarmed at the anticipated costs of
compensating forestry companies in Caithness, suggested to his Cabinet
colleague, Malcolm Rifkind, that Scotland should receive its own conservation agency and shoulder the burden itself. With the Conservative party's

popularity at an all-time low in Scotland, Rifkind must have seen political advantages in such a gesture, and ordered his Scottish Development Department to prepare a plan for detaching the Scottish part of the NCC and merging it with the Countryside Commission for Scotland. The case for Scotland automatically created a similar case for Wales. It seems, though, that Wales received its own devolved agency without ever having asked for one.

The secrecy in which all this took place is surprising, but it enabled ministers to rush the measure through before the inevitable opposition could get going – an early example of political 'spin'. The NCC had few influential friends north of the Border, where voluntary nature conservation bodies were weak. Moreover, the afforestation issue had encouraged separatist notions among the NCC's own Scotland Committee and staff. Broadly speaking they saw the future of wild nature in Scotland in terms of sustainable development and integrated land use, which in some vague way should reflect the value-judgements of the Scottish people. It made little sense to draw lines around 'sites' in the Highlands where wild land was more or less continuous. Hence they saw more merit in processes – making allies and finding common ground – than in site-based conservation, which, as they saw it, only served to entrench conflict. That, at least, is what I construe from the statement of the chairman of the NCC's Scotland Committee, Alexander Trotter, at the break-up, that 'It has been clear to me for some time that the existing system is cumbersome to operate and that decision making seemed remote from the people of Scotland'.

Some of the opposition to the break-up was blunted by the obvious appeal of combining nature and landscape conservation in Scotland and Wales. Many believed that the severance of wildlife and countryside matters back in 1949 had been a fundamental error, and that in a farmed environment like the British countryside they were inseparable. However, Ridley refused to contemplate their merger in England, arguing that the administrative costs would outweigh any possible advantages (a view the Parliamentary committee concurred with when the question was reopened in 1995). The main objection, apart from the well-founded fear that science-based nature conservation had suffered another tremendous, perhaps fatal, body blow, was the void that had opened up at the Great Britain level. Following a report by a House of Lords committee under Lord Carver, Ridley's successor, Chris Patten accepted the idea of a joint co-ordinating committee to advise the Government on matters with a nationwide or international dimension. This became the Joint Nature Conservation Committee or JNCC, a semiautonomous science rump whose budget would be 'ring-fenced' by contributions from three new country agencies. Some of the NCC's senior scientists ended up in the JNCC, only to find they were scientists no longer but 'managers'.

Creating the new agencies took many months, during which the enabling legislation, the (to some, grossly misnamed) Environment Protection Bill, passed through Parliament, and the NCC made its internal rearrangements. Separate arrangements were needed under Scottish law,

NCC's spending in 1988 (in £,000s)

Income

From government grant-in-aid	36,105	
Other income	2,461	(mainly from sales of publications rents and research undertaken on repayment terms)
	£38,566	

Expenditure

Staff salaries and overheads	14,310	
Management agreements	7,287	
Scientific support	4,992	(including 3,736 for research contracts)
Grants	2,510	(made up of 1,109 for land purchase, and the rest staff posts and projects, mainly to voluntary bodies)
Maintenance of NNRs	1,399	
Depreciation	1,089	
Other operating charges	7,280	(e.g. staff support, books and equipment, accommodation, phones)
	£38,867	

From NCC 15th Annual Report 1 April 1988–31 March 1989

and so an interim body, the Nature Conservancy Council for Scotland was set up before the Scottish Natural Heritage was established by Act of Parliament in 1992. From that point onwards, the history of official nature conservation in Britain diverges sharply. Because of the interest in the new country agencies' performance, I will present them in some detail. They form an interesting case study of conservation and politics in a devolved government. In Scotland and Wales particularly it has led to a much greater emphasis on popular 'countryside' issues, and less on wildlife as an exclusive activity. In England, too, there have been obvious attempts to trim one's sails to the prevailing wind, with an ostentatious use of business methods and a culture of confrontation-avoidance. Let us take a look at each of them, and the JNCC, starting with English Nature.

English Nature

Headquarters: Northminster House, Peterborough PE1 1UA.

Vision: '*To sustain and enrich the wildlife and natural features of England for everyone*'.

Slogan: '*Working today for nature tomorrow*'.

English Nature began its corporate life on 2 April 1991 (April Fool's Day was a public holiday that year) with a budget of £32 million to manage 141 National Nature Reserves, administer 3,500 SSSIs and pay the salaries of 724 permanent staff. Most of the latter were inherited from the NCC, including a disproportionate number of scientific administrators, and only 90 were new appointments. EN's Council was, as before, appointed on the basis of individual expertise, and intended to produce a balance of expertise across the range of its functions. However, they were now paid a modest salary and given specific jobs to do. From 1996, under the new rules established by the Nolan Report, new Council posts were advertised. All of them had to be approved by the chairman, a political appointee. What was noticeable about EN's first Council was that only one was a reputable scientist. None were prominently affiliated to a voluntary body, nor could any of them be described as even remotely radical. This Council was less grand than the NCC's: fewer big landowners, no wildlife celebrities, and no MPs. In 1995, at the request of Lord Cranbrook, EN's chief executive, Derek Langslow, became a full member, unlike his predecessors who just sat in on meetings and spoke when required. This made him a powerful figure in English Nature's affairs.

English Nature inherited the structure of the NCC, with its various administrative branches, regional offices and headquarters in Peterborough. Externally the change from NCC to English Nature was brought about simply by taking down one sign and erecting another. An agency designed to serve Great Britain could, with a little readjustment, easily be scaled down to England alone. English Nature could, if it wished, carry on with business as usual. Even its official title remained the Nature Conservancy Council (for England); the name 'English Nature' was only legalised in 2000.

In the event, it opted for a radical administrative shakedown. The new administration was keen to present a more businesslike face to the world with a strategic approach in which aims would be related to 'visions' and goals, and tied to performance indicators monitored in successive corporate plans. A deliberate attempt was made to break down the NCC's hermetic regions and branches into 'teams', each with their own budget and business plan. At Northminster House, partition walls were removed, and

American corporatism comes to nature conservation. This card, carried by English Nature staff in the late 1990s, borrows the language of big corporations ('strategic change', 'inside track', 'empower/accredit').

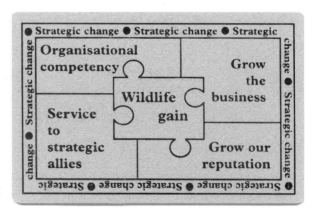

the warren of tiny offices replaced by big open plan rooms in which scientists, technicians and administrators worked cheek by jowl. There were also significant semantic changes. English Nature saw landowners and voluntary bodies as its 'customers'; its work as a 'service' – one of its motto-like phrases was that 'People's needs should be discovered and used as a guide to the service provided'. Its predecessors had considered themselves to be a *wildlife* service. English Nature was overjoyed to receive one of John Major's Citizen Charter marks for good customer service. Henceforward English Nature's publications bore the mark like a medal.

English Nature's tougher organisation was mirrored in its presentations. Its annual reports seemed more eager to talk up the achievements of English Nature as a business than to review broader events in nature conservation. Looking back at EN's first ten years, Michael Scott considered that the 'strategic approach' had engendered more bureaucracy along with tighter administrative control: 'Senior staff talk more about recruitment levels, philosophy statements, strategic management initiatives and rolling reviews than about practical policies on the ground' (Scott 1992). Nor was EN's much-vaunted 'philosophy statement' exactly inspiring to outsiders, with its talk of 'developing employee potential' and achieving 'efficient and effective use of resources through the operation of planning systems'. To those, like the postgraduates who listened in on EN's lectures on corporate strategy, it might have sounded impressively professional, but, with the best will in the world, it didn't sound much *fun*; and to some they seemed to have more to do with what happened behind the dark-glass windows of Northminster House than out there in the English countryside.

The internal changes were not as radical as they looked. English Nature's statutory responsibilities were much the same as the NCC's, and the focus was still on SSSIs, grants and nature reserves. But now that the SSSI notification treadmill had at last ceased to grind, staff could turn their attention towards more positive schemes and participate more in 'wider countryside' matters. English Nature reorganised its grant-aid projects into a Wildlife Enhancement Scheme for SSSIs and a Reserves Enhancement Scheme for nature reserves. Both were based on standard acreage payments, and every attempt was made to make them straightforward and prompt. They were intended to be incentives for wildlife-friendly management, for example, low-density, rough grazing on grasslands and heaths, or to fund management schemes on nature reserves. The take-up rate was good. The trouble was that they were never enough to cover more than a fraction of SSSIs. Meanwhile EN's grant-aid for land purchase virtually dried up. Country wildlife trusts turned to the more lucrative Heritage Lottery Fund instead.

English Nature also took the lead on a series of themed projects to address important conservation problems. In each, the idea was that EN would provide the administration and 'strategic framework' for work done mainly by its 'partners'. The first, a 'Species Recovery Programme' to save glamorous species such as the red squirrel and fen raft spider from extinction, was up and running within weeks. The following year, it introduced a Campaign for Living Coast, arguing that it was wiser in the long run to work with the grain of nature than against it. In 1993 came a Heathland Management Programme, the start of a serious effort to conserve biodiversity on lowland heaths by reintroducing grazing. In 1998, this swelled into an £18 million Tomorrow's Heathland Heritage programme, supported by the Heritage Lottery. In 1997, English Nature proposed an agenda for the sustainable management of fresh water, detailing the 'action required' on a range of wildlife habitats, and started another multimillion pound project on marine nature conservation, part-funded by the EU LIFE Programme. More controversial was EN's division of England into 120 'Natural Areas' based on distinctive scenery and characteristic wildlife. The basic idea was to show the importance of wildlife everywhere and emphasise its local character. Each area had its own characteristics and 'key issues' which, for the South Wessex Downs, included the restoration of 'degraded' downland and fine-tuning agri-environmental schemes to benefit downland wildlife. The critics of 'Natural Areas' were not against the idea as such (though some Areas were obviously more of a piece than others) but saw it as a long-winded way of stating the obvious, involving the production of scores of 'Natural Area Profiles' replete with long lists of species. As with the Biodiversity Action Plan, part of the underlying purpose seems to be to foster working relations with others, especially local authorities.

Like its sister agencies, English Nature wanted to present positive ideas for helping nature and avoid the wrangles of the 1980s. It did so with considerable success, helped by the fact that conservation was gradually becoming more consensual. But the awkward fact remained that, by EN's own figures, between a third and a half of SSSIs were in less than ideal management. Moreover, in its zeal to work positively with 'customers and partners', some found English Nature too willing to compromise and to seek solutions in terms of 'mitigation'. An early instance was the 'secret deal' with Fisons over the future of peatland SSSIs owned or operated by the company. Fisons had agreed to hand over 1,000 hectares of the best-preserved peatlands to English Nature in exchange for a promise not to oppose peat extraction on the remaining 4,000 hectares. Those campaigning actively to stop industrial peat cutting on SSSIs were excluded from the negotiations, and left waiting on the pavement outside the press conference. Whatever tactical merit there might have been in a compromise agreement, the protesters felt that EN had capsized their campaign. English Nature argued that to try and block all peat cutting on SSSIs, as the campaigners wanted, would have involved the Government in compensation payments costing millions, and put 200 people out of work. To which, the campaigners replied that that was the Government's business, not English Nature's. And who exactly were the 'partners' here – the peat industry or the voluntary bodies?

It was English Nature's misfortune to be seen to be less than zealous when an issue became headlines, such as the Newbury bypass (p. 217) or the great newt translocation at Orton brick-pits (p. 207). Of course, as a government body EN had to be careful when an issue became politically sensitive, but on such battlegrounds it was easy to see it as 'the Government' and bodies like the WWF or Friends of the Earth as the opposition; it contributed to the tense relationship between the agencies and the voluntary bodies at this time. The year 1997 was a particularly difficult one for English Nature. It failed to apply for a 'stop order' at Offham Down until prodded by its parent department (pp. 96–7). It wanted to denotify parts of Thorne and Hatfield Moors which would clearly enable the peat producers to market their product more widely. This ill-timed decision led to an embarrassing public meeting at Thorne, when chief executive Langslow was all but booed off the stage, followed by an enforced U-turn after the minister politely advised English Nature to think again. EN's latest strategy, 'Beyond 2000', was ill-received, despite its clumsy attempts to involve the voluntary bodies with questions like 'How can we improve our measurement of EN's contribution to overall wildlife gain' (uh?). On top of all that, in November WWF published a hostile critique of English Nature, *A Muzzled Watchdog?*, based on a longer report on all three agencies I had written for them. It was not so much what it had to say as the unwonted sight of one conservation body publicly attacking another that attracted attention. EN's refusal to comment, apart from some mutterings about 'inaccuracies', did not help its case.

And then, suddenly, all was sunshine again. New Labour had made a manifesto commitment to increase the protection of wildlife. It also lent a more friendly ear to the voluntary bodies, especially those with upwards of a hundred thousand members. English Nature's first chairman, the cau-

Thorne Moors SSSI was a bone of contention in the 1990s between English Nature, which sought a compromise deal with the developers, and campaigners who wanted to stop peat extraction altogether. (Peter Roworth/English Nature)

A fresh breeze. Barbara Young (Baroness Young of Old Scone), chairman of English Nature 1998–2000. (English Nature/ Paul Lacey)

tious and politically acute Lord Cranbrook, reached the end of his term and was replaced by the leftish-inclined late head of RSPB, Barbara Young, who also held a government job in the House of Lords. Council included more credible members. Parliament, investigating the work of English Nature and inviting voluntary bodies to participate as witnesses, kindly concluded that any lack of zealotry on the part of EN must have been due to insufficient money, and so increased its budget.

A friendlier minister and a more supportive social climate seem to have increased English Nature's confidence. Opposing harmful developments is back on the agenda. It dared to criticise the Government line on Genetically Modified Organisms. One particular case summed up the change in attitude. In 1999 EN prevented a proposal to tip ball-clay waste at Brocks Farm SSSI in Devon, having turned down the owner's offer to 'translocate' the grassland habitat. 'The first prerequisite for protecting an SSSI is to leave it as it is,' said EN's spokesman. Both the crispness of the language and the conviction behind it seemed a world away from the rather hapless appearance English Nature had created a few years earlier.

Scottish Natural Heritage (SNH)

Headquarters: 12 Hope Terrace, Edinburgh EH9 2AS

Mission: '*Working with Scotland's people to care for our natural heritage*'.

In 1992, Malcolm Rifkind, Secretary of State for Scotland, told his newly established natural heritage body that if it was not 'a thorn in his flesh from time to time' then it would not be doing its job properly. It was expected, however, to 'work with Scotland's people' more

successfully than its predecessor, which meant not running too far ahead
of public opinion. Scottish Natural Heritage was set up by Act of
Parliament in 1992. It combined the functions of the old NCC in Scotland
and the Countryside Commission for Scotland, a disproportionately small
body compared with England's Countryside Commission (for Scotland
had no National Parks), responsible for footpaths and non-statutory
'National Scenic Areas'. 'SNH' was given a generous first-year budget of
£34.6 million and inherited a combined staff of about 530. Its chairman,
the television personality Magnus Magnusson, was an unashamed populist
and 'aggressive moderate', professing to dislike 'the harsh voice of single-
minded pressure groups' quite as much as 'the honeyed tones of the developer'. The new chief executive, Roger Crofts, came fresh from the Scottish
Office, as did two of his senior directors.

Although the nature conservation responsibilities of SNH were similar to
its predecessor – new legislation had not changed the statutory instruments in Scotland, which were still SSSIs – the ground rules were different.
SNH's founding statute emphasised the magic word 'sustainable' for the
first time in British law, although exactly what was meant by the duty of
'having regard to the desirability of securing that anything done, whether
by SNH or any other person (*sic*) in relation to the natural heritage of
Scotland, is undertaken in a manner which is sustainable' – is open to
interpretation! It was plainly ridiculous to make sustainability a duty of a
minor government agency but not of the Government itself ('like giving a
wee boy a man's job'). SNH put on record its view that sustainable development in Scotland required serious changes in government policy and
the way public money was spent. But it, like English Nature, also espoused

Des Thompson, SNH's senior ornithologist, surveying Flow Country patterned bogs by
the Thurso River in Caithness. (Derek Ratcliffe)

a corporate ethos that sought consensus and partnership, which inevitably means doing things more slowly. Confrontation was the policy of the bad old days.

The second ground rule was accountability. To give at least the semblance of bringing SNH 'closer to its constituents', it was organised into four local boards, each with its own budget, work programme, and salaried board members, and responsible for three or more area 'teams'. Predictably enough, the regional boards proved expensive to run, sowed wasteful bureaucracy and duplication of effort, and set one local 'power base' against another. They were abandoned in 1997, and replaced by a new structure with 11 'areas' overseen by three 'Area Boards'. This was SNH's third administrative upheaval in five years.

Another significant change was what the former NCC's Scottish director Morton Boyd called 'the fall of science'. The minister in charge of environmental affairs at the Scottish Office was Sir Hector Monro (now Lord Monro of Langholm). He had served on the NCC's Council 'and had grown to dislike scientists' (Boyd 1999). The role of science must be advisory, he insisted, and should not be used as the basis of policy. Hence SNH's top scientist, Michael B. Usher, was not the 'Chief Scientist', as before, but the 'Chief Scientific Adviser', and he was eventually excluded from SNH's main management team. Nor were SNH's local boards particularly rich in scientific experience. The scientists sat on a separate research board under Professor George Dunnet, later named the Scientific Advisory Committee. It was rich in IQs but poor in influence, and, fed up with being repeatedly ignored, Dunnet resigned in 1995. As Boyd commented, the standing of scientists is not what it once was. Not only were they held

Humility? The NCC's scientific advisory committee dwarfed by the great beeches of the New Forest. (Derek Ratcliffe)

responsible for the disputes that had made the NCC unpopular in Scotland, scientists were also seen as an unacceptable 'élite'. The new approach had to be 'people-led'.

With the Scottish Office breathing down its neck, landowners asserting themselves and voluntary bodies inclined to be publicly critical, SNH was obliged to tiptoe over eggshells. Crofts kept in close touch with his minister and senior civil servants, and some saw SNH's new relationship with Government as one of servant and master. Rifkind's words, it seemed, were more to be honoured in the breach than the observance. When SNH tried to introduce notions of sustainability into transport policy, for instance, it was firmly put in its place by his successor, Ian Lang. The only thorns he would be prepared to tolerate, it seemed, were rubber ones.

All the same, SNH's reports give the impression of substantial progress in uncontroversial matters, with various initiatives carefully ticked off against Scottish Office targets. It has, for example, played a useful role in helping walkers and landowners to find common ground through an Access Forum. This has worked because landowners saw voluntary agreements on access as a way of staving off legislation, while the ramblers saw it as a means of 'trapping them into compromise on a matter of rights' (Smout 2000). The result was a grandly named 'Concordat on Access to Scotland's hills and mountains'. Though legislation is coming anyway, the talks have at least defused the situation by liberalising entrenched attitudes, and access is not now the contentious issue in Scotland that it became in England.

In terms of wildlife protection, SNH has kept a lower profile than the NCC, although it has experienced much the same problems. SNH's approach has been more tactful, and it has tried as far as possible to build bridges with bodies like the Crofter's Association, and with local communities. Local accountability was impressed upon it even more strongly by the new Scottish Parliament. In the early days, SNH inherited several outrageous claims for compensation by the owners of large SSSIs. It also had to cope with a statutory appeals system for SSSIs imposed on SNH by a group of landowners in the House of Lords led by Lord Pearson of Rannoch. Although in practice the appeals board was given little work to do, its existence tended to make the SNH cautious about notifying new SSSIs, and conservative about recommending Euro-sites. National Nature Reserves were also reviewed; those with weak agreements and no immediate prospect of stronger ones were struck off, or 'de-declared' (see Chapter 5). SNH was similarly cautious about acquiring land or helping others to acquire it. For example, SNH smiled benignly at the new owners of Glen Feshie, part of the Cairngorms National Nature Reserve, despite knowing nothing about them, and was not allowed to contribute so much as a penny towards the purchase price of Mar Lodge (only to its subsequent management). Like English Nature, it has stepped back from direct management into a more advisory role.

SNH are probably right that the future of Scotland's wildlife will benefit more from changing attitudes and shifting subsidies than from putting up barricades around special sites. While about 10 per cent of Scotland (and Wales) is SSSI, compared with 7 per cent in England, nearly three-quarters of the land is subject to the Common Agricultural Policy, while the equally profligate Common Fisheries Policy presides over Scottish inshore waters. Hence the Scottish Office's 1998 White Paper *People and Nature*, while voicing doubts about basing conservation policy on SSSIs, does at least contain a ray of hope by underlining the legitimate claims of 'the wider community' on the way land is managed; on what Smout has called 'the public nature of private property'. The forthcoming National Park at Loch Lomondside and The Trossachs may come to symbolise a new 'covenant' between land and people. SNH has also won plaudits for determinedly tackling wildlife crime, and for its leadership in trying to resolve the age-old conflict of raptors and game management. The Scottish Executive recently showed its appreciation of SNH, and the challenging nature of its work, by increasing its budget. It is difficult for outsiders to know to what extent SNH has helped to change hearts and minds in Scotland, but it can surely be given some of the credit.

Countryside Council for Wales (CCW)

CYNGOR CEFN GWLAD CYMRU
COUNTRYSIDE COUNCIL FOR WALES

Headquarters: Plas Penrhos, Ffordd Penrhos, Bangor, Gwynedd LL57 2LQ
Vision: under review (May 2001)

The Countryside Council for Wales was formed in 1991 by merging the Countryside Commission and the NCC within the Principality. Unlike Scottish Natural Heritage, 'CCW' had no custom-made legislation, just a ragbag of texts from Acts dating back to 1949. Unlike English Nature, it started with a serious staff imbalance. While over 100 staff from NCC took new jobs (or continued their old ones) in CCW, only four from the smaller Countryside Commission decided to stay on. And so CCW had to start with a recruitment drive. Having evolved in different ways, the NCC and the Commission were chalk and cheese, and welding them together was no easy task. The NCC had statutory powers, and enforced them. The Countryside Commission was more of a clap-happy, grant-aid body. Sir Derek Barber compared them with monks and gypsies, all right in their own way, but not natural partners.

    CCW was warned to be 'mindful of the culture and economy of rural Wales'. It would have to build on the Welsh NCC's relatively strong links with farmers and Welsh institutions. CCW inherited the NCC's headquarters at Bangor, and decided against a move to Cardiff. Apparently this was only because the minister responsible wanted the CCW and its job opportunities to lie in his own constituency, but to outsiders it seemed to signal

CCW's affiliation with the rural, Welsh-speaking heartland rather than the industrial south. Small, culturally homogeneous countries have advantages denied to larger ones. People know one another; there is a lot of cross-participation and a pervading sense of identity. It is important to 'belong', and to be seen to be 'people-centred'. CCW might have been straining a little too hard in describing its goal as '*a beautiful land washed by clean seas and streams, under a clear sky; supporting its full diversity of life, including our own, each species in its proper abundance, for the enjoyment of everybody and the contented work of its rural and sea-faring people*'. But behind this embarrassing guff there was an open-faced willingness to start afresh, and in a spirit of community.

CCW is much the smallest of the three country agencies, and began life with a relatively miserly budget of £14.5 million. With that it has to administer over 1,000 SSSIs covering about 10 per cent of the land surface of Wales, attend to all matters of rural access and carry out government policy on environment-sensitive farming. Its governing council was, like the others, well stuffed with farmers, businessmen and 'portfolio collectors', but scarcely anyone whom a conservationist would regard as a conservationist. Presumably CCW relied on their worldly wisdom more than their knowledge of the natural world. CCW's chairman for the first ten years, Michael Griffith, was a Welsh establishment figure with farming interests and, it is said, a gift for getting on with ministers of all hues and opinions. The present chairman is another prominent farmer, a former chairman of the NFU in Wales. CCW's first two chief executives both had a professional background in countryside planning rather than nature conservation, Ian Mercer in local government and National Parks, Paul Loveluck in the Welsh Office and the Welsh Tourist Board. Inevitably, therefore, it was the 'holistic' view of things that prevailed ('I work for the rural communities of Wales, not for wildlife,' was a phrase often heard on CCW corridors, perhaps to annoy the 'Victorian naturalists' from the former NCC). Senior posts were found for people with no background in nature conservation. People who ran processes were more highly valued than those who worked on the product. Some believed that core wildlife activities were being neglected at the expense of access work that overlapped with the remit of local authorities. Any blurring of functional boundaries held political dangers for a small, newly established body.

CCW went through much the same time-consuming reorganisations as its big sisters in Scotland and England. It organised its staff into Area Teams and Policy Groups, and delegated authority downwards while reserving all important decisions (and, it is said, many trivial ones also) to headquarters. Like English Nature, CCW was keener on mitigation than confrontation, especially where jobs were at stake. For example, it bent over backwards to accommodate the development of the 'Lucky Goldstar' electronics factory on part of the Gwent Levels SSSI. On the other hand a series of high-profile cases gave CCW a chance to make itself useful, such as the proposed orimulsion plant in Pembrokeshire, which it successfully opposed, and the wreck of the Sea Empress, from which it drew worthwhile

John Lloyd Jones, chairman of CCW. (CCW)

lessons. CCW's bilingual reports generally seem more down-to-earth and better written than the grammatically strained productions of English Nature and Scottish Natural Heritage, perhaps because they are concerned more with events and issues than with internal administration.

Among CCW's most distinctive policies are its championing of environment-friendly schemes such as *Coed Cymru*, introduced in 1985 to regenerate Wales' scattered natural woodlands, and its administration of *Tir Cymen* (now renamed *Tir Gofal*), Wales' integrated agri-environmental scheme. Judging by the desire of the Welsh Office, and later the Welsh Assembly, to take over *Tir Cymen*, it has been a success. Like SNH, CCW has also done its best to promote the Welsh countryside as 'a leisure resource', producing a stream of colourful publications, and devoting loving attention to matters like footpaths and signs. Some grumble that in its determined wooing of 'customers' and 'partners', CCW has been neglecting its statutory role of protecting wildlife. Possible signs of weakness are CCW's failure to publish comprehensive data on the condition of SSSIs (although it admits that most of the National Nature Reserves in its care are in unfavourable condition), and its slow progress on Biodiversity Action compared with its sister agencies, earning it a black mark in the review, *Biodiversity Counts*. It has had to struggle hard to retain its authority, and seems much less firmly entrenched in Welsh affairs than its English and Scottish sisters.

The relationship of CCW with the turbulent political climate of Wales in the 1990s is a story in itself, which I continue on p. 54.

Joint Nature Conservation Committee (JNCC)

Headquarters: Monkstone House, City Road, Peterborough PE1 1JY
Mission: it is not allowed to have one.

The JNCC is the forum through which the three country nature conservation agencies deliver their statutory responsibilities for

Great Britain as a whole, and internationally. These are primarily the drawing up of 'Euro-sites' for the Natura 2000 network (SPAs, SACs), the setting of common standards, and advising government on Great Britain-related nature conservation matters. Its committee, chaired by Sir Angus Stirling, formerly the National Trust's director, consists of three independent members, along with two representatives from each of the country agencies, and one each from the Countryside Agency and the 'Council for Nature Conservation and the Countryside' (CNCC) in Northern Ireland. The JNCC is based in Peterborough, with a small sub office in Aberdeen, specialising in seabirds and cetaceans. All members of its staff are assigned from one of the three country agencies. In 2000, it had 84 staff and a budget of £4,735,000. Among the Committee's projects were some grand-scale surveys inherited from the NCC, especially the Marine Nature Conservation Review, the Geological Conservation Review and the Seabirds at Sea project. JNCC also runs the National Biodiversity Network and publishes British Red Data Books, as well as a stream of scientific reports. Its most important task was co-ordinating the UK proposals for Special Areas of Conservation (SACs), based on submissions by the four country agencies (including Northern Ireland). Denied any real corporate identity, the JNCC is nonetheless the principal centre of scientific know-how in British nature conservation.

The JNCC has a problem: it lacks an independent budget and its own staff. Its annual grant has to be 'ring-fenced' from the three agencies, who, along with their control of the purse strings, also dominate its committee. Their influence has not been benign. From the start, the JNCC was seen as a refuge for reactionaries from the old NCC who refused to move with the times. Senior refugees from the NCC's scientific team quickly discovered how much they had lost influence. People with international reputations found themselves pitched into low status jobs, or dispensed with altogether once a Treasury review, brought at the request of English Nature, had scrapped half of the JNCC's senior posts and humiliatingly downgraded its director's post. The JNCC's first chairman, Sir Fred Holliday, a former NCC chairman, resigned after five months, complaining that he had been kept in the dark over the Scottish SSSI appeals pro-

Sir Angus Stirling, chairman of the JNCC. (JNCC)

cedure. In 1996, its new chairman, Lord Selborne, traded a leaner structure – downsizing its staff from 104 to 66 – for more autonomy within its
core responsibilities. Even so, the JNCC was visibly struggling against the
devolution tide. The four country agencies often failed to reach a consensus view, or indeed take much interest in matters of UK concern. As this
book went to press, a government review body has recommended that the
JNCC became a separate body within the newly organised government
department, DEFRA (the Department of the Environment, Food and
Rural Affairs).

The whip hand: the agencies and their budgets

As the smallest of the country agencies, the Countryside Council for Wales
might have expected a struggle to make its mark. It also had the bad luck
to receive a right-wing ideologue as Secretary of State in the person of John
Redwood. Towards the end of 1994, Redwood took a hard look at the role
of CCW. It is said that he was outraged to notice that a third of CCW's budget went on staff salaries. In fact this was normal for a nature conservation
agency, or, indeed, any government agency, but others had been cleverer
at disguising it. As far as Redwood was concerned, CCW was both over-
manned and overstretched. It should be 'encouraged to concentrate on its
core functions'. In May 1995, the Welsh Office produced an 'Action Plan
for CCW' which proposed to reduce its running costs over the next two
years by handing over supposedly peripheral activities, such as the funding
of Country Parks, to local authorities. It also proposed to 'privatise' some
National Nature Reserves and hand over CCW's flagship *Tir Cymen* scheme
to the Welsh Office. Furthermore, CCW was ordered to cut down its travelling and stay in more, with the help of computer technology. To encourage it in all these things, CCW's budget was cut by a third.

Redwood's attack was badly received, not just in nature conservation circles but also, much to his surprise, by parts of the Welsh establishment and
the media. This was linked to a related matter, Redwood's refusal to implement new, more environment-friendly planning guidelines, thus creating
an undesirable divergence of approach on planning matters between
England and Wales. John Redwood failed to find much empathy with the
Welsh; as John Major expressed it in his memoirs, Redwood did not take
to the Welsh people, 'nor they to him'.

Ironically, the Redwood fracas helped to put CCW on the map and
sparked a good deal of favourable publicity for its work. When Redwood
resigned in order to challenge John Major as Conservative Party leader,
William Hague, his more politically astute successor, demonstrated a
change of tack by visiting some of CCW's offices, and talking to staff in a
friendly spirit. There is a story that, on his visit to Snowdon, the fit young
Hague simply tore up the mountain, leaving CCW's warden, a heavy smok-
er, trailing far behind. CCW was able to stave off corporate starvation by
negotiating an EU Life fund to supplement its budget, thus pioneering a
rich and, until then, surprisingly neglected alternative source of income.
An ostentatious display of good housekeeping was rewarded in 1996 by a
20 per cent increase in grant-in-aid, bringing things more or less back to

normal. But that was not the end of CCW's financial tribulations. Its funding body passed from the Welsh Office to the Welsh Assembly in 1999. The architect of the Welsh Assembly, Ron Davies, had been a strong supporter of wildlife conservation in Wales, and his 'moment of madness' in Brixton was also a misfortune for CCW. Its Corporate Plan was rejected by the Assembly with the warning that the agency might have to muddle along for a while without a pay rise. Other warning signs were First Secretary Alun Michael's dismissal of CCW's request for the Assembly to debate its new 'vision', *A Living Environment for Wales*. There was talk about restructuring environmental activity in Wales, for example, by merging CCW with the Environment Agency, and having another look at the possibility of hiving off some of its functions to local authorities.

English Nature nurtured more constructive relations with its paymasters. In 1992 it was given an extra million pounds for restoring peatlands and to speed up the designation of EU Special Protection Areas for birds. The National Audit Office in 1994, and the Commons Public Accounts Committee in 1995, made critical comments about some aspects of its business, but on the whole supported EN's strategic approach to its tasks and wholehearted use of business language. EN endured a lean year in 1996, but fought off a further cut the year after. The incoming Labour Government's Comprehensive Spending Review increased EN's grant-in-aid by 16 per cent to £44.6 million, followed by another generous increase in 1998, coinciding with the appointment of Barbara Young as chairman.

Scottish Natural Heritage has had to tread carefully. The generous settlement it received in 1992 was tempered by an awareness that its every move was being shadowed by the Scottish Office, which expected SNH to be a 'people-friendly' body and avoid the controversies of the recent past. That it was as vulnerable as CCW to hostile trimming measures became clear in 1995, when the Scottish Secretary Ian Lang decided to carry out the dreaded 'high level review'. He was purportedly concerned about SNH's involvement in wider issues like agriculture and transport, and looked down his nose at the £800,000 it had spent fighting the proposed super-quarry at Lingerbay on Harris. His successor, Michael Forsyth, was similarly put out when he learned that SNH had spent £1.8 million buying out the peat-cutting rights at Flanders Moss, which, to make matters worse, lay in his own constituency (in his view, that sort of public money should be spent on schools and hospitals). Like Redwood, Lang wanted SNH to concentrate on its core activities and to trim what he saw as peripheral matters, such as public access to the countryside. But even if it had, the savings would have been insignificant. At the end of 1996, in which its budget had been cut by 10 per cent, SNH published its answer in *Natural Priorities*. This was a fairly defiant restatement of SNH's responsibilities over a broad range of heritage issues, and even hinted that it could do with a bit more co-operation from the all-powerful Scottish Office's environment, agriculture and fisheries departments. But the net was tightening. In 1998, chief executive Roger Crofts estimated that SNH's spending power had fallen by nearly a third since its establishment in 1992.

The publication of the Scottish Executive's 2001 policy statement, *The Nature of Scotland*, made it clear that Government intends to involve itself directly in the detail as well as the broad thrust of nature conservation north of the border. Increasingly, SNH and its sisters in England and Wales are becoming processing instruments, responsible for implementing legislation and as a conduit for government grants, but of diminishing importance as policy makers. By 2001, the dynamic of nature conservation was definitely moving from the state to the voluntary sector. In all the major recent events in nature conservation – biodiversity, the 'CROW' bill, SAC designation, devolution – the agencies have been either bystanders or supine instruments of government policy. This, some would say, is what comes of replacing scientists with bureaucrats. All the same, I think the agencies could win back some of the respect and influence that their predecessor, the NCC, enjoyed, if they showed more leadership, concentrated on outcomes rather than outputs, and spoke up fearlessly for the natural world. Or maybe I am just misreading the runes, and that it is the fate of the nature conservation world to complete the circle, back to the charities and pressure groups that nurtured it.

3

The Voluntary Army

This chapter is about the private sector of nature conservation, the voluntary nature conservation bodies – who they are and what they do. Perhaps few countries in the world have as many charities, trusts and associations active in the same broad field as Britain. Wildlife and Countryside Link, the forum where many of them meet and share ideas, serves 34 national bodies and many more local ones, varying from special-interest trusts (butterflies, reptiles, sharks) to international pressure groups (Greenpeace, Friends of the Earth) and world-famous charities (WWF, RSPB, National Trust). Every county in England and Wales has its own wildlife trust (Scotland and some of the smaller counties have federated trusts). Learned societies with small but enthusiastic memberships exist for practically every animal, plant or mineral that occurs in Britain: for example, water-beetles (the Balfour-Brown Club), microscopy (Quekett Microscopical Club), seaweeds (British Phycological Society) and molluscs (Conchological Society of Britain and Ireland). Hedgehogs, sharks and bats have their own societies. There is even a group busily recording the distribution of nematode worms. Some special-interest bodies have recently become active in nature conservation; for example, the venerable British Mycological Society (fungi) now has a part-time conservation officer, responsible for biodiversity projects and compiling a red data list.

In their glorious diversity, ranging from the National Trust to small groups that meet once a year to dine and reminisce, finding an adequate name to cover everyone is problematical. Government refers to them with statist disdain as non-governmental organisations (NGOs). Some prefer the term voluntary bodies, but this too, seems somewhat vague and reductionist (what is the alternative to a voluntary body – a compulsory body?). Besides, a voluntary body such as the RSPB has a membership larger than any political party and, if it is voluntary, it is every bit as professional as its official counterparts. Voluntary bodies now campaign successfully for new legislation and assist the Government in its statutory responsibilities, such as maintaining biodiversity. Perhaps the fact that they defy easy labelling says much about nature conservation in practice. Conservation is not, though it is sometimes portrayed as such, a homogeneous mass movement, working to a common programme. Although the 'vol. bods' do often pool their resources, as in the campaign to preserve peatlands, they have separate aims, and different sorts of members, ranging from committed activists to folk who simply enjoy wandering in pleasant countryside. They are united by a common interest in nature conservation, but that does not make them the same.

The influence of the voluntary bodies in the 1990s owed nearly every-thing to their mass memberships – no modern political party can afford to ignore a body with a million members. Their social base has obviously broadened. Nature conservation used to be caricatured as a concern of the urban middle classes, and there is still some truth in that. However, a mem-bership survey of the RSPB in 1982 suggested that a large proportion were in technical and clerical occupations, while 14 per cent were unskilled manual workers (Smout 2000). Today, perhaps one in ten people are mem-bers of an environmental pressure group of some sort. Many, of course, are members of more than one. Young people tend to gravitate towards envi-ronmental campaigning bodies, such as Greenpeace, where there are opportunities to join in the action. They think they can change the world. County trusts are traditionally the home base of older, reasonably well-off people, interested in wildlife and worried about the effect of developments on the local countryside. They think we are doing well if we manage to save just the best bits of our backyard.

The phenomenal growth of the voluntary bodies is very recent. In 1960, the RSPB had only 10,000 members, not many more than it had in 1945. Membership increased in the 1960s and 1970s, but really took off in the 1980s, when events propelled nature conservation from the hobby of a few to a mainstream issue. With power has come controversy. The assertiveness of some pressure groups has exhumed the old accusation of urban-based sentimentalists imposing their will on genuine countrymen; it is the *raison d'être* of the Countryside Alliance. There are also contrasts between places where conservation bodies are strong and others where they are weak. Donald MacKay (1995) observed that 'the more south-east England become agitated over conservation issues in Scotland, the stronger became the Scottish anti-conservation lobby, and the harder it became to recruit to the Scottish conservation cause'. It was not that the Scots man or woman was less keen on nature, but that they were Scots first, and wanted to do things in their own way. They now have their chance. Paradoxically, all this growth has not led to more field study or better-informed naturalists. Although birdwatching is more popular than ever, the expert amateur nat-uralist, and especially the all-rounder, is becoming an endangered species. Specialists in less popular groups belong to a small and ageing population. Love of wildlife is expressed differently in 2000 than it was in 1900. It has become less 'hands-on' (naturalists used to *collect* their subject), less based on knowledge-seeking, more of a personal lifestyle choice, more of a fash-ionable cause and less of a hobby.

For ease of reference, in what follows, I treat the main voluntary bodies one by one. For reasons of space I omit bodies whose interests are not pri-marily in the conservation of wildlife, such as the Ramblers Association and the Council for the Preservation of Rural England (CPRE), natural allies though they often are. Similarly, I have to exclude learned societies and clubs, such as the Ray Society, whose main interest lies in promoting field study and the advancement of science. Even so the number of players, each with a different focus or stance, is considerable, and perhaps baffling to some. Possibly if one started again with a clean slate, there would be far fewer 'vol. bods'. But today's 'conservationists' have a large range to choose from and can pick and mix. In this account of their background and activities, I emphasise the role of the county wildlife trusts, the one body that every naturalist should join, since they cater for what should matter most to most of us – the flora and fauna on our doorsteps.

The Big Three

Royal Society for the Protection of Birds (RSPB)

Britain's (and Europe's) largest wildlife and conservation society was formed in 1891 and acquired its Royal Charter in 1904. However, the RSPB's mass popularity and power are relatively recent. It broke the 100,000 tape only in 1972, but in the 1980s its growth was meteoric, reaching half a million members in 1989 and one million by 1997. The RSPB '*works for a healthy environment rich in birds and wildlife*'. It has good things to offer to its million members: free access to most of its 140 nature reserves and an excellent quarterly magazine, *Birds*. The RSPB has a grand UK office at Sandy Lodge, Beds, and separate headquarters in England, Scotland, Wales and Northern Ireland, as well as nine regional offices. It employs around 1,000 full, part-time and contract staff; its network of nature reserves throughout the UK covers some 111,500 hectares and receives over a million visitors a year. With in-house science expertise, RSPB investigates the impact of human activity on birds, as well as the needs of threatened species both at home and overseas. It has acquired matchless skill in presenting the conservation case, and in detecting and admonishing failures of policy. It has also successfully mounted legal challenges over conservation designations, and deals with an average of 350 planning cases per year. With birdwatching a popular hobby on both the Government and Opposition front benches, British birds receive far more sympathetic attention than any other forms of wildlife. The RSPB has been criticised in some quarters as exercising too much power; for example, in buying up a lot of land in Orkney or the Hebrides, where it is seen by some as an inappropriate outside influence. Gamekeepers have also fallen out with RSPB over raptors.

From the start, RSPB has been active in education, with special clubs for children (the Young Ornithologists' Club, recently renamed 'Wildlife Explorers', magazine *Bird Life*) and teenagers ('RSPB Phoenix', magazine *Wingboat*). It claims to have helped make the national curriculum more wildlife-conscious (though it would help to have more teachers who know their natural history). Internationally, RSPB represents the UK on Birdlife International, and contributes to bird protection overseas (for example, the publication *Important Bird Areas in Europe* was largely RSPB-funded). The RSPB is now rich: income in 2000 was £38 million, mainly from membership subscriptions and legacies, supplemented by grants, fund raising appeals and sales of goods. Today it often works in partnership with other conservation charities, and also with farmers and land owners. Increasingly RSPB champions wildlife more generally, as well as their habitats. Its slogan: '*for birds for people for ever*'. You can read a sympathetic account of the RSPB's eventful history in *For the Love of Birds*, written to celebrate its centenary (Samstag 1989). For hostility, try *Isles of the West* by Ian Mitchell (1999).

UK Headquarters: The Lodge, Sandy, Bedfordshire SG19 2DL.
Chief Executive: Graham Wynne.

The county wildlife trusts

Membership of the wildlife trust of one's home county is the logical first step for anyone interested in natural history. Nearly every county in England and Wales has a wildlife trust, many of them based on older natural history societies. Most of them were formed in the 1950s and 60s. Some, such as the trusts of North Wales or 'Bucks, Berks and Oxon', are federated, and Scotland has a federal system with different regions under a unified Scottish Wildlife Trust. The purpose of the trusts is to acquire land as nature reserves and encourage interest in wildlife. The founders of the Kent Naturalists Trust (now the Kent Wildlife Trust) spoke for many others who 'saw the speed of change of farming practice and urbanisation as a severe threat to our lovely county'.

The first county to receive its own wildlife trust (as opposed to a natural history society or field club) was Norfolk. The Norfolk Naturalists Trust was established by Dr Sidney Long in 1926 as a 'special non-profit paying company to hold and manage nature reserves'. Behind its formation lay a dissatisfaction with the National Trust, which came to the boil when the latter refused to take on Cley Marshes on the grounds that it was *only* of interest to naturalists. Norfolk had acquired several nature reserves by the 1950s, but although F.W. Oliver's prediction that one day every English county would have its own county trust proved right, it took a long time. It was not until 1946 that the Yorkshire Naturalists Trust was founded, on the Norfolk model, and again with the immediate purpose of looking after a nature reserve, Askham Bog. The Lincolnshire Naturalists Trust followed two years later, largely through the efforts of A.E. Smith, later Secretary of the Society for the Promotion of Nature Reserves

(SPNR). With the support of the Nature Conservancy, many more county trusts sprang up across England in the 1950s – Leicester and Cambridgeshire in 1956, the West Midlands and Kent in 1958, Surrey and Bucks, Berks and Oxon ('BBONT') in 1959, Essex and Hampshire in 1960, Cornwall and Wiltshire in 1962. The first Welsh trust, the West Wales Naturalists Trust, was formed in 1956, and the Scottish Wildlife Trust, covering the whole of Scotland, followed in 1964. Many of them emerged from the embers of an earlier natural history society, often through the efforts of a few dedicated local naturalists. For example, the Somerset Trust for Nature Conservation was formed in 1964 by members of the venerable Somerset Archaeological and Natural History Society, led by Ernest Neal and Peter Tolson. The Cornwall Naturalists Trust took over and much extended the activities of the Cornwall Bird Watching and Preservation Society. Many county trusts have changed their names (and acronyms) two or three times since. Originally they were naturalists trusts. Later some became trusts for nature conservation. Now they are nearly all wildlife trusts – and one rather dreads their possible future reincarnation as sustainability or biodiversity trusts.

Most trusts acquired a full-time conservation officer as soon as they were up and running, with the help of 'pump-priming' grants from the NCC and other bodies. During the 1980s, NCC grants helped the trusts to become more professional and to acquire a small corps of promotional, educational and marketing staff, as well as computer systems. In the 1990s, some trust nature reserves profited from English Nature's Reserve Enhancement Scheme, and still more by the Heritage Lottery Fund which, by 2000, had awarded a total of £50 million to buy land as nature reserves or fund capital improvements. A further £6 million worth of projects came from the Landfill Tax Credit Scheme. At the same time, increased public interest in nature conservation resulted in big increases in membership. For example, the medium-sized Somerset Trust, with 9,000 members, now has an annual income just over £1 million and assets of £3 million, together with about 30 full-time staff housed in beautiful surroundings at Fyne Court. Between them the county wildlife trusts now manage some 2,300 nature reserves, ranging in size from under a hectare to several square kilometres, and extending over nearly 70,000 hectares.

The activities of the county trusts have much in common, but they always reflect the nature of their constituencies. The Welsh Trusts have become adept at running seabird islands and restoring reed beds; the Scottish Wildlife Trust specialises in restoring peat bogs. Among their core activities are acquiring and managing nature reserves and campaigning against harmful developments. More recently, their work has become more inclusive, embracing ideas of sustainability enshrined in Agenda 21 and interpreting them on a local scale (see p. 78), or helping farmers to sell environment-friendly products, as in the Devon Wildlife Trust's 'Green Gateway' scheme. The nature of the membership is also changing. Twenty years ago, most trust members were keen naturalists. Today, many join out of a broader concern for the environment (that is, for our own quality of life), and often include whole families. Trust activities reflect such changes, with a greater emphasis these days on communities, education, and participatory activities.

The Wildlife Trusts partnership, formerly the Royal Society for Nature Conservation (RSNC), acts as a spokesman and administrative centre for the disparate county wildlife trusts. It had its distant origins in the SPNR, which was set up in 1912 for the purposes of 'securing' nature reserves and 'to encourage the love of Nature'. This Society struggled on for years on a shoestring budget without achieving very much (though its surveys are a valuable retrospective source for the state of wildlife in the first half of the twentieth century, see Rothschild & Marren 1997). It did, however, contribute organisation and expertise for the Nature Reserves Investigation Committee in 1942, which produced the original 'shopping list' for the subsequent selection of National Nature Reserves and other important sites. In the 1950s the SPNR assisted some of the fledgling county trusts with modest grants to set up their first nature reserves, along with advice on how to look after them. In 1957 the county trusts proposed that the SPNR should act as a co-ordinating body for their activities, in effect as their 'federal centre'. In the early 1970s a proposal to combine forces with the RSPB was briefly considered, but rejected, largely because the pair were mismatched: the RSPB was already too big. In 1976, the SPNR was granted a royal charter, becoming the RSPNR for a short period, before changing its name yet again to the Royal Society for Nature Conservation (RSNC) in 1981. In 1991, the RSNC joined with the 46 county trusts and 50 urban wildlife groups to form the Wildlife Trusts partnership, which now has a combined membership of nearly 300,000. All receive the wildlife trusts' quarterly magazine, *Natural World*, along with a copy of their local trust's magazine. There is also a junior arm, Wildlife Watch, founded in 1977 with young naturalists in mind.

The Wildlife Trusts partnership provides the local trusts with a common identity, promotes their common interests and campaigns on their behalf. On occasion it has gone too far down the centralising path, for example, when it tried to impose a common 'badger' logo (known as the raccoon by disparagers) on all the trusts. But in general the division of responsibility seems to work well enough, with each partner concentrating on its constituency strengths, leaving the umbrella body to organise training weekends, launch national appeals (for example '*Tomorrow Is Too Late*') and making its voice heard in the corridors of power. It has long had its head office somewhere in Lincolnshire for reasons lost in the mists of time, but the Trusts' director's office is in London. Its logo: the ubiquitous badger. Vision: '*the achievement of a United Kingdom that is richer in wildlife and managed on sustainable principles*'.

I cover the activities of a particular wildlife trust on pp. 75–9.

Head Office: The Kiln, Waterside, Mather Road, Newark NG24 1WT.
Wildlife Trusts partnership Director general: Simon Lyster.

The National Trust

THE NATIONAL TRUST

At the turn of the millennium, the National Trust's membership was just short of a stupendous three million. The public loves a bargain, and for the modest membership fee the whole of the Trust's vast estate is open to them. Moreover, to many, the Trust embodies all that is best in the countryside: beautiful scenery, benevolent stewardship and a good day out. However, until recently the National Trust was only on the margins of the nature conservation world. It is not a campaigning body, and much of its work is centred on maintaining stately homes and gardens. Its importance lies in the nature conservation work carried out on its own properties. The Trust is emerging as an important player mainly because, in common with other heritage bodies, it takes a greater interest in wildlife than in the past.

There are two separate National Trusts, one for England and Wales, the other for Scotland. The former, older Trust had its origins in the concern over the enclosure of commons in the nineteenth century. The desire of a few Victorian philanthropists to preserve 'all that still remained open, for the health and recreation of the people' led to the formation of the Commons Preservation Society, the first successful conservation pressure group in history. In 1885, the Society's solicitor, Robert (later Sir Robert) Hunter, proposed a 'Land Company' to buy and accept gifts of heritage land and buildings for the benefit of the nation. In 1893, joined by Canon Hardwicke Rawnsley's Lake District Defence Society, this became known as the 'National Trust for Places of Historic Interest or Natural Beauty', or National Trust for short. Its constitution was based on that of a similar American body founded two years previously. In 1906, Hunter drafted a private Bill that made the Trust a statutory body, and gave it the right to make bylaws and to declare its properties inalienable. This meant they could not be sold or taken away without the Trust's consent: a National Trust property is the Trust's for keeps. A separate National Trust for Scotland (see below) was established in 1931, and given similar powers to its sister body. A full account of the National Trust was published in 1995 (Newby 1995).

Although the National Trust acquired many places 'of special interest to the naturalist' in its early days, such as Wicken Fen, Cheddar Gorge and Box Hill, its management of them was for many years scarcely different to any other rural estate; modern farming and forestry methods that damaged wildlife often went through on the nod. Management of the Trust's de facto nature reserves, such as Wicken Fen or the tiny Ruskin Reserve near Oxford, was generally overseen by a keen but amateurish outside body. They tended to turn into thickets. The Trust's outlook began to change in the 1960s after it launched Enterprise Neptune to save the coastline from development, having found that a full third of our coast had been 'irretrievably spoiled'. By 1995, some 885 kilometres of attractive coast, much of it in south-west England, had been saved in this way.

Since the 1980s, the National Trust has developed in-house ecological expertise, and belatedly become a mainstream conservation body, managing its properties, especially those designated SSSIs, in broad sympathy with wildlife aims. Some of the basic maintenance is done by Trust volunteers in 'Acorn Workcamps'. Although public access remains a prime aim, some Trust properties are now in effect nature reserves, with the advantage of often being large, especially when integrated with other natural heritage sites. By its centenary year, 1993, the National Trust owned 240,000 hectares of countryside, visited by up to 11 million people every year. It owns large portions of Exmoor, The Lizard and the Lake District, and about 14,000 hectares of ancient woodland and parkland. Like the RSPB, its membership climbed steeply in the 1970s, breaching the million-member tape by 1981. The Trust is now Britain's largest registered charity, larger than any trades union or any political party. Members receive the annual Trust Handbook of properties, as well as three mailings a year of *National Trust Magazine*, and free admission to Trust properties (including those belonging to the National Trust for Scotland). It has 16 regional offices in England, Wales and Northern Ireland, as well as a head office in London.

Head Office: 36 Queen Anne's Gate, London SW1H 9AH [at the time of writing, the National Trust was set to move from its elegant Georgian house in SW1 to a graceless office block in Swindon, to the dismay of most of its staff.]

Director-general: Fiona Reynolds

National Trust for Scotland (NTS)

The National Trust for Scotland

The National Trust's sister body in Scotland was founded in 1931, and was made a statutory body with similar powers, including inalienability rights, seven years later. It has the same aim of preserving lands and property of historic interest or natural beauty, 'including the preservation (so far as is practicable) of their natural aspect and features, and animals and plant life'. The Trust acquired its first property, 600 hectares of moorland and cliff on the island of Mull, in 1932. It is now Scotland's second largest private landowner with nearly 73,000 hectares or about 1 per cent of rural Scotland in its care, including 400 kilometres of coastline. About half of this area consists of designated SSSIs, among them the isles of St Kilda, Fair Isle and Canna, and Highland estates such as Ben Lawers, Ben Lomond, Torridon and Glencoe. Perhaps its most important property is Mar Lodge estate in the Cairngorms, acquired in 1995, which is being managed as a kind of large-scale experiment in woodland regeneration and sustainable land use (pp. 240–41). Like the National Trust, the NTS was for many years more interested in access than habitat management; for example, it had no perma-

Afternoon sunshine sparkles the native pines of Derry Wood in Mar Lodge estate, now owned by the National Trust for Scotland. (Peter Wakely/SNH)

nent presence at Ben Lawers until 1972 and, apart from footpath maintenance, did no management to speak of until the 1990s. Though, with 240,000 members, relatively modest in size compared with the National Trust, the NTS is a mainstream and increasingly important partner in nature conservation in Scotland, all the more so since it is an exclusively Scottish body. It has four regional offices with a headquarters – a classic Georgian mansion – in Edinburgh.

Head Office: 5 Charlotte Square, Edinburgh EH2 4DU.
Chairman: Professor Roger J. Wheater.

International pressure groups

WWF-UK

WWF currently stands for the World Wide Fund for Nature. Until 1986 it was known (more memorably) as the World Wildlife Fund, 'the world's largest independent conservation organisation', with offices in 52 countries and some five million supporters worldwide. WWF was founded by Peter Scott and others in 1961, and is registered as a charity in Switzerland. Its mission: '*to stop the degradation of the planet's natural environment, and to*

Taking action for a living planet

build a future in which humans live in harmony with nature'. The UK branch, WWF's first national organisation, has funded over 3,000 conservation projects since 1961 (but especially since 1990), and itself campaigns to save endangered species and improve legal protection for wildlife. In the 1990s it produced a succession of valuable reports on the marine environment, wild salmon, translocations, SSSIs and other topics from a more independent viewpoint than one expects nowadays from government bodies. In a sense, it has taken over as the lead body reporting on the health of Britain's natural environment and the effectiveness of conservation measures. Among its most important contributions has been WWF's persistent prodding of the UK government over the EU Habitats Directive, which eventually led to a large increase in proposed SACs (Special Areas for Conservation) for the Natura 2000 network (see Chapter 4). All of WWF's work is supposed to have a global relevance. WWF-UK's work is currently organised into three programmes: 'Living Seas', 'Future Landscapes' (countryside, forest and fresh water) and 'Business and Consumption' ('our lifestyles and their impact on nature'). It also works with others overseas to promote sustainable development in ecologically rich parts of the world and good environmental behaviour by businesses. The organisation is funded mainly by voluntary donations (90 per cent), with the rest from state institutions. The UK branch has some 257,000 members and supporters, and 200 volunteer groups about the country. Its youth section is called 'Go Wild'. It has offices in Scotland, Wales and Northern Ireland with a headquarters at Godalming in Surrey.

Its logo: the famous panda, designed by Peter Scott. Its slogan: '*Taking action for a living planet*'.

Address: Panda House, Weyside Park, Godalming, Surrey GU7 1XR.
Chief Executive: Robert Napier

Friends of the Earth

FoE acts as a radical environmental ginger group, pressing for more environment-friendly policies, both at home and worldwide. It is careful to avoid alignment with any political party or to accept commercial sponsorship, and most of its funding comes from the membership. FoE is particularly effective at 'media management' and at shaming commercial interests into adopting more environmentally friendly policies. Founded in America, a British branch took root shortly afterwards in 1971. Its first newsworthy action was the dumping of thousands of non-returnable bottles on the doorstep of Schweppes, the soft drink manufacturers. On wildlife matters, FoE has taken the lead on major issues, such as the protests over the Newbury bypass, on peat products and GM crops, dumping in the North Sea and pressing for stronger measures to protect SSSIs.

FoE are a streetwise organisation with a youngish membership, and its language is characteristically urgent and emotional ('Our planet faces terrible dangers', 'We can't allow environmental vandals to lay waste the earth' etc.). By persistently hammering away at an issue in an outraged tone, whilst also seeming well informed, FoE builds up a momentum for change. Its quarter of a million UK members receive a highly professional quarterly magazine, *Earth Matters*, and plenty of appeals, stickers and campaign literature. Its slogan: '*for the planet* for people'.

Address: 26–28 Underwood Street, London N1 7JQ.
Executive director: Charles Secrett.

Greenpeace

The most headline-making of all respectable environmental organisations was founded in America in 1971. The public first heard about Greenpeace when a group of activists sailed into an atomic testing zone in a battered hire-boat. A UK branch was formed the same year. Greenpeace is an international environment-protection body, funded by individual donations. It specialises in non-violent, direct action: front page pictures of activists being hosed from whaling ships, or waving banners at the top of chimneys, or on derelict oil platforms, alerts public opinion and raises awareness of an issue. Famously, its ship, the Rainbow Warrior, was blown up in Auckland harbour by French agents in 1985. Behind the headlines lie quieter activities: producing reports, lobbying governments, talking to businesses and even conducting research. Greenpeace concentrates on international campaigns, such as nuclear test bans, or the banning of drift nets or mining in the Antarctic. Among recent activities that impinge on nature conservation in Britain are campaigns for renewable energy and against GM crops. It exploits 'consumer power' by dissuading companies from using products from ancient forests and peatlands. Greenpeace has 176,000 supporters in the UK and a claimed 2.5 million worldwide. Its slogan: '*Wanted. One person to change the world*'.

UK Office: Canonbury Villas, London N1 2PN
Executive director: Peter Melchett (until December 2001)

Lord Peter Melchett, former chair of Wildlife Link and Executive Director of Greenpeace until 2001.
(Greenpeace/Davison)

The link body

Wildlife and Countryside Link (and predecessors)

Wildlife and Countryside Link (originally Wildlife Link) is an umbrella body representing 34 voluntary bodies in the UK with a total of six million members. It is funded by the member organisations, which also take turns to chair its meetings, plus donations from WWF-UK, the Department of the Environment, Transport and the Regions (DETR), English Nature and the Countryside Agency. It functions through various working groups and 'task forces', as well as 'one-off initiatives' covering a wide range of environmental issues at home and abroad, including rural development, trading in wildlife, land-use planning and the marine environment. Wildlife Link has played an important co-ordinating role in shaping current wildlife protection policies, enabling the voluntary bodies to pool their resources and experience and present a common agenda. It has a small secretariat based in London. In keeping with the spirit of devolution, there are now separate Wildlife and Countryside Links in Scotland, Wales and Northern Ireland.

The need for an umbrella body to represent the proliferating voluntary societies was appreciated as early as 1958, when the Council for Nature was formed as a voice for some 450 societies and local institutions, ranging from specialist societies to local museums and local field clubs. Headed by the glitterati of the 1960s conservation world, it helped to establish nature conservation in hearts and minds with events like the two National Nature Weeks and the three Countryside in 1970 conferences. It also helped to set up the Conservation Corps, later to become the BTCV (p. 70), while the Council's Youth Committee, under Bruce Campbell, did its best to 'make people of all ages conscious of their responsibility for the natural environment' (Stamp 1969). Its publications were a monthly broadsheet, *Habitat*, and a twice-yearly *News for Naturalists*.

Despite its influence, the Council for Nature was always short of money. By the mid-1970s, it was ailing badly, and four years later had ceased to function. Its publication *Habitat* was continued by the Council for Environmental Conservation (CoEnCo, now the Environment Council) while its function as an umbrella body was taken on by the newly founded Wildlife Link, then a committee of CoEnCo under Lord (Peter) Melchett. Wildlife Link scored an early success with the Wildlife and Countryside Act (see next chapter). In 1993 the name was changed to Wildlife and Countryside Link to emphasise its wider remit.

Secretariat: 89 Albert Embankment, London SE1 7TP.
Chair: Tony Burton (2001).

Members of Wildlife and Countryside Link

Bat Conservation Trust
British Association of Nature Conservationists
British Ecological Society
British Mountaineering Council
British Trust for Conservation Volunteers
Butterfly Conservation
Council for British Archaeology
Council for National Parks
Council for the Protection of Rural England (CPRE)
Earthkind
Environmental Investigation Agency
Friends of the Earth
Greenpeace
Herpetological Conservation Trust
International Fund for Animal Welfare
Mammal Society
Marine Conservation Society
National Trust
Open Spaces Society
Plantlife
Ramblers' Association
Royal Society for the Prevention of Cruelty to Animals (RSPCA)
Royal Society for the Protection of Birds (RSPB)
The Shark Trust
Universities Federation for Animal Welfare
Whale & Dolphin ConservationSociety
The Wildfowl & Wetlands Trust
The wildlife trusts
Woodland Trust
World Conservation Monitoring Centre
Worldwide Fund for Nature – UK
Young People's Trust for the Environment& Nature Conservation
Youth Hostels Association (England & Wales)
Zoological Society of London

The special interest groups

The British Association of Nature Conservationists (BANC)

BANC was founded in 1979, and acts mainly as a forum for practising conservationists and planners through its influential journal, *Ecos*. Something of a trade journal, *Ecos* usually contains short articles on a wide range of conservation-related subjects, as well as news and reviews. BANC also holds conferences on particular topics, and publishes pamphlets on issues ranging from conservation ethics to feminism.

The British Trust for Conservation Volunteers (BTCV)

Established as the Conservation Corps in 1959, the BTCV organises practical tasks for people who wish 'to roll up their sleeves and get involved'. A sister organisation was formed in Scotland in 1984. It runs some 200 courses each year for up to 130,000 volunteers on habitat management, such as footpath maintenance, fencing, hedge laying and dry-stone walling, along with wildlife gardening and developing leadership skills. It also works in partnership with local authorities and with government schemes such as the Millennium Volunteers. BTCV organises weekend residential projects and 'Natural Break' working holidays; for example, 29,000 volunteers assisted the National Trust to the tune of 1.7 million hours in 1994–95, 'the equivalent of one thousand full-time staff'. Among its publications is *The Urban Handbook*, a guide to community environmental work. Its hands-on, open air, communal approach appeals particularly to the young. Its quarterly newsletter is *Greenwork*. Mission: '*Our vision is of a world where people value their environment and take practical action to improve it.*'

Head office: 36 St Mary's Street, Wallingford, Oxfordshire OX10 0EU.

British Trust for Ornithology (BTO)

The BTO was established in 1933 as a research and advisory body on wild birds. Its main task is the long-term study and monitoring of British bird populations and their relationship with the environment. From just one full-time administrator in the 1960s, it has grown into a leading scientific institution with a staff of over 50 and an annual income of nearly £2 million, mainly from funds and appeals. The Trust's work is a fusion of 'amateur enthusiasm and professional dedication', its members acting as a skilled but unpaid workforce. Among its many schemes are the long-running Common Bird Census and its replacement, the Breeding Bird Survey, which it runs jointly with the RSPB and JNCC. Other projects include a Nest Record Scheme, the Seabird Monitoring Programme, special surveys of wetland, grassland and garden birds, the census of special species such as skylark and nightingale, and the administration of the National Bird Ringing Scheme. The Trust also helps organise special events such as the recent Norfolk Birdwatching Festival, and contributes to bird study internationally. It publishes the world-renowned journal *Bird Study*, currently reaching its 48th volume, as well as a bimonthly newsletter, *BTO News*, and a quarterly magazine, *Bird Table*. Membership: 11,490. Mission: to '*promote and encourage the wider understanding, appreciation and conservation of birds through scientific studies using the combined skills and enthusiasm of its members, other birdwatchers and staff*.'

Address: The Nunnery, Thetford, Norfolk IP24 2PU.
Director: Dr Jeremy Greenwood.

Butterfly Conservation

Although we have so few species compared with most other European countries, butterflies are next to birds in popularity. In 1968 the British Butterfly Conservation Society (BBCS) was formed by Thomas Frankland and Julian Gibbs with the purpose of saving rare species from extinction and promoting research and public interest in butterflies. Growth was slow at first, but the Society took on the responsibility for the Butterfly Monitoring Scheme in 1983, and acquired its first nature reserve three years later. Since shortening its name to 'Butterfly Conservation' in 1990, the society has acquired considerable in-house expertise. With 10,000 members, it is said to be the largest conservation body devoted to insects in all Europe.

With an office in Dorset, probably today's richest county for butterflies, Butterfly Conservation has a network of 31 branches throughout Britain and runs 25 nature reserves. It has also opened an office in Scotland. It is funded mainly by grants, corporate sponsorship and legacies. The Society's most substantial achievement to date is its 'Butterflies for the New Millennium' project, a comprehensive survey of British butterflies involving thousands of recorders in Britain and Ireland, culminating in the publication of a *Millennium Atlas* in 2001 (Asher *et al.* 2001). Butterfly Conservation is the lead partner for several Species Action Plans, and administers some 30 other projects under its 'Action for Butterflies' banner. The Society also helps to monitor butterfly numbers using fixed transects (*'every butterfly counts, so please count every butterfly'*), and contributes to butterfly conservation internationally; Martin Warren, its conservation director, co-authored the European Red Data Book. The Society publishes a quarterly magazine, *Butterfly Conservation*, and a range of booklets. In 1997, it helped to launch a new *Journal of Insect Conservation*. Aim: *'Working to restore a balanced countryside, rich in butterflies, moths and other wildlife'*.

Address: Manor Yard, East Lulworth, Wareham, Dorset BH20 5QP.
Chairman: Stephen Jeffcoate. Head of Conservation: Dr Martin Warren.

The Farming and Wildlife Advisory Group (FWAG)

FWAG was the brainchild of an informal gathering of farmers and ecologists at Silsoe College, Bedfordshire, in 1969 to work out how to fit conservation into a busy, modern farm. (For a full account see Moore 1987.) Under the auspices of FWAG a network of local farm advisers was established, generally of youngish people with a degree in ecology but with a background in, or at least knowledge of farming. By 1984, some 30 advisers had been appointed, and a Farming and Wildlife Trust was launched to fund the local FWAGs, supported by grants from the Countryside Commissions and other bodies, as well as by appeals. The FWAG idea has helped to break down stereotypes and change farming attitudes. Its advisers became skilled at spotting ingenious ways of preserving wild corners, and planting copses and hedges without harming the farmer's pocket. A notable achievement was the creation or restoration of thousands of farm ponds. An external review indicated that FWAG is strongly supported by farmers, with a high rate of take-up of advice.

Some saw the FWAG project as essentially a public relations exercise, and criticised it for being too much under the thumb of the National Farmers' Union and hence reluctant to criticise modern farming methods. However, it has undoubtedly helped to find a little more space for wildlife on innumerable farms – up to 100 farms per county per year – and has built bridges with the farming community by organising farm walks and training visits, and appearing at the agricultural shows.

Head office: The National Agricultural Centre, Stoneleigh, Kenilworth, Warwickshire CV8 2RX.

Marine Conservation Society

Formed in 1978 ('Underwater Conservation Year'), this small but active national charity with 4,000 members is dedicated to protecting the marine environment and its wildlife, especially in the offshore waters of the British Isles. It publishes an annual *Good Beach Guide*, an *Action Guide* to marine conservation and a 'species directory' of all 16,000 species of flora and fauna found in British waters. It campaigned successfully for protection for the basking shark, and is the 'lead partner' for the UK marine turtles Species Action Plan. Among its multifarious activities for volunteer divers and beachcombers are 'Seasearch', a project to map UK marine habitats, a schools project called 'Oceanwatch', and an adopt-a-beach project for communal cleaning up. Another project, 'Ocean Vigil', records cetaceans and sharks. Fund raising is called 'Splash for Cash'. Internationally, the Society surveys coral reefs in the Red Sea and Sri Lanka, and is helping the Malaysian government to establish a marine wildlife park at the Semporna Islands. Members receive a quarterly news magazine, *Marine Conservation*. Its slogan: '*Seas fit for life*'.

Office: 9 Gloucester Road, Ross-on-Wye, Herefordshire HR9 5BU.

Plantlife

'Britain's only national membership charity dedicated to saving wild plants' was established in 1989, and now has some 12,000 members and a permanent staff of 18, based in London. Plantlife aims to achieve for plants what the RSPB has done for birds, that is, to improve their lot through a programme of campaigning, practical conservation work and public education. As a small charity with big ideas, it often works in tandem with bodies with similar aims, for example, as a member of the campaign to save peatlands, and has formal links with botanical societies and institutions nationally and internationally. Under its 'Back from the Brink' campaign, Plantlife is the 'lead partner' for Species Action Plans on a range of rare flowers, bryophytes and fungi. It also runs 22 nature reserves across 17 counties, mainly meadows, heaths and bogs with an outstanding flora. It contributes to plant conservation Europe-wide via the newly founded Planta Europa network, and commissions research reports on matters of current concern, such as bulb theft, controlling the sale of invasive plants ('At war with aliens') and managing woods for wild plants ('Flowers of the forest'). Its magazine, *Plantlife*, was recently voted the pick of the bunch. Its goal: '*A world in which the riches of our wild plant inheritance are not diminished by human activity or indifference but are recognised, cherished and enhanced*'.

Address: 21 Elizabeth Street, London SW1W 9RP.
Director: Dr Jane Smart.

Plantlife 'flora guardians' clearing invasive 'parrot's feather' weed from a plant-rich waterway.
(Tim Wilkins/ Plantlife)

Wildfowl and Wetlands Trust (WWT)

WWT is 'the only charity concerned solely with wild-fowl and the wetland habitats they rely on'. In 1946, Peter Scott leased 7 hectares of land at Slimbridge, Gloucestershire to establish the Severn Wildfowl Trust, renamed the Wildfowl Trust in 1954. Slimbridge has become home to the most comprehensive collection of ducks, geese and swans in the world. The Trust orig-inally specialised in breeding endangered species, most famously the Nene or Hawaiian Goose, and Slimbridge later became a major research and 'discovery' centre. Nine more autonomous 'Centres' were established at Peakirk, Walney, Arundel, Martin Mere and Washington in England, Caerlaverock in Scotland, Llanelli in Wales and Castle Espie in Northern Ireland, all but the first being designed as refuges for wild birds (with excellent viewing facilities) rather than as captive breeding centres. The Trust also helped Thames Water to set up the Wetland Centre, on the site of Barn Elms Reservoir in London. It organises wildfowl surveys and advises on conservation world-wide, for example, on the design of a new wetland reserve in Singapore and on reed-bed filtration systems in Hong Kong. It changed its name in 1989 to reflect the Trust's wider interest in wetland habitats. In 1992, WWT produced a global review on the conservation and management of wild-fowl, and played a leading role on the first international conference devot-ed to these birds. Today it has some 70,000 members, while up to 750,000 people visit its Centres each year. Its newsletter is called *The Egg*. There is also a biannual research newsletter, *Wetland News*.

Main Centre: Slimbridge, Gloucestershire GL2 7BT.

The Woodland Trust

Founded by Kenneth Watkins in 1972, the Woodland Trust has been one of the voluntary move-ment's surprise successes, strik-ing a chord with our British love of trees and woods. It acquired its first property, Avon Valley Woods in Devon, near Watkins' home, in 1972, and its 1000th, Coed Maesmelin, near Port Talbot, in 1999. The Trust's straightforward purpose is to acquire woods of historic, scientific or ameni-ty value, open them to the public and manage them in sympathy with their character. Although still only a medium-sized charity, with 63,000 mem-bers, the Trust has a sizeable income – £16.3 million in 1999 – from lega-cies, landfill tax credits, corporate sponsorship and appeals. It also sells timber on the Internet. It has offices in Scotland, Wales and Northern Ireland, and a headquarters at Grantham, Lincolnshire. An admirable proportion of the Trust's budget goes straight into conservation; for example, in 1999, it spent £5 million on acquiring woods and £6.3 on managing them. Many Trust properties are SSSIs, managed by agreement with one of the conservation agencies, and nearly all of them are de facto nature reserves, with nature conservation a primary aim. Today it owns or manages 1,080 sites covering 17,700 hectares.

With its open house policy, the Woodland Trust aims to promote public enjoyment ('Wild about Woods') and to 'engage local communities in creating, nurturing and enjoying woodland'. It publishes an attractive quarterly newsletter, *Broadleaf*, and many of its properties have their own leaflets. The Trust also contributes towards the national Millennium Forest project. Though on occasion a little too anxious to plant trees where no trees are needed, the Woodland Trust has saved many fine woods from oblivion, and its overall influence on British woodland management has been benign, and considerable. On ancient woods, its aim is 'no further losses'.

Address: Autumn Park, Grantham, Lincolnshire NG31 6LL.
Chief Executive: Mike Townsend.

Membership of the large conservation societies (in thousands)

	1971	1981	1990	1995	2001
National Trust	278	1,046	2,032	2,323	2,700
RSPB	98	441	844	890	1,011
Wildlife Trusts partnership	64	143	250	252	300
National Trust for Scotland	37	110	218	234	240
WWF	12	60	247	200	257
Friends of the Earth (England and Wales)	1	18	110	200	250
Woodland Trust	–	20	66	62	63

Afterword: my own county trust

The county of Wiltshire has been the domain of great naturalists ever since John Aubrey wrote the first county natural history (*Memoires of Naturall Remarques*) in 1685. Richard Jefferies lived at Coate, near Swindon, and the location of his Bevis stories is preserved today as a country park. The county boasts one of the classic floras – Donald Grose's 1957 *Flora of Wiltshire*. Its lepidopterists include Baron de Worms, who was in charge of a chemical laboratory at Porton Down, and the Marlborough schoolmaster Edward Meyrick, perhaps the greatest microlepidopterist that ever lived, who lies in my parish churchyard. For 150 years we have had a flourishing Archaeological and Natural History Society based at the county museum, with its own journal. Even the county's coat of arms commemorates its most characteristic, certainly its most spectacular, species, the now nationally extinct great bustard, standing back to back, in their proper colours.

The Wiltshire Wildlife Trust, then the Wiltshire Trust for Nature Conservation, was formed in 1962. At that time nature conservation was scarcely more than the hobby of a few hundred local naturalists. As one of the founders, Lady Radnor, recalled, 'We thought then, very innocently, that if we could stop egg-collecting and the men with butterfly-nets, if we could persuade government to ban some of the more deadly pesticides then all would be well'. Agricultural pesticides were the big issue then. A Trowbridge farmer recalled how, walking through the town centre in the early 60s, there was often a lingering niff in the air: not the emissions of some factory but the agricultural sprays from farmers' fields.

The transformation of the Wiltshire Trust from a modest local charity to a business with over 50 full or part-time staff and an annual turnover of well over a million pounds has taken place quite recently – mostly since 1990. Financially the Trust's main benefactors have been the Heritage Lottery Fund (HLF) and the Landfill Tax Credit Scheme, both created in the mid-1990s. Lottery grants have provided many trusts with their biggest windfall, enabling them to buy those long-needed fences or set about restoring wetlands by ambitious damming and drainage schemes. Typically the Fund would provide three-quarters of the costs, leaving the Trust to make up the rest from other donations or its own resources. The Wiltshire Trust was among the first to see the opportunities this presented. As it happened, the HLF's administrative body, the National Heritage Memorial Fund, was already a good friend of the Trust, having helped it to acquire three nature reserves, including Ravensroost Wood, the Trust's showcase reserve near Swindon. The opening for business of the Heritage Lottery Fund coincided with the sale of a traditional farm at Jones's Mill near Pewsey, which the Trust was anxious to save. Its director dashed over to take pictures and filed an application that same day. Two weeks later it had enough money to purchase the important part of the site, which is now a well-loved nature reserve. The HLF has since helped the Trust to purchase sites of national importance, including Clattinger Farm, a 'time warp' vista of flower meads untouched by the plough or agricultural chemicals, Coombe Bissett Down, one of the best British sites for burnt-tip orchid, and, most recently, a 235-hectare property at Blakehill Farm for restoration to its former flowery glory. The Wiltshire Trust passes on its experience of working with the Lottery by chairing the Trust partnership's working group on Lottery funding. Gary Mantle, the Trust's director since 1990, attributes part of its success in this field in part to the Trust's relatively modest size: 'We're lean and hungry, fleet of foot'. It is also, as I am able to attest (wearing my other hat as a Lottery assessor), impressively businesslike. When it says it will do such and such, it does it. The Lottery appreciates bodies that demonstrate value for money.

James Power and Gareth Morgan of the Wiltshire Wildlife Trust at Clattinger Farm, which became a Trust reserve in 1997.

Landfill Tax Credit, introduced in the mid-1990s, is the great unsung windfall for voluntary nature conservation. Essentially it is a tax levied on every ton of rubbish buried. Twenty per cent of the tax collected can be retained by the operator and given away to a registered environmental body of its choice. The rules are complicated, but the potential largesse is enormous, with £100 million becoming available for good causes in the first year of operation alone. Wildlife bodies have often been a little slow to spot a potential winner, but the Wiltshire Trust sniffed the air like an emergent vole and pricked up its ears. For several years it had worked with the Hills Group, a large aggregates and landfill operator, on the Braydon Forest countryside management project in the north of the county, in an effort to preserve some glorious countryside close to Swindon and open it to the public. The Trust got itself on the list of eligible charities, and brought along its shopping list of activities. The upshot was that it received a present of £200,000 towards a range of activities. There is, however, a potential conflict of interest between landfill operators and environmental bodies since the latter would really prefer rubbish to be burnt or recycled rather than buried. A shake-up of how the landfill tax is spent seems imminent.

Like other county trusts, Wiltshire runs a network of nature reserves. Among the 40-odd examples are Blackmoor Copse, famous for its woodland butterflies, which it took over in 1963, and several fine sweeps of downland, including Morgan's Hill, Great Cheverell Hill and Middleton Down. Two, Ramsbury Meadow and High Clear Down, lie within walking distance of my home. All the Trust's reserves are run by volunteers; unlike some trusts, Wiltshire has no full-time paid wardens. The Heritage Lottery and other donors have enabled the Trust to specialise in grassland management – an obvious choice since Wiltshire has more chalk downland than any other county, and also a fine series of unimproved neutral grassland meadows. However, the Trust no longer regards nature reserves simply as an end in themselves, but as demonstration sites, and as kernels within a wider area where sustainable and wildlife-friendly land management is the aim. The Wiltshire Trust is 'farmer friendly' and many landowners have served it in one way or another. 'Farmers appreciate a pat on the back,' says Gary Mantle. 'It's nice for them to hear a conservationist say "what a fantastic bit of land", instead of being criticised all the time, especially when times are hard.' The trouble nowadays is that managing almost any wildlife habitat has become uneconomic unless it is subsidised in some way, and the kind of stock farmer the trusts rely on most is going out of business.

Most county wildlife trusts contribute to the local planning process by providing details of places of local importance for wildlife which are not quite important enough to be SSSIs. In Wiltshire, these places are called, simply, 'Wildlife Sites'. They are generally good examples of diminishing habitats, such as coppiced woodland or chalk downs, but also include sites for rare species. Their protection depends on the local authority, generally the district, but in the case of roadsides the county council. 'Wildlife Sites' are non-statutory, but in Wiltshire they appear in local plans with a presumption against development. The Trust is given an opportunity to object to unfavourable development, and if necessary defend its stance at a public inquiry. Broadly speaking these places receive about the same level of protection as SSSIs did in the 1970s – perhaps more so, given that local authorities are much more environmentally friendly than they were then.

In other traditional areas of trust activity, the focus has broadened. The county's local biological records centre, long based at the county museum in Devizes just across the road, is now under the Trust's wing. This in turn is now part of the National Biodiversity Network, a computerised Wildlife Sites system and public information service. Like all trusts, the Wiltshire one stands up for wildlife at public inquiries, 'fighting hard to stop the destruction of important wildlife habitats'. But it also joins in the wider struggle to find acceptable policies that would avoid such destruction in the first place, both with ideas and by involving its members and the wider public in local and national campaigns.

Most recently, the Wiltshire Trust has become interested in broader environmental issues. The underlying premise is that it is no longer possible to separate wildlife issues from our own future. The Trust wants to demonstrate that it is doing its bit to promote ideas of energy saving, fair-trading and 'sustainable gardening', and practising what it preaches. Behind it lies a conviction that the Trust ought to have the support of at least 100,000 residents in the county, not just its current 10,000 members. To achieve that it needs to come to grips with issues that concern a lot of people, not just those who are keen on natural history. In 1994, the Trust took on the role of managing the county's local Agenda 21 process. At the time, Mantle went on record as saying he was uncertain whether this would be a complete waste of time or the most important thing they could do – though it would be one or the other. Six years on, having seen the impact of Agenda 21 on tackling issues such as global warming by energy efficiency advice, minimising waste and working for fair trade at home and abroad, he believes they made the right decision. The Trust's strategy for 2000–2005, headed 'a sustainable future for wildlife and people', is upbeat about 'presenting a positive, hopeful face to the world': 'Working to a common purpose we can make a real difference'.

The Trust's founder, Lady Radnor, recalled a line by Rudyard Kipling: '*And gardens are not made, By saying Oh how beautiful And sitting in the shade*'. Today life seems more complicated than it was back then: 'the tunnel has grown longer and darker, and taken some very nasty turns'. Wildlife trusts are richer, which enables them to do more, and also to rethink the ground rules about what a local trust is for and what it has to say to the world. Gardens are indeed made by hard work, but they also need creativity and hope, as well as clean rainwater and sustaining soil.

Director: Gary Mantle OBE
Office: Elm Tree Court, Long Street, Devizes, Wiltshire SN10 1NJ.

Growth of a county wildlife trust: Wiltshire Wildlife Trust 1985–2000

Year	Staff	Members	Turnover (in £)
1985	5	2,293	54,636
1986	5	2,233	69,504
1987	5	2,825	107,598
1988	6	3,238	124,622
1989	9	< 4,000	247,118
1990	10	< 4,000	217,322
1991	12	< 4,000	235,268
1992	18	> 4,440	240,589
1993	20	> 4,500	264,396
1994	22	> 6,000	536,924
1995	24	> 7,000	686,104
1996	24	9,204	1,426,269
1997	35	9,924	1,099,939
1998	39	10,204	1,271,821
1999	46	10,720	1,576,784
2000	49	10,774	2,405,597

4

Conservation Politics:
SSSIs and the Law

I have in front of me a thick volume published by the then Department of the Environment and titled *Wildlife Crime: A guide to wildlife law enforcement in the UK* (Taylor 1996). Its purpose is to try and sort out the legal labyrinth of wildlife law as it stood in 1996, mainly for the benefit of policemen and other law enforcers. More than half of it is devoted to species – birds and their eggs, badgers, deer, seals and salmon, as well as the trade in endangered species. Lists of protected birds in their various grades and schedules take up seven pages, non-avian protected animals and plants another four. On the face of it, Britain's wildlife looks well protected. But although protection laws may look like nature conservation, much of them are about animal welfare issues. Kindness to animals is an issue for the RSPCA. Conservationists are more concerned with the survival of populations and species than with individuals. However, the legal benefit enjoyed by our wild animals is decidedly mixed. The law has evolved, rather like the landscape, in an ad hoc way, and the result is chock-full of anomalies. Pat Morris (1993) has pointed out that while it is technically illegal to shine a torch at a hedgehog, you can squash one flat with your car without worrying about prosecution. An antique dealer risks a heavy fine for selling an old coat trimmed with pine marten fur, but the law does not help living martens very much. The badger is an exceptionally well-protected animal, but the Ministry of Agriculture slaughters thousands of them. Contrariwise, in the interests of the environment more deer need to be culled, but no one insists on it and so deer continue to multiply. In practice, the government nature conservation agencies spend remarkably little time on species protection. Unlike the US Fish and Wildlife Service, who carry guns and have the power to arrest, Britain's wildlife agencies have no special powers of enforcement. They devote far more time to managing species under the Biodiversity Action Plan (see Chapter 11), but until 2000 the Plan had no basis in British law. The really important species laws boil down to two: the Wild Birds Protection Act of 1954 and its subsequent amendments, which protect virtually all wild birds, and the Wildlife and Countryside Act of 1981, which protects a lot of other rare animals and plants, mostly from imaginary threats, like collecting, and, more importantly, protects all species of bat, as if they were honorary birds. It is easy enough to protect a species on paper – you simply declare it protected – but quite another thing to bring a successful prosecution. In practice, most prosecutions are to do with birds and bats. Egg collectors, errant gamekeepers and careless timber treatment companies are the principal targets.

Animal smugglers are dealt with under international codes enforced under EU regulations.

Protecting a species is pointless unless its habitat is protected too. As the law stood until recently, you could convict someone picking rare orchids in a meadow, but do nothing to prevent a developer or farmer from destroying both the meadow and its orchids. Hence, nature conservation in practice is directed at saving the habitat. Most land in Britain is privately owned and dedicated to some sort of productive use that is usually not nature conservation. In 1949, Government was persuaded by the principle that some portion of the land should be set aside for nature in the interests of at least that part of the population which cares about such things. The principle was not new. The Norman kings set aside land for game for their own selfish purposes. And as Professor Smout reminds us, the eighteenth-century enlightenment took the view that while most of the land is destined for agricultural improvement, some of it should be set aside to delight rather than for productive use – 'most for man, a little for nature' (Smout 2000). The contribution of twentieth-century conservationists has been to work out where the best spots for nature lie. The next stage is to see to it that these valuable areas are looked after in a way that ensures they stay valuable.

The Nature Conservancy and its successors evolved methodologies for grading semi-natural land according to its value for wildlife. These were based on attributes such as size (generally the bigger the better), diversity and 'naturalness' – based that is, on the quality of the habitat rather than on rare species in isolation. The original idea had been to preserve all the very best examples of woods, heaths, chalk downs and so on as nature reserves. However, that was never going to be enough, so an alternative to direct ownership (or at least proxy management) was needed. From 1949, the Nature Conservancy was allowed to 'schedule' any area of land of special interest 'by reason of its flora and fauna, or geological or physiographical features'. These were (and are) called SSSIs or Sites of Special Scientific Interest. This clumsy term has caused much head-scratching. Here 'Scientific' really means 'nature conservation' in an adjectival sense. Every now and then someone suggests changing the name, but nothing has ever come of it for fear of adding to the confusion. In any case, until 1981, SSSIs did not amount to very much. The job of the Conservancy was to identify SSSIs, say why they were important and notify them to the local planning authority. In the case of a development requiring planning permission, the local authority would then decide whether to allow the development or refuse it. Of course the local authorities had their own plans and guidelines that were broadly in favour of SSSIs – but 'the national interest' always came first, and this could be interpreted in all sorts of ways. Moreover, the most common threats to SSSIs – agricultural improvement or tree planting – did not normally require planning permission. Altogether the Conservancy had been given a very poor hand. It could offer only pennies in compensation, while the 'improvers', by way of the Ministry of Agriculture and the Forestry Commission, could offer pounds.

The shrinking of a
wildlife site. The Wye
and Crundale Downs
was recommended by
the Wildlife Special
Committee as a
1,500-acre (607-
hectare) 'National
Reserve' in 1947 as
'a first-class example
of typical Kentish
chalk communities
with many character-
istic and rare plants
and insects'. By 1970
the site had been
reduced by plough-
ing to 415 hectares.
Today it measures
about 257 hectares
(based on W.M.
Adams, *Nature's
Place*).

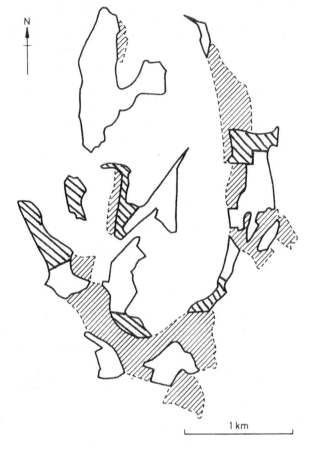

 Area removed
from SSSI
1963–1981

 Area removed
from SSSI
1981–1990

Remaining area
of SSSI

1 km

It is not surprising then, that, before 1981, many SSSIs went under the
plough or turned into spruce plantations. For example, in 1963 a farmer
received a ploughing grant for the destruction of Waddingham Common
SSSI, one of the best natural grassland sites in Lincolnshire. The farmer
offered to leave a token acre unploughed. Representatives of the Nature
Conservancy and the local wildlife trust insisted on five acres as a bare minimum. The farmer laughed; the entire site was ploughed (Sheail 1998). In
Wiltshire, 15 out of 27 chalk grassland SSSIs were ploughed out of existence between 1950 and 1965. In Kent, most of Crundale Downs, proposed
in 'Cmd. 7122' as a 'National Reserve', went the same way (see above). The
Nature Conservancy was urged to do more to obtain the co-operation of
owners and occupiers through moral persuasion backed by 'suitable annual payments'. Unfortunately the cash-strapped Conservancy was unable to
pay anybody very much, and certainly could not compete with grants for
agriculture and forestry.

If anything, the situation worsened after Britain's entry to the European
Community in 1973. In 1980, the NCC's chief advisory officer, Norman

Moore, estimated that 8 per cent of all SSSIs had suffered damage *during the past twelve months*, of which the main causes were agricultural improvements and the 'cessation of traditional practices' (due mainly to agricultural improvements). There had, in previous years, been attempts to strengthen SSSI protection, but they had all failed. In 1964, Marcus Kimball MP had presented a Private Member's Bill that would have imposed a mandatory period for negotiations over the fate of an SSSI, during which agricultural grant-aid would be withheld. The agriculture departments quashed that idea on the grounds that no one knew how many SSSIs would be designated, and that it would cause 'unnecessary disruption to farm businesses'. The Conservancy seemed to agree, stating that it preferred doing things by voluntary means. Four years later, the 1968 Countryside Act offered another opportunity to strengthen SSSIs, but again it was lost, largely through resistance by the agriculture lobby. The Act did enable the Nature Conservancy to enter formal management agreements with owners, but gave it no extra cash to do so. Moreover it effectively restricted the incentive on offer to a laughable one pound per acre. Until this limit was waived in 1973, not a single agreement was made (Sheail 1998).

It was clear to many that something needed to be done if Britain were to have a system of nature conservation worthy of the name. In 1977–78, the Amberley Wild Brooks case (see Chapter 9), decided that, on occasion, conservation and amenity aims should outweigh agricultural production. The Government's own review body advised that legislation would be necessary to reconcile agricultural production and countryside conservation. By then there was growing public concern about the diminishing quality of the countryside, epitomised by the piecemeal loss of moorland on Exmoor, despite its National Park status.

The Wildlife and Countryside Act – origins and arguments

When the Conservatives under Margaret Thatcher came to power in May 1979, they inherited the bare bones of a Wildlife Bill. A new law was needed to implement an uncontentious Euro-directive for Special Protection Areas for certain wild birds, and also to ratify the Council of Europe's Berne Convention on wildlife and natural habitats. Beyond that, the review committee had advised that something had to be done about strengthening SSSIs, but no one was sure what. Enter Michael Heseltine, the flamboyant new Secretary of State for the Environment, who combined a 'managerial' broadbrush approach to government business with a genuine interest in wildlife. In his memoirs, Heseltine mentioned that one of his first acts had been to summon the chief executive of the RSPB – the most important figure in nature conservation by virtue of his then 400,000 members – and ask him what he would like Heseltine to do. 'If I may say so,' added the Secretary of State, encouragingly, 'you are unlikely ever to find a minister more sympathetic than me.' It seems to have been Michael Heseltine's influence that assured the eventual passage of the Wildlife and Countryside Bill, despite opposition within Cabinet (which at that time contained several country landowners). However, after laying down broad principles on involving landowners in safeguarding SSSIs, he characteristically left the details to his officials, and most of the Parliamentary business to his deputy, Tom King.

Hurriedly, Department officials drew up a consultation paper on 'the conservation of habitats', based on the recommendations of the review committee and setting out what the Government proposed to do about SSSIs. The answer, it seemed, was not much. Great faith was placed in the beneficence of landowners and in 'the voluntary process'. In the anticipated rare cases when this was not enough, the NCC could in future apply to the minister for an order, not to save the site but to purchase breathing space in which to continue negotiating. However, not all SSSIs would qualify for the order; indeed it seemed that it would apply only to an unspecified number of sites 'of national or international importance'. This was probably not what the chief executive of the RSPB had asked for – it was more likely that it represented what the barons of the CLA and NFU were prepared to accept. At this stage, the NCC had not been asked at all.

It might seem surprising, therefore, that the NCC's new chairman, Sir Ralph Verney, expressed himself broadly satisfied with this Lenten fare. His main complaint was a technical quibble over who should draw up the list of 'sites of national or international importance', his Council or the minister. On this point, Heseltine told him that it was politically impossible to impose further restraints on private landowners unless matters were handled directly by government. The NCC meekly replied that it hoped the minister would at least consult his statutory wildlife advisory body about all this, since only the NCC had the appropriate knowledge and experience to advise him. It pointed out that the NCC had spent nearly ten years reviewing and categorising Britain's wild places, and really knew a lot about it.

From the standpoint of progressive, early twenty-first-century green legislation, this must sound like the Dark Ages. However, back in 1979, nature conservation had not yet assumed the public importance it has today. The Conservative Government was reluctant to tie the hands of private landowners with more regulatory red tape. The late Sir Ralph Verney was himself a Buckinghamshire landowner of ancient lineage, a past president of the Country Landowners' Association and sometime chairman of the Forestry Commission, with a long record of meritorious service on government royal commissions and advisory committees. He listed his recreation in Who's Who as shooting. Verney's deputy was Viscount Arbuthnott, a Scottish aristocrat, Lord Lieutenant of Grampian Region, past president of the Scottish Landowners Federation, as well as a past chairman of the Red Deer Commission and an influential figure in rural matters generally north of the border. It would be hard to hold all these important jobs without an underlying philosophy, and inevitably their views on the countryside were a landowner's perspective, imbued with concepts of stewardship and 'balance'. NCC's Council contained others that had interests in farming and forestry, soon to be joined by a right-wing MP, Sir Hector Monro, close to the levers of power. Verney saw the NCC's role as ensuring that conservation was 'properly integrated' into a balanced rural land use. Some saw such views as an acknowledgement of the lowly place of nature conservation in the rural pecking order. Promotion of a single interest seemed respectable when it came from the Ministry of Agriculture or the Forestry Commission, but unacceptable in a less powerful quango like the NCC. Nor, as the NCC's staff were well aware, did these comfortable notions of integration and stewardship sit easily with the well-documented losses of natural habitats.

The NCC's acquiescence was also in part tactical. As Verney and most of his Council saw it, an important principle had already been established – that, as a result of a recent streamlining of capital grants on farms, farmers already had to give the NCC advance notice of agricultural improvements on SSSIs. That meant that changes in farm practice could at last be made subject to regulation. The door had at least opened, and, with the right kind of positive encouragement, the NCC might find itself in 'a good position to orchestrate the proper evolution of the SSSI system', they hoped.

By 1980, the voluntary bodies, or 'non-governmental organisations' (NGOs) as they were called, most significantly the RSPB and the Council for the Preservation of Rural England (CPRE), were becoming more assertive. Though hardly militant, they were learning how to apply pressure on the Government. As the Bill began its passage, the voluntary bodies thrashed out a common platform via a hitherto invisible focus group called Wildlife Link. An effective leader emerged in the radical and persuasive Labour peer, Peter Melchett. As one of the participants put it, 'hours of gruelling discussion' on Wildlife Link, each party representing a sizeable client group, ensured that 'the right people knew what to ask for, what to fight for, and what to accept' (Vittery 1982). Wildlife Link provided, among other things, a crash course in the art of political lobbying.

Outside events lent a helping hand. In October 1980, Marion Shoard's *The Theft of the Countryside* was published. Although Shoard's advocacy of planning controls on farming held little appeal outside the far left, her analysis of the 'engine of destruction' and its complete lack of public accountability, plus the piling up of ever larger mountains of surplus food, opened many people's eyes to what was happening behind the keep-out signs. 'SSSI' became an everyday acronym (though few seemed to know what the hissing letters stood for). *The Common Ground* by Richard Mabey, also published that year – Mabey was soon to become a member of the NCC – was less angry in tone, but argued persuasively for a 'land ethic' that recognised the legitimate interest of the wider community in preserving a healthy, attractive countryside. He voiced the public demand for fields of wild flowers as well as fields of wheat. Yet another useful nudge came from within the NCC itself with a now famous article in *New Scientist* by David Goode (number two in NCC's scientific team) demonstrating just how many downs, heaths and commons had vanished under the plough since the War. One of the latest had been Horton Common in Dorset, enclosed and rotovated in front of the television cameras.

It was these undeniable statistics of habitat loss that obliged the NCC to reconsider its position. In October 1980, Verney and his Council listened to a forceful presentation by Peter Melchett arguing that ministerial orders and 'reciprocal notification' should be available to all SSSIs, and not just to a selected few. The latter phrase meant that the landowner and the NCC would form what was, in effect, a contract. The NCC would inform each owner about the site's special interest, and advise how this should be conserved, while the owner would agree to give advance notice of any intended operation that might damage it. In that event, the NCC would be expected to offer the owner a formal agreement, including compensation for any lost income incurred. The NCC was won round to Wildlife Link's position. Within the week, Verney wrote to Heseltine to inform him of his Council's U-turn on policy, and issued perhaps the bravest press release in conservation history. The NCC was now committed to the principle of 'reciprocal notification'. Reportedly, the Department's only reply was an acknowledgement postcard.

More than a year after discussions began, the scene moved from behind the closed doors of government offices to the theatre of Parliament. The Wildlife and Countryside Bill was introduced on 25 November 1980. But with an already crowded programme of business in the House of Commons, Government decided to send the Bill first to the House of Lords. From its point of view, this was a serious tactical mistake. Many hereditary peers were landowners, with a vested interest in the Bill's outcome as well as much practical experience of matters in it. They had a great deal to say about it all. Over several weeks of debate, a thousand amendments were tabled, apparently a parliamentary record. Most of them were of course ephemeral, but provision for Marine Nature Reserves was added to the Bill by Lord Craigton, while Lord Sandford moved a key amendment to enable agriculture ministers to make grants to farmers for conservation purposes (he had in mind the recent ploughings and reseedings in Exmoor National Park). The most crucial amendment to the Bill, tabled by Lords Buxton and Onslow, called for reciprocal notification on all SSSIs. It failed – but by only two votes.

The second reading in the House of Commons on 27 April 1981 was introduced by Michael Heseltine in his first and, as it turned out, only speech on the Bill. The Government's watering-down of the Lords amendments was badly received, and it seemed that this supposedly non-partisan Bill was now at risk of being 'talked out' by the Opposition. The main stumbling block was the Ministry of Agriculture, which had placed the NCC under considerable pressure to agree to a purely voluntary code, whereby a farmer would give the NCC due notice only if he felt like it ('otherwise bad things might happen', suggested the NCC's parliamentary monitor). A second blizzard of amendments followed before the Bill entered its all-important Committee stage (no Conservative member who had spoken up for conservation during the second reading was allowed to serve on Committee). But with Labour MPs such as Tam Dalyell threatening to ramble on for hours about the colour of a goldeneye's eye, the Committee began to run out of time, while the 'queue here' signs outside the Committee Room attested to the unusual public interest in this Bill. At the last minute, Government began to waver. In Cabinet, Heseltine won approval for reciprocal notification against reported opposition from the landowners William Whitelaw and Francis Pym, as well as from the Minister of Agriculture, Peter Walker. At the subsequent Report stage, concessions suddenly spilled out of Tom King's hitherto tightly closed sack. Limestone pavements would be protected. Marine nature reserves could be set up. Above all, the Bill would now contain a clause requiring landowners to give advance notice of any 'potentially damaging operation' on an SSSI – the crucial change everyone had been pressing for. The original proposal of a Nature Conservation Order for 'Super SSSIs' would remain as a back-up power to facilitate negotiations, but, since the Bill did not define what a Super SSSI was, no one knew how many sites would qualify for it. And there was another catch. A new amendment introduced by the Government required all the costs of compensating landowners for 'profit foregone' to be met by the conservation agencies – within National Parks by the relevant park authority, otherwise by the NCC, whose annual income at that time was only £6 million. Not one penny of the then £176 million of annual grant-aid for farm improvement could, it seemed, be spared. As the Secretary of State for Agriculture explained, it was not the business of his Ministry to pay for nature conservation.

The Wildlife and Countryside Bill received royal assent on 30 October 1981. As drafted, it left a number of questions hanging, most notably on how management agreements on SSSIs would be paid for (Government's financial guidelines on the subject were not forthcoming until early in 1983). What mattered, though, was that the rules had changed. Politicians could no longer afford to ignore public opinion. The dormant, mainly urban, electorate had woken up. The preservation of wildlife was henceforth a matter that concerned everyone.

The Act in practice

Nature conservation in Britain really came of age with the Wildlife and Countryside Act of 1981. In the usual way of modern legislation it was a portmanteau chunk of law, catching up previous piecemeal laws and laying down rules on everything from the tethering of bulls to the close season for snipe. The important bits were tucked away in the small print. Although the Act, with its numerous Parts, Sections and Schedules, takes up 128 printed pages, the vital sections on wildlife habitats (Sections 28 and 29) are covered in just four. They set out the new ground rules for 'notifying' SSSIs, bringing in government and the local authority, as well as the owner and occupier. They laid out the rights and responsibilities of the interested parties and allowed three months for negotiations before the site would be notified. Section 29 also introduced 'special protection' for certain areas of special scientific interest, but this amounted only to extended negotiating time, not to protection per se. By introducing a statutory mechanism on SSSIs, the Act did at least determine that SSSIs were now the main instrument of nature conservation in Britain. It did not halt the destruction and damage to SSSIs, but it did at least ensure that

Winter-flooded fields on the Somerset Levels, a tranquil landscape bitterly fought over by conservationists and drainage men in the 1980s. (Natural Image/Bob Gibbons)

some SSSIs were looked after better. To a large extent it all hinged on the attitude of the owner or occupier, and the ability of the NCC to persuade him. Some farmers remained unpersuaded, especially in wetland areas such as the Somerset Levels and poor hill farms in Wales and Scotland. These quarrels were well publicised, and raised the level of public debate. Widespread interest and concern about our wild places led to the unprecedented growth of the voluntary conservation sector in the 1980s.

The Wildlife and Countryside Act extended the protection of wildlife habitats, but created more bureaucracy and regulation. Rather than attack the fundamental cause of wildlife damage – subsidised agriculture – the Act only sought to protect parcels of land 'of special interest'. Even so, it may be doubted whether those who drafted and debated the Bill were fully aware of the scale of the task. In 1982, there were 2,670 SSSIs covering 1,367,000 hectares; by 2000 the number had grown to over 4,000 SSSIs in England alone, covering over a million hectares (6 per cent of the land surface), and involving 32,000 owners and occupiers. To read the debates in Hansard you would be forgiven for assuming that most SSSIs were field corners and small copses of negligible economic potential that any landowner would be happy to set aside for wildlife. In fact, they include entire river estuaries and vast tracts of upland. The Wash and the New Forest are SSSIs; so is 44,000 hectares of the North York Moors, much of the North Pennines and most of the Cairngorms. The first people whose lives were changed by the Act were civil servants, most notably the staff of the NCC, who were expected to make it all work. Any sense of exhilaration within the NCC was soon followed by the dawning horror that all *existing* SSSIs would need to be renotified under the new Act (Treasury lawyers insisted on it) as well as new ones designated under new guidelines. At first it was thought the job might take two years. In fact, it took nearly ten. As Morton Boyd put it in his memoirs, 'a single notification was a piece of precise teamwork between scientists, land-use assessors, regional staff and council secretariat' (Boyd 1999). It seems to be a uniquely British affliction to need to dot every i and cross every t when implementing the rules. Those of us charged with 'renotifying' SSSIs had 14 separate stages to cover for each SSSI, each one zealously vetted by officials at headquarters. Guidance papers on SSSI selection and procedure flew about from headquarters to the regional offices, the daddy of them all being a 288-page tome on SSSI selection, delivered rather late in the day in 1989 (most of the job had been done by then).

When the 'SSSI notification' papers arrived at the farmhouse or agent's office, reactions varied. Owners of large country estates and big farms, who had followed events closely and knew what to expect, broadly accepted the system and knew how to exploit it. Some people, especially in Scotland, expressed indignation at the language used. 'Damaging activities' implied misconduct, while 'consent' smacked of condescension, if not arrogance. Further resentment stemmed from the lack of any appeals procedure. The Government took the view that an SSSI statement was scientific fact enshrined in law, and therefore not open to a contrary opinion. Besides, if everybody appealed, the system would grind to a halt. Many farmers feared that SSSI designation would lower the land value, and saw themselves being deprived unjustly of their freedom of action. Farmers from the Somerset Levels were particularly angry, clubbing together to make a bonfire of their notification papers, along with an effigy of the NCC's chairman and regional officer, and the local RSPB man. Their local MP, John Peyton, complained that Sir Ralph Verney had been unable 'to keep his zealots and minions in any sort of check'. It took a visit from the minister and promises of compensation to calm things down. A scapegoat was found in the person of Sir Ralph who was effectively sacked for doing his job. A few landowners hurriedly sprayed or ploughed their special sites before they could be notified – small meadows being particularly vulnerable in this respect, since they were so easy to destroy. But these were not typical reactions. On most SSSIs, there was a measure of co-operation. 'Loss and damage' continued, but more often as a product of neglect or less than ideal practice than of wilful mismanagement. The Act worked passably well because the British are fundamentally law-abiding people. It would not have worked in a country such as France where small farmers are more militant, nor in a larger, wilder country such as the United States, where farming and nature conservation are geographically separated.

In the following years, the Wildlife and Countryside Act was tested in the courts through public inquiries and prosecutions. Test cases helped to resolve a matter that had been kept deliberately vague: that is, which SSSIs would qualify for a Nature Conservation Order under Section 29. Government, it seemed, preferred to take each case on its merits, while the NCC wanted a ministerial order made available to *all* SSSIs. The main benefit of such orders is that they could be extended until an agreement was reached, and, while they are in effect, deliberate damage to an SSSI meant the perpetrator could face conviction and a hefty fine. However, civil servants know better than to bother ministers too often with awkward matters, and it was tacitly accepted that Nature Conservation Orders would be served only in the most obdurate circumstances.

The first one was made in 1983, two years after the passage of the Act. Baddesley Common SSSI is one of the last remaining remnants of natural grassland, heath and bog on the coastal plain of Hampshire outside the New Forest. It came under threat when the estate of which it was a part was broken up and sold at auction. The local wildlife trust managed to purchase part of the Common, but the rest went to a former builder with farming aspirations. These had been encouraged by the land auctioneer who 'painted encouraging pictures of a waving sea of corn' (Tubbs, pers. comm.). The new owner forthwith gave notice of his intent to clear, drain and plough, and refused to negotiate. To gain time, the NCC applied for an order under Section 29, which was served by 'the largest available' ministry official. The builder appealed, and four months later a public inquiry was held. In July 1983, on the advice of the inspector, the Secretary of State upheld the order.

Baddesley Common near Southampton, rescued from the builders by the first Nature Conservation Order, served in 1983. Now a 50-hectare nature reserve owned by the Hampshire Wildlife Trust. (Hants and Isle of Wight Wildlife Trust)

That the Government was willing to invoke Section 29 on an 'ordinary' – that is patently non-Super – SSSI was encouraging. Less so was the amount of time taken up by this single case. Behind the order lay a lengthy succession of meetings, within NCC, with the landowner or his legal representative, with the Department, with the District Valuer, with legal counsel, amounting to some 55 man-days for the NCC. And Baddesley Common was but one SSSI among more than a hundred in Hampshire alone. Clearly, it would not take many Baddesley Commons to bring the system to a grinding halt, and perhaps bankrupt the NCC in the process.

Moreover, subsequent events suggested that Baddesley Common might have been an exceptional case, and that not all SSSIs, after all, would qualify for a Section 29 Order. Out of a total of 17 applications made by the NCC between 1983 and March 1985, three were refused on the grounds that the sites were 'not of national importance'. Hence, short of compulsory purchase – which was theoretically available but ruled out on political grounds – there was nothing more the NCC could do to stop Brimham Rocks SSSI being turned into a pig farm, or Sherburn Willows SSSI being dug up to create ponds. Refusal to serve an order implied that the Secretary of State had a better understanding of what constituted a 'nationally important' SSSI than his official advisers (and, if that was the case, one might wonder why he bothered having advisers). The NCC stated bluntly in its annual report that, as far as it was concerned, '*all* SSSIs serve a national function and they are selected on that basis'. It also reminded Secretary of State that the legislation made no distinction between categories of SSSI. Legally they were all equally deserving. (Does the more forthright tone of NCC's annual reports at this time have something to do with the fact that it was I that drafted them? I do not know the answer to this question.)

A further twist in the story concerned the case of West Mersea Meadow in Essex. Learning that the NCC intended to notify this small hay meadow, noted for its green-winged orchids, its owner, who had plans of his own to sell the land for housing, sprayed it with herbicide. To prevent further damage, the minister made a Section 29 Order just as soon as the NCC had notified the site, damaged as it was, as an SSSI. The owner objected, as was his right under planning law, and another public inquiry was held. This time it was the inquiry inspector who thought he knew what 'national importance' meant. In his view, West Mersea Meadows was not nationally important. The minister felt bound to concur, and withdrew the order. Perhaps it was all rather academic, since the orchids were dead in any case. The meadow was bulldozed away a few years later.

Other ambiguities in the Act were revealed during the handful of prosecutions held for wrongful damage to SSSIs during the 1980s. The first concerned a farmer who had damaged some natural grassland in Ulverscroft Valley SSSI in Leicestershire by spreading lime on it. He pleaded guilty, and was fined a hardly exemplary £200 (the maximum allowed under the Act was £2,000). Had he chosen to defend himself, he would probably have got off. Subsequent test cases showed that the Act as it stood was effective-

Frequent victim: green-winged orchid, a signature species of old, unimproved meadows.

ly toothless. One avenue of escape was that the NCC could not prove that an owner had been served with documents. In consequence, Gwynnydd yr Afon Fach SSSI was ploughed, Broadstone Meadow SSSI sown with swedes, and part of Aberlady Bay SSSI sprayed with weedkiller. The NCC held a moratorium on SSSI notification while its lawyers put the Act under the microscope. Their opinion was that the legislation was deficient, and the NCC's only recourse was to ask for amending legislation. The minister found he was too busy, but Government gave tacit support to a Private Member's Bill introduced by David Clark, the Opposition spokesman on the environment. This resulted in the Wildlife and Countryside (Service of Notice) Bill, which became law on 25 July 1985. It saved the NCC from having to serve every document by hand by a new procedure in which the service of SSSIs was linked with that of planning applications. This required a register to be maintained by the local authority and made available for public inspection. Furthermore, the landowner was now obliged to confirm receipt of SSSI notification papers. Of course the amending legislation did not apply to any SSSI notified between 1982 and 1985, but, to their eternal credit, this opportunity to torch them all without fear of prosecution was not taken by Britain's landowners.

Another weakness in the Act was the three-month interval allowed between serving notice of an SSSI and its actual notification. The notorious 'three month loophole', designed to allow an owner time to comment and, if he wished to, to object, was abused by a few spiteful people who used the time to render the site 'no longer of scientific interest'. By March 1985, at least 21 proposed SSSIs had been 'deliberately destroyed or seriously damaged' in this way. For example, Northampton Golf Club literally stripped Kingsthorpe Field of its special interest by hiring bulldozers to remove turf and topsoil. At Rhos Derlwyn-Fawr in Dyfed, a few busy days with a tractor and deep plough were enough to damage beyond reasonable hope of recovery this ten-hectare island of bog and natural grassland.

The time it took for an order to be served offered yet another loophole. For each ministerial order, the NCC needed to submit a detailed case. Where agricultural or forestry interests were affected, the 'Min. of Ag.' might need to be brought in, and all this could take weeks. The NCC could, in theory, offset the delay by applying for an order early in the process, but that could easily be presented in court as sharp practice and evidence of a cynical lack of faith in an owner's intentions. Nor would it bode well for future relations. As with the service of documents, it took amending legislation, following the recommendation of a House of Commons Select Committee, to close the loophole. So hard was this particular legal knot to untie that the Bill's sponsor, David Clark again, was obliged to use diagrams to explain his point.

The most serious drawback of all was the time it took to notify SSSIs to more than 20,000 owners and tenants. Bill Adams (1984) memorably likened the process to 'walking to the moon'. Our legislators had either overlooked or misunderstood the historical nature of land tenure in Britain; for example, on commons, farmers may have rights over land without owning an inch of it. This made the system difficult to operate in places where ownership was shared, such as upland commons and river banks. Understandably, the NCC tended to leave the more complex cases till last, but until an SSSI had been notified under the Act, it lacked protection, with only the planning laws to fall back on. Despite the quadrupling of its budget between 1980 and 1989 from £9 million to £38 million (for how the money was spent, see p. 41), the NCC's work was dominated by SSSIs throughout the 1980s.

Progress of SSSI notification in the 1980s

Year	Number of SSSIs notified under 1981 Act	Total area (hectares) of SSSIs so notified
1982–83	35	18,487
1983–84	1,079	229,823
1984–85	1,906	415,465
1985–86	2,828	690,158
1986–87	3,956	1,021,958
1987–88	4,398	1,190,183
1988–89	4,846	1,414,335
1989–90	5,264	1,618,641
1990–91	5,576	1,721,502

Source: Appendix 5a of NCC annual reports 1982–1991. The figures take account of all SSSI denotifications made during that time.

Total SSSIs in 2000

	Number	Area (hectares)
England	4,088	1,053,796
Scotland	1,458	990,809
Wales	1,010	225,454

Source: Agency annual reports 1999–2000.

The Act in retrospect

The Wildlife and Countryside Act formed the legal basis of nature conservation in Britain for 18 years. Was it a good Act? Did it help to save our wildlife and countryside? Any assessment must be relative. In 1982, the Act's measures to increase protection for SSSIs seemed wonderful; in 2001 they seem pathetic. Critics included Friends of the Earth who spotted right away that the Act was addressing only the symptoms, not the cause, and condemned it from the start as 'a wretched and dishevelled piece of legislation' (Rose & Secrett 1982). It imposed regulatory machinery on conservation bodies while doing nothing to pay for it. Ann and Malcolm MacEwan claimed that 'Ministers have sold Parliament a pig-in-a-poke' by refusing to say how much money Government would provide as compensation for 'profits foregone' (MacEwan & MacEwan 1982). The Act was a dead end, leaving agriculture and conservation on a collision course, with no way of regulating the conflict except 'by pouring money into a bottomless pit'. Oliver Rackham (1986) considered that the Act had only made things worse by neutralising the NCC and 'bogging it down in the paperwork of administration. Its provisions for compensation are scandalously open to abuse ...'.

By contrast, Government repeatedly insisted that the Act was working very well, and indeed, that it was the best, most forward-looking, wildlife legislation in the world. Sir Hector Monro, then at the Department of Environment, suggested that the 'one or two' problem cases should be set against 'thousands of SSSIs all managed extremely well'. His colleague, Lord Caithness, claimed, against plenty of evidence to the contrary, that very few SSSIs had in fact suffered any damage. Neither of these statements was in fact true, but of course it depends on what you mean by 'few', 'well' and 'damage'. The NCC made a reasonable job of making the legislation work, that is by managing to keep a lid on difficult cases, and keeping inquiries, prosecutions and other public embarrassments to a minimum. Farmers and landowners, while not liking it much, by and large went along with the Act, and by no means all of them demanded full payment for profits foregone (those that did tended to be wealthy 'businessman-farmers' who would have made a pile of money in any walk of life). The most notorious example of blatant profiteering was the Glen Lochay case, in which the new landowner, a wealthy businessman, ran rings around NCC, forcing it to pay heavy compensation for foregoing afforestation and agricultural improvement. Conservation bodies that owned land, such as the National Trusts and the county wildlife trusts, tended to be more supportive of the Act – and the NCC – than those that did not, such as Friends of the Earth and WWF-UK. The critics exaggerated their case. FoE's claims that the Act would soon cost the taxpayer upwards of £20 million per year in compensation payments proved well wide of the mark; the average at that time was closer to £4 million. However, they were correct in pointing out that, since the details of management agreements were kept strictly confidential, neither they nor the taxpayer had any means of knowing whether or not they were getting value for money.

The Wildlife and Countryside Act did not prevent 'damage' to SSSIs. Much of this came from forces beyond the NCC's control, notably agricultural subsidies, and more generally from the CAP's encouragement to farmers to drag every last ounce of productivity out of the land. What was almost equally damaging was the neglect of land no longer in agricultural production, causing coarse grass and scrub to spread on heathland and natural grassland. Even protected land could be managed ignorantly or inappropriately; some of the National Trust's properties, such as Hatfield Forest and Wimpole Hall, form classic case studies in weathervane management. The Act could not prevent local authorities from granting planning permission for developments that damaged or even destroyed SSSIs. Indeed, there the Act had done little to change the situation, and no compensation for 'profits forgone' was available for developments dealt with under planning law. Nor did the Act do anything about Interim Development Orders, open-ended planning permissions that enabled companies like Fisons to dig peat on SSSIs, or drainage cabals to suck the water out of wetlands, or aggregate companies to blast away chunks of the Mendips.

Where it did have a positive and lasting effect was in creating a new land ethic based on co-operation between conservation agency and owner/occupier. Much of our wildlife depends for its survival on grazing animals, regular, low-key maintenance and the harvest, the thousand and one activities of a traditional farm. It was easy for desk-bound conservationists and civil service mandarins to think of SSSIs as being cared for by the NCC, but in fact the NCC looked after nothing except its office buildings and a minority of its National Nature Reserves. Its arts were advice and persuasion, backed by science, a small amount of practical expertise and a pocket full of small change. In 1980, as in 1880, the quality of Britain's wildlife depended in the last analysis on what farmers and landowners did. That the 1880s were a much better time for wildlife than a century later was nobody's fault. The system has changed from one dependent on labourers and carthorses to another ruled by the agro-chemical industry, crop breeding and machines. Turning the clock back is not an option. For one thing there are hardly any labourers or carthorses left.

In retrospect, the Wildlife and Countryside Act could indeed be pilloried as a leaky, badly draughted piece of legal gobbledegook, easily undermined by any competent lawyer. But this did not matter particularly, since it was intended to work through persuasion, not compulsion. What undid it more was the slow progress on notifying SSSIs – a defect of process rather than law – and the lack of sufficient resources to tempt farmers into entering long-term, positive agreements. On occasion, the Act was used in ways that were never intended, for example as the basis for a land-use strategy in some upland areas. There, unsurprisingly, it failed. To suggest that the Act's overall effect was to assure that things got worse more slowly sounds cynical, but it was still something. However, it also increased the polarity between SSSI land and 'wider countryside'. By the 1990s, it was possible, in the English lowlands at least, for a trained eye to identify with a fair degree of certainty which land is an SSSI and which is not. This is the gauge both of the Act's success and of its limitations.

Offham Down and after

Pressure to plug the many remaining loopholes in the Wildlife and Countryside Act built up during the 1990s, as depressing (if confusing) statistics of SSSI loss and damage continued year on year. In 1997, nearly half (45 per cent) of the SSSIs in England were in 'less than ideal' condition, according to English Nature. Evidently some progress was made, for by 2000 this proportion had fallen to a third. Some 300 SSSIs were being damaged every year, claimed Friends of the Earth. Despite this (and as before), the wildlife agencies seemed less than enthusiastic about asking for new legislation, preferring instead to pin their faith on good relations with owners and occupiers. There were a number of possible reasons for this. Agency culture in the 1990s was deeply non-confrontational. Prosecutions were time-consuming and, even if successful, there was little 'wildlife gain'. English Nature's chief executive, for one, saw no point in them. In his view, better results were obtained by working in 'partnership' with landowners than by shaking a worm-riddled stick at them. When, in 1995, Friends of the Earth drafted a wildlife bill to strengthen the protection of SSSIs – doing the agencies' job for them as they saw it – English Nature's ambivalent response was to offer a minimum of necessary tacit support while publicly doubting whether it would serve any useful purpose.

By the mid-1990s, some of the voluntary bodies had become dedicated students of wildlife law. The Worldwide Fund for Nature (WWF-UK) commissioned a report by Terry Rowell, *SSSIs: A Health Check* (Rowell 1993), reviewing the effectiveness of SSSI protection, and analysing its limitations. WWF summarised its recommendations in a ten-point plan to strengthen SSSIs, which was later worked up into a draft Wildlife Bill by Friends of the Earth. Launching the Bill in 1994, the former NCC chairman, Sir William Wilkinson, reminded the press that SSSIs were not what they seemed: 'I should like to set to rest the serious and widely held misconception that SSSI designation confers actual protection. In fact it does nothing of the sort. It acts as a feeble tripwire. It sets up a negotiating process, and the final decision rests with Government'. Also at issue was the billions spent through the CAP each year on supporting activities that damaged wildlife, compared with the scant few millions available to help wildlife and protect SSSIs. FoE's Wildlife Bill aimed to safeguard SSSIs by improving incentives for positive management and restoration, while making it more difficult for wrongdoers to escape prosecution. It wanted to extend the Section 29 'Stop Order' to all SSSIs, to scrap the landowner's automatic right to compensation for 'profits foregone' and to extend protection to include third-party damage. A Conservative MP, James Couchman, agreed to sponsor it as a private member's bill. Of course, the bill, like most private member's bills, ran out of time after passing through the Commons in 1997, but it formed a useful marker, and enabled the FoE to cry 'I told you so' every time another SSSI was lost. Even so, the prospect of such a bill reaching the statute book seemed as far off as ever. It was at this point that, in May 1997, as a general election approached, Justin Harmer, a Sussex farmer, decided to plough up an SSSI.

In 1996, the European Union had announced a subsidy of £591 per hectare for planting flax, grown to produce one of those mysterious vegetable oils for which the modern world seems to have an insatiable appetite. Perhaps through some oversight or crossed wire, the subsidy was available for any land suitable for growing flax, not excluding protected land such as SSSIs. At Lewes, that historic town sitting snugly in a declivity of the South Downs, the local office of English Nature had been trying to negotiate a management agreement with Mr Harmer over the future of Offham Down, part of the chalk escarpment running west of the town from Offham to Clayton, and designated as an SSSI because of its rare plants, birds and insects. But nothing English Nature was able to offer him could equal the EU's generous bounty for flax-growing. The four months allowed by law for negotiations passed without agreement, and reluctantly English Nature had to let him plant his flax. It could have asked the minister to make a 'stop order' under the familiar Section 29. For some reason, it did not do so. Publicly, EN claimed that this particular bit of the SSSI did not meet the necessary national standard. Privately it admitted that, since the owner flatly refused to negotiate, it saw no purpose in prolonging negotiations. Compulsory purchase was ruled out on the grounds that, since there was no public access to the site, management to conserve it would be impossible. And so the farmer climbed into his tractor and began ploughing the virgin chalk turf with its cowslips and early butterflies. Most of the work was done at night to avoid confrontation with protesters. Soon Offham Down was striped with muddy plough lines, 'as if a giant clawed beast had swiped at the hill' (Phillips 1998).

There the matter might have rested, without much publicity or fuss, had it not coincided with the 1997 general election campaign. Ever quick to spot an opportunity, Friends of the Earth, with the help of Sussex Wildlife Trust, held a press conference on the public bridleway overlooking the mutilated SSSI, to show the media how Britain's wonderful wildlife laws worked in practice. The outgoing Secretary of State, John Gummer, found himself staring at news headlines, and a mountain of mail full of words like 'ludicrous' and 'crazy' about a place he had probably never heard of. By all accounts, Gummer practically ordered English Nature to ask him for a stop order, which he duly granted. Mysteriously, it turned out that Offham Down *was* of national importance after all, on the basis of certain plants, such as bastard toadflax and round-headed rampion, which EN had suddenly realised were there (round-headed rampion is in fact found on practically every chalk hillside in the county – its local name is 'Pride of Sussex'!).

Meanwhile, even before the stop order arrived, protesters had invaded the site, set up camp and had begun to turn back the turves, to which downland plants still clung, thereby coining a new word – 'unploughing' – in the lexicon of land management. Soon it turned into a media festival of

'Giant claws scraping at the hill'. The 'unploughing' commences at Offham Down.
(FOE)

unploughing, involving hundreds of volunteers, working in long lines to restore the down with their bare hands. Once the stop order had been served, the farmer himself good-humouredly joined in. The Opposition environment spokesman, Michael Meacher, turned up to claim that none of this would have happened under a Labour government. Two days later, Tony Blair, asked about the Offham incident at a fortuitous public meeting in East Sussex, scented the public mood: 'You asked me about SSSIs and the crazy situation just near here where you've got a farmer being paid European Union money to effectively tear up a place that's of particular scientific and natural interest. This is completely crazy. It's a crazy situation ...' It would not happen under a Labour government, he added. Strengthening SSSIs had been Labour policy since 1994, and 'ensuring greater-protection for wildlife' now became a manifesto pledge. Blair's words undoubtedly helped to make legislation a more urgent priority once he was installed in Downing Street. Most conservationists now see this impressive example of 'people power' as the event that led to the Countryside Act and a significant strengthening of the protection afforded to SSSIs. Meacher was appointed Environment Minister in John Prescott's restructured Department of the Environment, Transport and the Regions (DETR). One of his first acts was to instruct English Nature to withdraw its permission for Justin Harmer to plough Offham Down and part of the nearby Offham Marshes. Meacher also promised a comprehensive review of wildlife legislation since 1982, and a consultative green paper. A green paper sounded like a green light.

Breakthrough: the 'CROW Bill'

To help Michael Meacher with his green paper, and to draw public attention to what they saw as the main issues, the 22 principal voluntary wildlife bodies, acting through Wildlife and Countryside Link, produced a document they called the Wildlife Charter. Published on 17 November 1997, it outlined the kinds of legal measures they considered necessary if the new Government was to meet its manifesto commitment 'to improve protection for wildlife'. The Charter covered a lot of ground, with the deft use of magic words such as 'biodiversity' and 'sustainability', but more or less repeated the proposals for SSSIs made in FoE's Wildlife Bill. The House of Commons obligingly tabled an early day motion in which 292 thoroughly lobbied MPs of all parties expressed support for the Charter. Meanwhile, FoE broke another taboo by launching an interactive website on SSSIs, Wild Places!, an illustrated public database of Britain's 6,000 plus SSSIs, complete with details of over 2,000 incidents of loss and damage in England and Wales. Thousands of electronically aware people were thereby able to study the latest SSSI casualties: Selar Farm, in South Wales, obliterated by a coal mine; Porth Ceiriad Dunes, also in Wales, partly levelled by the owner's bulldozers; another wetland site ploughed and reseeded by a farm tenant, the owner, Severn-Trent Water, having apparently forgotten to tell him the land was an SSSI. Some 2,000 people logged on to FoE's website during its first month's operation.

The promised consultative paper, *SSSIs – Better Protection and Management*, was eventually published in September 1998. It showed a Government receptive to the ideas in the Wildlife Charter, if offering rather less than was demanded, and at an uncertain date. There was agreement on the need to strengthen SSSIs by revising planning guidelines, closing legal loopholes and increasing penalties; and, more importantly, by offering better incentives for 'positive' management. The 'ransom money' of profits foregone would be scrapped. Views were invited on a range of other issues, including responsibility for limestone pavements and whether upland commoners should be regarded as tenants.

These proposals applied only to England and Wales. The parallel green paper in Scotland, *People and Nature: a new approach to SSSI designations in Scotland*, took a broader line, emphasising the claimed differences between Scotland's wilder countryside and England's (the differences are not in fact national but physical: between upland and lowland). Scotland is even more keen to stress local accountability, and to demand a much greater say by local communities in their own affairs. In 2001, the Scottish Executive set out the devolved government's proposals in a glossy consultation document, *The Nature of Scotland: a policy statement*, which proposed 'substantial reform to the way in which we protect and manage our most special natural places'. There would be 'less bureaucracy' and an appeals procedure. The document also held out the possibility of a shake-up of the system of protection in favour of the European system of SPAs and SACs, although this would require new legislation. Introducing the Government's proposals to the new Scottish Parliament in 1999, the late Donald Dewar doubt-

Sites of Special
Scientific Interest
(SSSIs) in Scotland
(from Scottish Natural
Heritage. *Facts and
Figures 1999/2000*

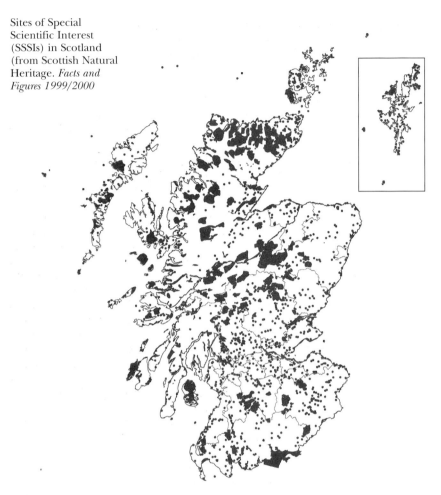

ed whether the sites designated by the European Union need all be notified as SSSIs. He didn't like the term SSSI anyway, and thought it should
be replaced by a more 'people-friendly' name.

In England and Wales, the voluntary bodies responded to the green
paper with a ten-point 'Wildlife Challenge', calling on Tony Blair to introduce comprehensive new wildlife laws in the next Queen's Speech. It contained the by now familiar demands on SSSIs in a context that included
common standards across the UK (a worried reference to Scotland's separatist tendencies) and a clearly defined *purpose* for nature conservation law
(whose absence hitherto is a fine example of how the trees have managed
to disguise the wood these past 30 years). It also reminded Government of
the absence of similar proposals for that perpetual Cinderella of nature
conservation, Northern Ireland.

Late in 1999, Government announced that it would introduce a
Countryside Bill to Parliament. But with so many Bills crowding

Parliamentary time, it decided to combine wildlife protection measures with the introduction of a statutory right to roam. This linked proposals that had become uncontentious with a radical measure that could expect a much bumpier ride, especially in the House of Lords. Many feared this could still scupper the Bill. Moreover, some backbench Labour MPs were bent on making the mix still more controversial by trying to graft onto the Bill their pet issue, the banning of hunting with hounds. Eventually they were bought off by being promised a free vote on a Hunting Bill in the next session.

The attempted hijack by anti-hunting campaigners was the most exciting moment in what, compared with the passage of the Wildlife and Countryside Bill 20 years earlier, was a rather uneventful parliamentary journey. The immensely complicated measures on SSSIs in the Countryside and Rights of Way Bill ('CROW' Bill) were tucked away near the back in Part 3 on *Nature conservation and wildlife protection*. The Bill was first read in March 2000, and, to the surprise of many, duly reached the statute books six months later. In the Commons, there was little opposition to the proposals on SSSIs since the Conservatives chose to direct their fire on the right-to-roam part. The reformed House of Lords had rather more to say, but mostly on the detail rather than the principle. Lord Buxton was against imposing prison sentences on gamekeepers caught poisoning protected species. Lord Ferrers expressed astonishment that anyone should want to protect noxious creatures such as stag beetles and adders. About 300 amendments were tabled, mainly about access. There was, however, one surprise gift – a new clause that gave statutory backing for Biodiversity Action Plans, thanks entirely to pressure from the voluntary bodies.

Under the new legislation, the agencies can at last refuse consent for any operation likely to damage an SSSI. The discredited 'profits foregone' regime was replaced by one emphasising positive management (by 1998, 84 per cent of management agreements issued by English Nature were of this kind, with only 16 per cent based on compensation – though they absorb 30 per cent of the total costs). On problem sites, the agencies can now issue a 'management notice', requiring remedial action on pain of prosecution and bringing their powers in line with those of the Environment Agency. Maximum fines have been increased to £20,000 in a Magistrates Court and an unlimited fine in a Crown Court. The CROW Act also irons out most of the remaining ambiguities and nonsenses over SSSIs. Agency staff at last have the right to enter and inspect SSSIs (although they are expected to ask nicely first). They are required to make a simple statement about preferred site management to owners. The legal power to denotify SSSIs is made explicit, and the process defined. Public bodies and statutory undertakers such as water companies and port authorities henceforth have a 'duty to take reasonable steps' to protect SSSIs 'consistent with the proper exercise' of their primary functions, and are encouraged to consult English Nature over their management. The CROW Act does not override the planning laws; local authorities can still allow someone to build a shopping mall on an SSSI, but they are supposed to consult the wildlife agency first, and to inform it if they intend to ignore the agency's advice (at which point, the agency would, one hopes, be having an urgent word with the minister). In keeping with the contemporary preoccupation with human rights and 'natural justice', everybody has a right to appeal about everything. However, the Act does its best to outline the respective duties of owners and occupiers, local authorities, statutory undertakers, wildlife agencies and other public bodies, in the hope that ignorance will no longer form an excuse for inaction. 'The thrust of the Act,' it explains, 'is on co-operation, not confrontation.' 'We continue to place a high priority on building effective partnerships with owners and occupiers' commented English Nature, and 'seek to manage SSSIs by agreement.' The law still does nothing to override previous planning permissions given for mineral and peat extraction – the loophole that enabled peat companies to dig up Thorne Moor. It has little to say about marine conservation other than initiating another review. Even so, it represents a huge improvement.

The CROW Act applies equally to all SSSIs, big or small. Hence the old Section 29 Order, together with the still undefined 'super sites', is now redundant and has been scrapped. But while SSSIs now have equal status as sites of *national* conservation importance, some sites will nonetheless be more equal than others because SSSIs are the chosen legal instrument in Britain for enabling European and international decrees, notably Ramsar sites (q.v.) and European SPAs and SACs. Hence, an SSSI that is also an SPA, SAC or Ramsar site (see below) will inevitably count for more than a plain SSSI. Moreover, it seems probable that SSSIs in Scotland may come to mean something different to what they mean in England and Wales, and probably under a different name. Never mind. The Government now dares to set itself the target of 95 per cent of SSSIs in a favourable condition by the year 2010 – and to achieve this it will have to do something about overgrazing and peat exploitation, currently the biggest causes of SSSI damage. What really matters, is that, for the first time in their half-century existence, Sites of Special Scientific Interest have become protected sites.

Euro-conservation

At the start of the new Millennium, habitat protection in Britain was in a state of transition between national and European systems. This section will deal briefly with the European Union's system, called Natura 2000, and what it means in practice. First, however, let us touch on an international convention that requires Britain to establish protected sites. This is the Ramsar Convention, named after the town in Iran where the parties first met, or the Convention on Wetlands of International Importance Especially as Waterfowl Habitat. It requires signatory nations to compile a list of 'Ramsar sites', get them designated, and report back to a special Ramsar Bureau. The Convention defines wetland widely, as meaning almost any wild wet place, natural or artificial, including estuaries, lakes, washes, reservoirs and bogs, so long as it has a rich wildlife or assemblages of rare species, especially birds. In Britain, we deal with this by designating Ramsar sites as SSSIs. This causes no particular problems since they are

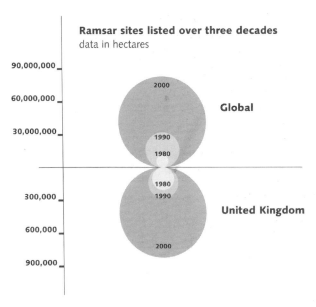

Ramsar sites listed over three decades
data in hectares

90,000,000

60,000,000 — Global

30,000,000

2000

1990

1980

1980

300,000 — 1990 United Kingdom

600,000

2000

900,000

Slow train to Ramsar 1970–2000. Britain has about one per cent of the world's notified wetland sites. (JNCC)

selected on the same basis. Even so, progress was slow, probably because no one could see much point in the exercise. Most of our 'Ramsars' are already protected by something else. By 2000, there were 149 UK sites on the Ramsar list covering 676,000 hectares, plus another 59,000 hectares in the UK's Overseas Territories and Dependencies. Another, more recent, international agreement is the World Heritage Convention, which runs a fund to establish World Heritage sites of cultural or natural importance. So far (2001) only St Kilda and the Giant's Causeway have made the grade in the UK as natural sites, though the Caithness and Sutherland peat flows and the Cairngorms have been put forward as candidates. This designation is more likely to produce better public facilities than better conservation.

Much more important is the developing European system of protected site designation, the Natura 2000 network. This is based on two Directives, the Conservation of Wild Birds, made in 1979, and the Conservation of Natural Habitats and Wild Flora and Fauna ('the Habitats Directive') in 1994. The first of these requires member states to designate 'Special Protection Areas' (SPAs) to conserve the habitats of certain birds, such as nightjar or stone curlew, that are rare or declining in Europe, or of migratory species of birds. Since British estuaries are used by large numbers of passage waders and geese, many of them have been made SSSIs *and* Ramsar sites *and* SPAs. Altogether some 230 SPAs have been designated, covering 11,165 square kilometres, or 4.6 per cent of the land surface. The Habitats Directive requires Britain to set up a more comprehensive series of 'Special Areas for Conservation' to protect a long list of habitats and species (other than birds) of special interest, both on land and in the sea. These are known as SACs. There will be about 500 SACs, covering 21,658 square kilometres, or 8.4 per cent of the land surface. Together, SPAs and SACs will form part of the trans-European network of sites, Natura 2000, chosen on similar criteria. In theory the same system of site selection and protection will operate throughout the European Union, from Britain to Portugal and Greece. The Habitats Directive requires SACs to be maintained in 'favourable conservation status', and protected against harmful developments except in cases of 'overriding public interest'. Under British law, all Natura 2000 sites have first to be notified as SSSIs. The increased protection for SSSIs under the CROW Act will, it is hoped, cover the EU requirements, but positive management will have to be achieved by negotiation and agreement on the individual sites – another job for the country agencies.

In 1999, the European Commission began a process known as 'moderation' to consider whether the candidate SACs submitted by member states, including Britain, would meet the requirements of the Directive adequately. For many habitats and species, Britain's candidate list, drawn up by the JNCC, fell well short. For example, there were not enough designated rivers to protect otter and Atlantic salmon. Fortunately the Regulation was amended in 2000 so that candidate SACs (cSACs) are now protected as if they were already SACs. But there is a lot of work to be done even to reach candidate stage, and public consultation on SACs has dominated the recent site protection work of all three British nature conservation agencies. By the time the process is complete, there are likely to be at least 300 SACs, many of them very large. The very biggest are National Park-sized, notably 'Caithness and Sutherland Peatlands' at 143,539 hectares and the Moray Firth at 151,352 hectares. Some 'lucky' sites may end up being Ramsars *and* SPAs *and* SACs *and* SSSIs – and, in some cases, National Nature Reserves as well! To outsiders, at least, this looks like gross bureaucratic overkill. It is one of the prices we pay for the ambiguities of modern life: a nation state which is also part of a European Union, allied to the twist of changing circumstances.

What matters, of course, is not the labelling but what the labels achieve. SACs, in particular, form a potentially powerful conservation tool because they must be maintained favourably, which in turn requires adequate incentives for occupiers and managers. Through its projected targets for SSSIs, the Government is now committed to this process, which will require reforms to the CAP (for example, on stock-rearing subsidies) and national legislation (for example, removing old planning permissions for minerals, with or without compensation). A widespread fear is that SACs may strip away resources from 'ordinary' SSSIs, and this does indeed seem likely to happen in Scotland. This would be highly undesirable. Not only do SACs cater for a smaller range of habitats and species than the SSSI series, but they are chosen on European criteria, and hence omit a lot of wildlife and wild places that happen to matter to us (for example, most of my county's downs and woods). Concentrating entirely on the European system would produce a grotesque distortion of nature conservation in practice. This will surely become an issue for the near future. The story, it seems, is never over.

PART II

Wildlife Habitats

5

Nature Reserves

Places of peace, havens of delight

In a world that so often seems to confuse the shadow with the substance, nature reserves are easily taken for granted. Today's culture values the new and sensational, but neglects things of solid, permanent worth. So let's say it. It may sound corny, but nature reserves are the greatest achievement of half a century of nature conservation in Britain. The reason is simple. It is only on land specifically set aside for it that nature comes first. There are other forms of protective designation, but in practice all of them entail a compromise between conservation and production. Nature reserves are different. We may need to borrow traditional forms of land husbandry to maintain habitats in their desired state, we may misunderstand things and make terrible mistakes, but on reserves it is wildlife that are the stakeholders, we the onlookers. We have been remarkably successful at preserving the best examples of natural habitats as nature reserves. Of the 284 places in Britain and Ireland identified in 1915 by Charles Rothschild and his helpers as the 'best spots', getting on for half are now nature reserves of some sort – a proportion that would probably have amazed Rothschild himself. All of Britain's wild diversity is represented in the nature reserve series: pristine limestone pavements, wild moors and mountain tops, fern-fringed Atlantic woods (Britain's rain forest), heaths and commons that have never been ploughed, wild estuaries and islands teeming with birds. A large proportion of our rarest plants and insects occur in nature reserves; some are almost restricted to them.

The acquisition of a network of over 200 National Nature Reserves and ten times that many other nature reserves has absorbed much of the time and energy of conservation bodies since 1950. The need for them is stated constantly in conservation literature. Wild places are under threat through technology-aided human progress. Naturally we all want to preserve the best of what remains. Some commentators have discerned more anthropocentric motives that may operate at a half-conscious level. In his perceptive study of nature conservation, *The Common Ground*, Richard Mabey saw something deeper in our zeal for nature reserves – in 'the planning, the fund-raising, the planting and building, the digging of ponds, the erecting of nest-barrels, the officiation by wardens, the public gatherings and the private vigils' (Mabey 1980). In all this frenetic activity is there not 'something sacramental ... a kind of temple-building (albeit to some very fleshly and familiar spirits)? We make these sanctuaries as acts of celebration and charitableness ... We are preserving something in ourselves as much as in the outside world'. There are some who would snort at such notions, but that the appeal of nature reserves goes beyond purely scientific considerations is undeniable. Even the unsentimental scientists that sat on the founding Wild Life Conservation Special Committee in 1947 recognised their 'cultural value' and compared them with ancient monuments and national museums, that is, as artefacts of the human spirit as well as a means to a desired end.

I have been lucky enough to have seen a lot of National Nature Reserves. My earliest memory of practical conservation work was cutting the invading sycamores and banging in fence posts at Yarner Wood in Devon. I later took part in some scientific monitoring of woodland vegetation at Wistman's Wood, that curious collection of dwarf oaks in the middle of Dartmoor, and investigating the life cycle of rare flowers high on the bleak fells of Upper Teesdale. In the late 1970s I found myself working for the Nature Conservancy Council in north-east Scotland as one of the team that looked after some wonderful wild places – the Sands of Forvie, that great duneland wilderness by the mouth of the Ythan, the Muir of Dinnet with its bearberry moors and pair of lily-studded lochs, the alpine Morrone Birkwood, and one of the great native pine woods, at Glen Tanar. My formal job was drafting their respective management plans, but I found myself so much in love with them that one of these plans turned into a publication, *The Muir of Dinnet. Portrait of a National Nature Reserve* (Marren 1979), and I helped organise seminars and edit collections of papers on two more, St Cyrus and Glen Tanar. Had I stayed there, I would probably have written a lyrical tome about the grand, wild, nature reserves of Scotland. As it was, I moved south and, in 1993, wrote a book about England's instead, about generally softer landscapes where the sound of larks and water birds was often drowned by a chugging tractor trailing a cutting machine or weed-wipe, or a dredger squirting out accumulating mud from the bottom of a lake.

Nature reserves come in all kinds of guises, from unique one-offs, such as the bouncing bog or *schwingmoor* at Chartley Moss in Staffordshire, or the isolated limestone hill at Stanner Rocks in Powys with its own suite of wild flowers, to places more typical of the wild landscape, such as the chalk downs at Wye, with their rare orchids and moths, or the series of pristine bogs at Silver Flowe in Galloway, a place whose very name has magic. If you live in the lowlands, nature reserves may come to mean relatively small places, often run by the local authority or the county wildlife trust. However, to see National Nature Reserves at their best and as they were originally envisaged, do visit some of the bigger, wilder ones, such as Invernaver or Inchnadamph in Sutherland, Beinn Eighe in Wester Ross, Roudsea Woods and Mosses in Cumbria, the cluster of Atlantic oak woods in Snowdonia National Park, the pine woods at Abernethy in Speyside or the Lizard Heaths in Cornwall. These places are important because they are big enough to support a high natural biodiversity, and have the human resources to ensure they are looked after properly without compromising their natural character.

If we lived in a pre-industrial society in which farming and wildlife existed

Highland fastness: Traligill Burn and Conival in Inchnadamph National Nature Reserve, Sutherland. (Derek Ratcliffe)

in harmony, we might not need nature reserves at all. In today's circumstances we do. Beyond question, many a pleasant down or heath would now be a field of rippling wheat or gloomy plantations of spruce if someone had not had the foresight to acquire them with the intention of leaving them as they are – beautiful, wild and transcending mere commodity value.

A short review of our nature reserves

For many, nature reserves are the embodiment of nature conservation in practice. They secure, for choice areas, a security of tenure under a form of management that benefits wildlife and preserves diversity. Nature reserves do not necessarily have to be set apart exclusively for wildlife. Many continue to be grazed by livestock, and not a few provide products for sale, from firewood and barbecue charcoal to premium quality hay or thatching reed. A few look the way farmland used to be – meadows, hedges, even cornfields. However, few, if any, British nature reserves provide enough rents or sales to pay for their upkeep. Admission is free for all but a few, like Wicken Fen in Cambridgeshire. It could hardly be otherwise with rights of way running through many of them. Instead, nature reserve maintenance is funded mainly by grants and carried out partly in-house, partly by volunteers, partly by contractors. Large capital grants often come from the Heritage Lottery Fund and conservation agencies, while many reserves are acquired by appeals, or by grant-aid from the nature conservation agenices or the local authority. One way or another, they are paid for by public funds.

What is the oldest nature reserve in Britain? Probably the private one set up by the Yorkshire squire and explorer, Charles Waterton, in 1821. The oldest formal one is the bird sanctuary at Breydon Water in Norfolk, purchased by a local society and declared a bird reserve in 1888. The earliest to protect a natural habitat is probably Wicken Fen, the first few acres of which were purchased by Charles Rothschild in 1899 and handed over to the National Trust for safekeeping. But although nature reserves are relatively recent arrivals on the rural scene, the idea of setting land aside for wild animals is almost as old as history. From Norman times, large areas of relatively wild country were preserved under special laws designed to safeguard wild deer and other game. However unjust and corrupt forest law might have been, it certainly preserved many areas of woodland and heath, most notably, of course, in the New Forest, where the system, in modified form, survives to this day. Another characteristic British habitat, the deer park full of ancient trees, is also the product of aristocratic fashion. Though maintained as status symbols, some were also nature reserves in all but name. Commons, relics of peasant agriculture, are another godsend for wildlife. We owe more to pre-conservation ages than is sometimes admitted. Nature reserves are sometimes only the latest manifestation of a long conservation history.

Most nature reserves are designed, naturally enough, to protect wildlife habitats. But a large minority are run by special interest bodies, such as the RSPB, the Wildfowl Trust or Butterfly Conservation, and managed in the interests of their client groups. Some habitats, such as limestone grassland and lowland heath, have a disproportionate number of nature reserves. This is partly because they are rich in rare species, but also because they are passing out of the economic system, and hence are under threat. Today, about the only conceivable future for habitats such as heaths and peat bogs is under some form of protective agreement, of which freehold ownership by a conservation body is the safest. Other reasons for acquiring land as nature reserves are to save it from imminent destruction, to protect rare species (especially birds, orchids and butterflies), to secure undisturbed areas for ecological experiments, and sheer sentiment. As Richard Mabey (1980) shrewdly observed, the act of creating them can seem as important as their effectiveness. Acquiring a nature reserve looks like an achievement, but it will come to naught unless the place is looked after properly.

Who looks after nature reserves in Britain? Acquiring a nationwide network of National Nature Reserves (NNRs), representing the best examples of natural habitats and geological formations, was one of the core tasks of the Nature Conservancy, set up in 1949. Over the next 21 years, the Conservancy acquired over 100 NNRs. Some were test-beds for long-term research and experiments in scientific management in the 1960s; some became well known, nationally and internationally, and were much visited by foreign scientists setting up their own nature conservation institutions. It probably helped that, by chance, so many NNRs had such memorable names: Castor Hanglands, Kingley Vale, Rodney Stoke, Tregaron Bog,

Braunton Burrows, Craigellachie, as well as a gaggle of Welsh Coeds and Cwms (I can still hear Morton Boyd describing 'the jewels in our crown from Caerlaverock to Muckle Flugga!'). They were famous once; Sir Fred Holliday claimed that the name Monks Wood resounded around the world, and everyone in ecology and conservation circles knew at least the names of Moor House, Upper Teesdale and Rum (formerly spelt Rhum). Each of the older ones has a scientific and conservation history, and I maintain that you can learn more about nature conservation in practice from the story of a single nature reserve than from a whole shelf of the latest agendas and strategies. The portfolio of NNRs has grown since 1970 to nearly 300 (200 in England alone), but although they do their job as well as – and, as far as management is concerned, better than – they did in the past, they seem to have lost their onetime significance. From being newsworthy and exciting, they have gradually moved to the margins of things.

Nature reserves are also run by local authorities. Some of the wild green spaces in and around London, such as Wimbledon Common, Richmond Park, Burnham Beeches and Epping Forest were the product of a far-sighted realisation that Londoners needed nearby havens of calm and beauty. The authorities had the good sense to preserve at least some of their natural character. Powers to designate 'Local Nature Reserves' (LNRs) were given to all local authorities in 1949. After a promising start, when wild seaside places such as Gibraltar Point, Ravenglass Dunes and Aberlady Bay were designated as LNRs, most local authorities found they had more urgent priorities. Local Nature Reserves did not really take off until nature conservation entered the mainstream national agenda in the 1990s. By

Part of Wye National Nature Reserve in Kent: the deep chalk coombe known as the Devil's Kneading-trough. (Natural Image/Bob Gibbons)

1998, however, there were 598 LNRs covering 29,032 hectares, representing more than a doubling by number since 1991. Most are in densely populated areas and are managed as public amenities. Some are SSSIs, and a few, such as Castle Eden Dene, Chobham Common and Kenfig Burrows have even been made National Nature Reserves. The first nature reserve to be run by a parish council, Coppice Leasowes in Shropshire, was designated in 1999. Country Parks are another form of rural amenity run by local authorities since 1968, which in some cases are de facto nature reserves. A well-known example is Sherwood Forest, an SSSI-designated fragment of old forest containing 600 'veteran' oak trees, some of which might have sheltered Robin Hood.

Another idea dating from 1949 was the Forest Nature Reserve (FNRs), which could be set up by agreement between the Conservancy and the Forestry Commission or another Crown body. They did not achieve very much. At Bernwood Forest near Oxford, some marginal areas were left for rare butterflies, but nature reserve status did not spare the rest from normal forestry practice, which in the 1960s meant clear-felling most of the wood and restocking it with conifers. Among the Forest Nature Reserves, however, is a very important one – Lady Park Wood in the Wye Valley, possibly the wildest wood in England. Exceptionally, it has been left to develop naturally – though not without a struggle with the FC, which, as usual, wanted to cut down the oldest trees to encourage regeneration. Scientists have learned more about natural woodland from studying Lady Park Wood than from any number of managed nature reserves. The concept of FNRs was revived in 1988 by the Forestry Commission, mainly for native pine woods in its ownership (they are now called Caledonian Forest Reserves).

Lady Park Wood NNR in the Wye Valley – near natural woodland preserved unmanaged for ecological study. (Derek Ratcliffe)

Another form of nature reserve was the statutory bird sanctuary, created by the Home Secretary after consulting the local authority. Most were on estuaries or other soft shores used by wildfowl and waders, such as the Ribble, the Humber or Caerlaverock on the Solway. Designation did not involve a change in ownership. The main purpose was to protect the roosting grounds of geese and other birds by a voluntary ban on shooting. The Wildfowl Trust also retained the word 'sanctuary' for its small but significant network of wetland nature reserves in Britain and Ireland (see Chapter 3).

The majority of nature reserves in Britain are run by voluntary conservation bodies, most notably the county wildlife trusts and the RSPB. A number of National Trust and National Trust for Scotland properties are also managed as nature reserves, either by name or in practice. Between them, the voluntary bodies now look after getting on for 3,000 nature reserves, ranging from large, famous places such as the Nene Washes and Dungeness to pocket-handkerchief reserves of less than a hectare. County trust reserves were often run on a shoestring, but, at least for those of SSSI status, much greater resources became available in the 1990s through agency and lottery grants and tax credits. The RSPB's reserves are larger on average and have better facilities, including trails and hides. Some such as Minsmere, Abernethy and the Insh Marshes are the equal of any National Nature Reserve. In the early days the RSPB did a lot of digging to create 'scrapes' and ponds for water birds. Today their purpose has broadened and they are managed for wildlife in general – if still birds in particular. The management of nature reserves designated as SSSIs must be agreed by the country agency, and this has achieved a kind of standardisation. Today, it matters less and less who runs a nature reserve, so long as the resources are there and the managers know what they are doing.

What do we hope to achieve by having nature reserves? Broadly speaking, of course, the aim is the protection of wildlife, but the ways in which the reserves support that aim are surprisingly varied. In the very early days, the accent was on pure protection. Nature reserves protected birds from being shot or otherwise disturbed, or wild flowers and butterflies from being collected. The main need was therefore for a fence and keep out sign (for the previous New Naturalist volume on nature conservation, the jacket designers incorporated a huge red no-entry sign). Unfortunately, simply fencing a site is counter-productive, for it often denies it the use that sustained it. The classic example is at Badgeworth in Gloucestershire, a pocket-handkerchief nature reserve fenced in to protect that very rare buttercup relative, the adder's-tongue spearwort. Inside the fence, the ungrazed vegetation grew tall and rank, and the spearworts temporarily disappeared. What they really needed was not fences but cows. The only places where unmodified protectionism actually works is on the seashore or a few mountain tops, cliffs and ravines. Practically everywhere else depends on livestock grazing (on heaths burning was a bad substitute) or forms of wild harvest, such as coppicing and hay-cropping.

The architects of the National Nature Reserves realised that natural vegetation is dynamic, and that habitats need maintenance. To develop tech-

An experimental plot high on the North Pennine fells at Moor House NNR. (Derek Ratcliffe)

niques in wildlife management, the Nature Conservancy acquired a few properties that were representative rather than outstanding, such as Moor House in the North Pennines or Monks Wood in what was then Huntingdonshire. They were intended to be 'outdoor laboratories' for experiments in grazing and tree planting (hence these places, at least, had to be owned to provide security for long-term experiments). As nature reserves, their value has increased over time. Moor House now contains the only unshot, unmanaged grouse moor in England. Some 400 scientific papers have been based on this particular outdoor laboratory, on subjects varying from vegetation production and geology to climate monitoring and tree establishment. In 1991 it was chosen as a 'lead site' for the Environmental Change Network, for remote monitoring of global warming and atmospheric chemistry. Moor House and Monks Wood are perhaps Britain's best known *scientific* nature reserves. But the people that set them up in the 1950s would have been disappointed there were so few of them.

Most nature reserves have the more direct aim of conserving and, if possible, enhancing whatever is inside them. Their management has generally been aimed at maintaining diversity, through a mosaic of habitats or a series of 'management zones', and habitat quality, achieved by a combination of grazing, mowing and cropping. Scrub-cutting is often necessary to prevent heaths or downs from turning into thickets, and regular coppicing preserves the open structure of a wood so necessary for carpets of wild flowers and visiting insects. However, until the 1980s, probably most nature reserves were under-managed. The development of 'recipe' handbooks for habitat management retrieved the situation a little, but what really saved the situation was more money, through agency enhancement schemes for nature reserves of SSSI quality, and the sudden fall of manna from the Heritage Lottery Fund. Improved technology also made a big difference, such as broad-wheeled vehicles that could cross a peat bog without leaving a mark, and powerful tractors capable of towing a variety of swipes and mowers through rough country. On wetlands, wildlife managers became adept at channelling water using sluices, pipes and dams. On the Broads, 20-tonne JCBs excavated ponds to recreate lost habitats for fen orchids and stoneworts; indeed JCBs have now become part of the conservation armoury, along with the universal four-wheel quad-bike (didn't someone declare there is little in the English countryside that cannot be improved by the creative use of a JCB?). All this investment, along with a new determination to tackle invasive herbage such as bracken and gorse, has made it possible to undo decades of neglect. The NCC and its successors tackled square miles of thicket using swipes, non-persistent herbicides and the full might of late twentieth-century hydraulic power. Before myxomatosis, Lullington Heath in East Sussex contained the biggest and best example of chalk heath, a rare and unusual habitat in which heather and other plants of dry, acid soil alternate with lime-loving plants such as wild thyme and rockrose. Unfortunately, having acquired it by lease from a water company

Lullington Heath NNR in East Sussex, a rare example of 'chalk heath' reclaimed from scrub and tangle by nature reserve managers using chain saws and hydraulic diggers. (Natural Image/Bob Gibbons)

in 1955, the Nature Conservancy was refused permission to introduce sheep to replace the lost rabbits in case they contaminated Eastbourne's water supply. By the time the company changed its mind, most of Lullington Heath had scrubbed over. Today, after an epic struggle led by tracked vehicles with hydraulic buckets, much of the reserve is once again open grassland and heath, maintained by grazing and mowing. Places like Lullington Heath have value not only in what they are, but by showing that, however hopeless it may look at first, restoration is possible. In the 1990s and today a struggle on a grander scale is underway to tackle invasive scrub on southern heathlands ('Tomorrow's Heathland Heritage'). The aim is to double the amount of open heath over the next 20 years. In Wales, the conservation bodies are faced by an even more difficult problem: invasive rhododendron – more difficult, because rhododendron toxins sterilise the soil, requiring surface treatment as well as chain saws and bonfires.

Hence, the practical management of nature reserves has changed over the years from almost none to a level that in some cases is comparable with a park or wild garden. Some argue that the pendulum has swung a little too far, and that it can be hard to find wilderness on English nature reserves; management is too conspicuous, man's imprint too obvious. There should be places that we simply leave alone, for better or worse. In Scotland, the retention of wilderness seems a more conscious aim. Of course, much of the burst of management on England's National Nature Reserves from the 1980s was aimed at undoing the neglect of previous decades, or the activities of the previous owner.

How are nature reserves selected? Before 1970, even National Nature Reserves were acquired in an ad hoc way, though most of them were recognised as being in some way special for wildlife. By 1965 the Treasury had started asking awkward questions. How many nationally important sites were there? How many nature reserves were needed to complete the series? How much would it all cost? The Nature Conservancy had no ready answer, and so it embarked on a project to provide a factual basis for nature reserve acquisition that could be defended scientifically. Under Derek Ratcliffe's leadership, this 'reserves review' became a fundamental review of Britain's wildlife resources that identified 735 'key sites' on a range of criteria, including size, rarity, representativeness and 'naturalness'. This was the *Nature Conservation Review*. By the time it was published, in 1977, however, the review had been decoupled from its original purpose as a shopping list. The Government never did get a direct answer to what might have seemed simple and logical questions. Each new reserve is justified on its merits, as before. There are, in practice, only two tiers of nature reserves – those that are designated SSSIs (or SACs) and the rest.

Why some nature reserves are better than others

Older nature reserves are generally better than recent ones. Their wardens or managers have had time to get to grips with the place, and to see the fruits of their labours. Big reserves are better than small ones. They are more stable, have more wildlife and are more cost-effective. Fortunately it

Morrone Birkwood near Braemar, a rare alpine birch wood with an understorey of juniper and a rich flora dependent on base-rich flushes and rills. Recently 'struck off' as a nature reserve because of differences between SNH and the private owner.

has sometimes been possible to link several small reserves and quasi-reserves and manage them as a unit (see below). Nature reserve management has improved over the years. In the early days, a warden might spend a lot of time fiddling about by planting trees or marram grass, or seeing off poachers. Today we think we have a better idea of what we need to do to benefit wildlife, and have developed the technical skills to do it. Even so, it is surprising how much nature reserves continue to reflect fads and fashions in conservation, rather than the unchanging bedrock of some scientific imperative.

Some have argued that, as an idea, nature reserves, good or bad, big or small, are conceptually flawed. 'Reserving' land for a quality that is in fact universal threatens to reduce nature to the status of a resource or tribe. In confinement, nature is apt to be treated with less than awe, farmed pragmatically as one crops a field of carrots. For a nature reserve to be successful implies an acceptance that conservation outside the reserve will be unsuccessful (if not, who needs a nature reserve?). By their existence, therefore, they reinforce a polarity between the protected countryside and the rest. Not all wildlife can be accommodated by nature reserves anyway. While they can support rare, relatively non-mobile species, such as a pasqueflower or an Adonis blue, there is little they can do to prevent the decline of widespread species, such as skylarks or harvest mice, or wide-ranging ones such as golden eagles or otters. Nature reserves exist partly because they are places where we can do something, immediately and on home ground. They give us something tangible to show for our efforts. They also exist because conservation bodies exist. Apart from purely campaigning bodies such as WWF, everybody seems to want a portfolio of nature reserves to show off. They are more solid than a policy, more permanent than a pamphlet. And they offer a service to the membership.

Conservation bodies do not always own their nature reserves. Among National Nature Reserves, fewer than a quarter are owned by the country agency; the rest are either leased or managed by agreement with a private owner over a fixed period. It was certainly the intention of the founders of official nature conservation in Britain that the bulk of the national reserves would be purchased. However, the Nature Conservancy's budget was limited, and not all owners were willing to sell. Furthermore, under the Conservatives in the 1950s, and, periodically, later on, there was political resistance to the Conservancy acquiring too much land lest it be seen as 'land sterilisation', and, worse, 'backdoor nationalisation' (National Nature Reserves had been conceived of during the postwar reconstruction when nationalisation was all the rage). In practice, NNRs were acquired as opportunity allowed, often by agreement with a private owner. At their worst, these agreements allowed the Conservancy to do little more than put up a sign. While all National Nature Reserves were of 'national importance for nature conservation', the management of some of them fell a good way short. By contrast, the RSPB owns most of its reserves, as, increasingly, do the county wildlife trusts. Another trend is for multiple ownership, with a conservation body contributing most, but other parties, notably the local authority, holding a stake.

The National Nature Reserves have (or at least had) more resources than most voluntary bodies can call on, but they are also more subject to outside interference. The unsatisfactory tenure of so many NNRs has been a major limitation on what they can hope to achieve. Some NNR agreements seem to have been made between owner and estate agent over a glass of sherry, without much involvement by scientists. In Scotland, SNH has begun a weeding-out process. Of the 71 NNRs in its care, only 31 met the three necessary qualifications: that nature conservation should be the *primary* purpose, that they should exemplify the best available standards of management, and be open to the public. Given that SNH owns only 17 of them (it part owns another 17) it is not all that surprising. Some 14 Scottish NNRs have been, or are about to be, 'de-declared', that is, struck off, on the grounds that their NNR status has become irrelevant. More rarely, this has happened in England and Wales, too. In 1994, a famous nature reserve, Braunton Burrows in Devon, was struck off after English Nature was prevented from introducing livestock there to prevent scrub invasion. National Nature Reserves have, rather ludicrously, come under political fire from right-wing ideologues opposed to anything with 'national' in the name. In 1988, the Environment Minister, Nicholas Ridley, sensing nature reserves might be making life difficult for field sports, 'invited' the NCC to consider selling off some of them, and in the meantime put an embargo on further land purchase. Since the NCC owned less than a third of its nature reserves, there was not much to sell. It took refuge in the time-honoured tactic of prevarication. Ridley expressed himself disappointed, but fortunately he was himself removed before the NCC could gauge the extent of his disappointment in its next annual grant. Ridley's successor, Chris Patten, promptly lifted the embargo, but since then the agencies have been reluctant to take on more National Nature Reserves unless someone else can be found to run them.

Another limitation of nature reserves is that nature is – by its nature, so to speak – not fully under our control. What happens on the ground may not relate very closely to the management plan. Monks Wood NNR offers an object lesson in humility. As the 'back garden' to the Nature Conservancy's field station specialising in nature reserve management, Monks Wood was something of a nature reserve showpiece. Its original management plan, published in book form in 1973 (Steele & Schofield 1973), partitioned the reserve into various 'management categories', notably coppice, high forest, glades and 'non-interference', each with its own set of prescriptions. This was partly to create a diversity of woodland habitats consistent with the wood's history, but also to hedge one's bets; the wood's scientific managers admitted they were 'not completely certain that man can manage more effectively than nature', and were therefore keen to leave parts of the wood to develop naturally. Within the zone earmarked as coppice, the warden planted nursery hazel cuttings, and 'layered' existing hazel shoots in a technique familiar to generations of woodmen. Planting was thought to be necessary because grey squirrels were eating most of the hazelnuts; acorns and oak saplings were also planted for the same reason. In the area designated as high forest, trees were thinned and old coppice stools singled to promote stands of tall oak and ash trees. The sides of the main rides were to be cut every eight years to create a fringe of blackthorn, the foodplant of the rare black hairstreak butterfly, and other shrubs. The glades, first formed by wartime potato fields, were cut annually with a tractor and swipe. Hopefully they would fill up with wild flowers.

That is how its managers thought Monks Wood should be. But, in fact, Monks Wood is not like that at all. The plan could not have taken into account Dutch elm disease, which created some wholly unplanned glades, nor the increase in muntjac deer, which created havoc with the coppice plots. Because of the deer, the wood's primrose paths have been replaced by tussocks of pendulous sedge, which seems to be the only plant the deer will not eat. The high forest developed willy-nilly, not from deliberate management but through the abandonment of it. In effect the whole wood has become 'non-intervention', except in the rides and glades, and they have become lawns. If we lose the battle with deer – and we are losing it – Monks Wood may become open and park-like, and all the earlier work will have been in vain (though we may rediscover some virtues in parks). This does not mean that management plans are a waste of time, or that the original Monks Wood one was wrong, only that they are inevitably based on experience and value-judgements made at that particular time. Complete control is an illusion. In practice, nature reserves are shaped not only by planning, but by natural events, such as gales, diseases, rising sea levels and non-native invaders, as well as by financial shortages, changing fashions and different neighbours.

Changing concepts of conservation

Recently the distinction between what is a nature reserve and what is not has become blurred. What sort of sense does it make to call the Cairngorms a nature reserve, but not the New Forest? Areas such as the Stanford Practical Training Area in the Breckland, or Windsor Great Park, contribute as much to nature conservation as any formally named nature reserve. Conversely, some nature reserves contribute little, if anything. Besides, in Britain, man and nature are indivisible. Multi-purpose use – blending amenity, access, conservation and production – or 'resting' the land as long-term set-aside, is becoming more characteristic of the modern landscape than sanctuaries set apart for wildlife. As conservation gains ground in national policies, nature reserves tend to lose significance. Some might argue that woodland nature reserves, for example, are no longer necessary, or that 'nature reserve' is not the right name for the increasingly large tracts of the Highlands and Islands owned by conservation and heritage bodies. The Woodland Trust does not call its properties nature reserves. Even the RSPB now owns and manages land where the term may not be appropriate.

In truth, 'nature reserve' always had a rather vague meaning. As we have seen, reserves cover a range of activities, from 'quiet enjoyment' to scientific experiments, biodiversity management and even non-intensive forms of farming and silviculture. In the case of The Wash, the National Nature Reserve label confers not exclusion, but a kind of sustainability plan, a forum in which conflicts can be resolved by discussion and a common strategy worked out. Conceptually, this sort of nature reserve resembles a medieval common more than the idea of a sanctuary set apart from use. In others, such as the Cairngorms, most of the activity is not about safeguarding wildlife in the narrow sense, but in trying to reconcile what Christopher Smout describes as 'use and delight' – the contending claims on our wild countryside (Smout 2000).

If one had to define the moment when nature reserves started to lose their traditional meaning, I would choose 1973, when the NCC took over responsibility for the North Meadow at Cricklade in Wiltshire. The original lists of nature reserves were made up of relatively wild places, but North Meadow is just a big field, cut for hay in summer and afterwards grazed by cattle and horses. Fields are first and foremost farm workplaces, and if they are also havens for wildlife, that is incidental. In taking on North Meadow, the NCC was really standing in for the parish stockman, and before him the medieval reeve. The reason it became a National Nature Reserve was North Meadow's large population of fritillary lilies, but the flowers required no special scientific management, only the slow rhythms of the farming calendar: flooding, mowing, grazing, pollarding and ditching. In harvesting its crop of lilies, the NCC was acknowledging that husbandry is as important as science, and that it is stability within a traditional farming

Hope springs eternal. This protected verge, in what is now a housing estate in Thetford, marks a colony of the endangered fingered speedwell.

system that sustains a place like North Meadow, Cricklade. It is the measure of the opening gap between agriculture and nature that such places are made nature reserves.

One could say the same about protected road verges, which form a kind of nationwide linear nature reserve. A network of them has been established, generally on the advice of the county wildlife trust and with the co-operation of the local highway authority. Most are sections of road bank with a particularly rich flora or one or more rare species. The system works best when there is someone nearby to keep an eye on things, such as Eric Simms and the beautiful, flowery verges he looks after at North Witham, Lincolnshire. Unfortunately the local authority sometimes forgets to tell their contractor or interprets protection as meaning no management, with the result that the verges become overgrown. In Cumbria, important verges have been 'improved' by farmers to supplement their rye-grass fields. Worse, the protection signs cannot stop eutrophication as the result of traffic fumes or the runoff from fields, which turn flowery waysides into monocultures of stinging nettle or cow parsley. Once again, protected verges are not a sign of 'wildlife gain': at their most successful they remind us of what has been lost elsewhere.

An acknowledged defect of Britain's nature reserves is that most of them are too small. Small reserves are vulnerable to change from surrounding land uses; a small reserve in the middle of an arable field, for example (and there are some), is probably doomed. Moreover small reserves are often awkward to manage. For this reason, it is desirable to link nearby nature reserves with corridors of 'wildlife-friendly' countryside wherever possible. For example, a series of nearby sites in the Chilterns of Bedfordshire are now managed as a unity by a forum composed of English Nature, the local wildlife trust, the National Trust and the local authority. At its core lies Barton Hills NNR, but this once isolated reserve is now buffered by a mosaic of set-aside and restored grassland with other 'oases' of natural chalk grassland which, with the right encouragement, should gradually expand and perhaps attract target species such as the stone curlew. The National Nature Reserve at Loch Lomond is today part of a wider nature conservation area with the National Trust's property at Ben Lomond and the RSPB's reserve at Inversnaid, the 'bonny banks'. Events there began to move rapidly in the late 1990s, with the opening of Ben Lomond National Memorial Park (allied to Forest Enterprise's plans to replace blocks of conifers with native trees) and the first tree plantings and woodland restoration work for the Millennium Forest for Scotland at the nearby Cashell estate. This whole area is set to become Scotland's first National Park. The preservation of its beauty and wildlife is the primary aim, and, because so much of the area is already under protective management, its realisation may be nearer than in most of England's National Parks.

One of the best-tested examples of land-integration for conservation and amenity is the Sefton Coast, a 27-kilometre curve of wide beaches and high dunes between the Mersey and the Ribble. Though broken by the sprawling towns of Crosby, Formby and Southport, much of this coast is still open and surprisingly wild, despite its proximity to Liverpool and ease of access from road and rail – so accessible, indeed, that they held car races here in the 1930s. More or less fortuitously, the wilder bits came into benign ownership between 1956, when a National Wildfowl Refuge was established at Southport, and 1978, when the integrated Sefton Coast Management Scheme started. The NCC acquired Ainsdale Sand Dunes, and later the Ribble Estuary and Cabin Hill, as National Nature Reserves. The National Trust acquired Formby Point and its famous red squirrels as part of its Enterprise Neptune scheme. Birkdale Sandhills and the nearby Ravensmeols Sandhills are Local Nature Reserves, run by Sefton Metropolitan Council. Together they own some 2,100 hectares of sand dunes, making it the largest protected area of this habitat in England. Moreover, this coast is particularly rich in wildlife, from the croaking chorus of natterjack toads in the spring to a variety of rare plants and insects, such as the sandhill rustic moth and the petalwort, a liverwort resembling a tiny crisp of lettuce.

Mobile dunes at Ainsdale
NNR, part of the Sefton
Coast conservation scheme.
(English Nature/
Peter Wakely)

The Sefton Coast scheme, begun by the local authorities with the backing of the Countryside Commission, but soon joined by other parties such as the NCC and the National Trust, attempts to find solutions to the many problems in keeping this area wild and attractive and yet cater for thousands of visitors. The problems include erosion – much of the sand had been removed by beach-cleaning machines, depriving the dunes of their necessary diet of loose sand – and the spread of scrub, especially sea buckthorn. Successful conservation demands co-operation between neighbours – for example, there is little point in digging ponds for natterjacks if the water is being piped away to a new housing estate. The tendency for the dunes to become overgrown with coarse grass and brambles has been overcome by winter-grazing them with hardy Herdwick sheep from the Lake District. They were introduced first to the National Nature Reserve, and, after signs of success there, are being used more widely as 'woolly mowers'. Some residents now consider there is too *much* management, and that in particular the shelterbelts of pine taking up space for perfectly good sand dunes should be left alone. Since the 1990s, the Sefton Coast scheme has been underpinned by candidate SAC status covering much of the area, and hence Euro-funding is available for an integrated nature conservation strategy. Its success owes much to an early start and to the broadly compatible aims of the main players. Integrated conservation areas like the Sefton Coast point towards a new, twenty-first-century role for nature reserves: as core areas of wild nature within multi-purpose 'parks' managed for outdoor recreation and tourism.

Wardening

The peculiarity of Britain's landscape is that practically all of it has an agricultural or silvicultural history. We are constantly told that Britain has no virgin wilderness comparable to, say Yellowstone or Serengeti. Even our wildest, remotest places, such as the blanket bogs of northern Scotland or the Cairngorms plateau, have been grazed by sheep or deer, scorched by fire or polluted from the air. Interestingly, when land becomes neglected and grows wild the wildlife value does not necessarily go up. Indeed, in formerly stable, traditionally managed landscapes, such as water-meadows or chalk hills, it goes down with a bump. Scrub has its value, but it is generally lower in biodiversity and less heterogeneous than the ancient grassland or heath it displaced. A farmed meadow such as North Meadow, Cricklade, manured, cut for hay and after-grazed by horses and cows, will contain more wild flowers than one which is ring-fenced, called a nature reserve, and left alone. The challenge faced by nature reserve managers in the1950s and 60s was in maintaining this inheritance of wildlife without the benefit of a reaper or shepherd, or a skilled woodman.

From the start, all National Nature Reserves had a warden. Sometimes the warden lived on the reserve in a tied house, but, as more and more NNRs were set up, he was often given several reserves to look after. The first generation of wardens were often from a background in forestry, gamekeeping or estate work. Several were retired servicemen (resourceful, self-disciplined, cool under fire). Traditionally, wardens divided their time between recording and monitoring wildlife, and care and maintenance: repairing fences and bridges, filling in potholes in the track, building culverts and dams. In the Cairngorms area, they shot deer and brought the carcasses down from the hill on the back of a pony. At Beinn Eighe, the warden spent much of his time trying to catch deer poachers. The first generation of wardens did a surprising amount of planting, of trees or marram grass. They also helped the scientists to carry out their experiments, which generally meant putting up more fences to keep animals out, or, alternatively, in. One ex-keeper inherited from the previous estate 'couldn't abide the sight of boggy ground, and took a disproportionate delight in digging small drainage schemes' (Boyd 1999).

Management activities on nature reserves are based on a management plan, which, in the NCC, was traditionally written by a scientist. Its purpose is to make a clear statement about what the reserve was for, to set management objectives within a time frame, and to focus resources in the most efficient and cost-effective way. A standard plan covers the next five or ten

Field study at Lathkill Dale, part of the Derbyshire Dales NNR. (English Nature/ Peter Wakely)

years in detail, while setting out broad aims and detailed tasks. In many cases, especially in low-maintenance places such as offshore islets or peat bogs, the aim is to keep the place more or less as it is. However, that does not necessarily imply lack of activity. Seabirds must be monitored, bogs need firebreaks and systems of water control, grassland needs a grazer. Just maintaining the status quo might entail a perpetual battle with birch, bracken and rhododendron. Such tasks were routinely itemised in the last section of the plan, called 'Prescriptions', following an analysis of the nature of the site and potential threats to its wellbeing.

Perhaps the greatest effort of all has gone into restoring peat bogs. From the late 1980s, the NCC and its successors inherited a large acreage of peat bogs devastated by mechanised peat cutting as a result of a controversial deal with Fisons. Since peat bogs are a living resource, formed by the bog-moss *Sphagnum*'s unusual ability to grow indefinitely and retain water like a sponge, even devastated bogs are renewable – providing they are wet enough. To ensure this the agencies embarked on a marathon exercise in dam-building to prevent the precious water from draining away. An object lesson in what happens to bogs when it does drain away is the National Nature Reserve at Thorne Moors, where fire took hold in August 1989 and consumed some 405 hectares of crumbling peat before it could be brought under control. Damming ditches with peat is a low-tech activity that takes a long time. By 1997, English Nature had built hundreds of dams at Thorne Moors, with the help of a wide-track Hymac excavator borrowed from the peat-diggers. Along with this work it has had to control invading bracken and birch scrub, which sucks water out of the peat almost as surely as a ditch drains it. Once the bog is sufficiently wet again, the scrub invasion should come to a natural stop. However, bogs must not be made too wet too quickly in case the natural acidity of the peat is softened by silt, which turns it into a fen. This has become a problem at Woodwalton Fen, where there is now plenty of water, but not pure water, as in the original fen, but water polluted with agricultural chemicals. For the same reason, it is considered unwise to build bonfires on very acid bogs in case the ash fertilises the peat. Altogether, restoring peat bogs requires a lot of care and patience, and is probably impractical unless the site is already a nature reserve, complete with an experienced person on the spot, such as Frank Mawby at Glasson Moss, or Peter Roworth at Thorne Moor, or Joan Daniels at the Shropshire mires.

Trust reserves

National Nature Reserves are chosen on scientific grounds as the best examples of wild habitats, in terms of size, naturalness and biodiversity. Almost by definition they are exceptional places. Many of the nature reserves run by the county wildlife trusts, on the other hand, sound like that place you know just down the lane. Among those run by the Kent Wildlife Trust is the bank near his home where Darwin studied wild orchids ('Downe Bank') and the wild garden of the astronomer Sir John Herschel. Others include a storage reservoir constructed in the 1960s that now attracts migratory birds and nesting little ringed plovers, a dump of chalk spoil that has since become another wild garden, and a mine that once produced Kentish ragstone and is now home to several species of bat. The very names of trust reserves seem ingrained in the landscape: The Mens, Chickengrove Bottom, Avery's Pightle, Dancers End, Oxey Mead, Coulters Dean. While a few cover 100 or more hectares, more are between one and 20 hectares: a stretch of bank too steep to plough, an overgrown common the trust has agreed to take over (in the absence of any remaining active commoners), or a wooded dingle above the remains of an old tramway.

Whether or not they are considered to be of SSSI quality, trust reserves often express the local character of the landscape, and are well-loved places to those who know them. An 'average' county might be Dorset, which has 22 wildlife trust reserves covering 760 hectares – which represents only 10 per cent of conservation and heritage land in the county, though more than half of its nature reserves. The 'jewel in the crown' – most counties have one – is the Kingcombe Estate, 152 hectares of traditional meadows, permanent pasture and woods with a field studies centre near Maiden Newtown. Another fine reserve, Powerstock Common, lies nearby. The Dorset Wildlife Trust has attempted to acquire examples of all the main natural habitats in the county, notably woodland (Bracket's Coppice 23 hectares), chalk grassland (Townsend 16 hectares), heathland (Winfrith Heath 103 hectares), fen (East Stoke 5 hectares) and meadow (Loscombe 10 hectares). Other nature reserve owners and managers in the county are English Nature (1,909 hectares), National Trust (5,500 hectares), RSPB (881 hectares), Woodland Trust (157 hectares), Plantlife (15 hectares) and several local authorities (383 hectares). With its beautiful wild coastline and extensive heathland, Dorset has a disproportionate number of National Nature Reserves and National Trust properties, but even so, if you live in that county the chances are that your backdoor nature reserve will belong to the Dorset Wildlife Trust.

Some trusts own considerably more properties than the Dorset one. BBONT, the trust for Bucks, Berks and Oxon, looks after more than 100, as does the Scottish Wildlife Trust, although within a much larger constituency. Typically a trust will own a large number of low-key nature reserves looked after by a farming neighbour or by volunteers, and a few high-profile ones with public facilities, a resident warden, and usually something exciting to see. It helps, for example, that ospreys nest at the Scottish Wildlife Trust's showcase reserve at the Loch of the Lowes, where they can be viewed from a hide. BBONT's equivalent is its Warburg Reserve, named after a celebrated Oxford botanist, a 100-hectare slice of Chiltern downland and forest, famous for its wild orchids and comparable in quality, scale and complexity with English Nature's Aston Rowant nature reserve not far away.

Some trusts own a flock of hardy sheep, which are taken from reserve to reserve as they are needed, rather like a touring party of conservation volunteers. (Conservationists are just as unsentimental about farm animals as farmers, often referring to them as 'management tools' or 'mowing machines'.) Woodland management presents special difficulties, since the

Exmoor ponies are widely used by the National Trust, county wildlife trusts and English Nature to keep rough grassland and heathland free of scrub. This also helps to conserve the breed, believed to be closely related to prehistoric wild horses. (Natural Image/Bob Gibbons)

skills that maintained our ancient woods have fallen into disuse. Years ago I had need to visit a lot of woodland nature reserves, and the contrast between the work of a professional woodman and a band of weekend volunteers was sadly obvious, and the former made a better conservationist. Fortunately skills are being relearnt, and the renewed demand for coppice products, especially for barbecues, is producing a new generation of professional woodmen.

My own trust, the Wiltshire Wildlife Trust (p. 75) specialises in grassland management. The Trust's 310 hectares of grassland nature reserves are let for livestock grazing to up to 20 different farmers. However, in the recurrent farming crises of the 1990s, it has become more difficult to find suitable grazers. Fortunately for some reserves a new type of farmer has stepped into the breach, smallholders who specialise in rare and traditional breeds of farm animals. There is now an eager market for such animals among discerning diners, which is lucky because they are generally hardier and better adapted to rough grazing than modern breeds. They look better too – Belted Galloways and Aberdeen Angus cattle on Wiltshire reserves look as though they have been part of the landscape since the Iron Age. By cropping the grass, and munching unwanted invasive plants, the animals maintain the grass in the optimum condition for wildlife. For its downland reserve at Morgans Hill, the Trust had the luck to find a shepherd with a small herd of miniature Dexter cattle, tough enough to withstand the harsh winter winds of the Marlborough Downs, light enough to use wet ground without poaching it too badly. The coarse grass that has overrun the Green Lane Meadow reserve may soon be tackled by Exmoor

Nature reserve signs are often 'dumbed down' with minimal information and no map. This is one of the better ones, for Overhall Grove, Cambridgeshire.

1 Cothill
2 High Halstow
3 Cavenham Heath
4 Moor House - Upper Teesdale
5 Ham Street Woods
6 Kingley Vale
7 Holme Fen
8 Monks Wood
9 Blean Woods
10 Orfordness - Havergate
11 Hartland Moor
12 Scott Head Island
13 Old Winchester Hill
14 Swanscombe Skull Site
15 Woodwalton Fen
16 Castor Hanglands
17 Bridgwater Bay
18 Bielham Bog
19 North Fen
20 Axmouth-Lyme Regis Undercliffs
21 Wychwood
22 Hales Wood
23 Wybunbury Moss
24 Westleton Heath
25 Morden Bog
26 Fyfield Down
27 Lullington Heath
28 Wren's Nest
29 Winterton Dunes
30 Rusland Moss
31 Rodney Stoke
32 Ling Gill
33 Weeting Heath
34 Aston Rowant
35 Thetford Heath
36 Knocking Hoe
37 Hickling Broad
38 Bure Marshes
39 Shapwick Heath
40 Rostherne Mere
41 Chippenham Fen
42 Chartley Moss
43 Lindisfarne
44 Ainsdale Sand Dunes
45 Dendles Wood
46 Ebbor Gorge
47 Holkham
48 Saltfleetby - Theddlethorpe Dunes
49 Pewsey Downs
50 Stodmarsh
51 Walberswick
52 Derbyshire Dales
53 North Meadows, Cricklade
54 Chaddesley Woods
55 Leigh
56 The Lizard
57 Swanton Novers
58 Castle Hill
59 The Swale
60 Barnack Hills & Holes
61 Great Asby Scar
62 Clawthorpe Fell
63 Ashford Hill
64 Gait Barrows
65 Forge Valley Woods
66 Wyre Forest
67 Thursley
68 Moccas Park
69 North Solent
70 Wye
71 Studland & Godlingston Heath
72 Parsonage Down
73 Morley Meadows
74 Roudsea Wood & Mosses
75 Park Wood

76 Upwood Meadows
77 Cotswold Commons & Beechwoods
78 Wylye Down
79 Ribble Estuary
80 Prescombe Down
81 Stiperstones
82 Blackwater Estuary
83 Martin Down
84 Brettenham Heath
85 Colne Estuary
86 Hamford Water
87 Dengie
88 Gibralter Point
89 Holt Heath
90 Holton Heath
91 Arne Reedbeds
92 Lady Park Wood
93 Stoborough Heath
94 Bredon Hill
95 Pevensey Levels
96 Castle Eden Dene
97 Muston Meadows
98 Barton Hills
99 Newham Bog
100 Lewes Downs (Mount Caburn)
101 Downton Gorge
102 Highbury Wood
103 Beacon Hill
104 The Flits
105 The Wash
106 Lundy Marine Nature Reserve
107 Hog Cliff
108 Gordano Valley
109 Barrington Hill
110 Lower Derwent Valley
111 Buckingham Thick Copse
112 Somerset Levels
113 Goss Moor
114 Golitha Falls
115 Hallsenna Moor
116 Tarn Moss
117 Dersingham Bog
118 Gowk Bank
119 Cabin Hill
120 North Walney
121 High Leys
122 Collyweston Great Wood & Easton Hornstocks
123 Ingleborough
124 Scoska Wood
125 Aqualate Mere
126 Thornhill Moss & Meadows
127 Derwent Gorge & Muggleswick Woods
128 Finglandrigg Woods
129 Malham Tarn
130 Dunsdon Farm
131 Greenlee Lough
132 Thrislington
133 Slapton Ley
134 Farne Islands

Key
Spotlight NNR
NNR

135 Wicken Fen
136 Redgrave & Lopham Fen
137 Benacre
138 Bassenthwaite Lake
139 Muckle Moss
140 Chimney Meadow
141 King's Wood, Heath & Reach
142 Burnham Beeches
143 Kingston Great Common
144 Duncombe Park
145 Foster's Green Meadows
146 Wem Moss
147 Blakeney
148 Hambledon Hill
149 Bradfield Woods
150 Hatfield Forest
151 Ant Broads & Marshes
152 South Solway Mosses
153 Roydon Common
154 Holme Moor
155 Teesmouth
156 Chobham Common
157 Ludham - Potter Heigham
158 Humberhead Peatlands
159 Martham Broad
160 Hardington Moor
161 Westhay Moor
162 Ham Wall
163 Ashtead Common
164 Newtown Harbour - Isle of Wight

165 Dunkery & Horner Wood
166 Ashford Hangers
167 Castle Bottom
168 Titchfield Haven
169 Wistman's Wood
170 Black-a-Tor Copse
171 Fenn's, Whixall & Bettisfield Mosses
172 Broxbourne Woods
173 Cassop Vale
174 Drumburgh Moss
175 Spurn
176 Sandy Beck Meadows
177 Walton Moss
178 Cliburn Moss
179 Duddon Mosses
180 Sutton Park
181 Calthorpe Broad
182 Ruislip Woods
183 Bardney Limewoods
184 Smardale Gill
185 Mid-Yare
186 Whitbarrow
187 Elmley
188 Langley Wood
189 Kielderhead
190 Dungeness
191 East Dartmoor Woods & Heaths
192 Kielder Mires
193 Durham Coast
194 Lydden Temple Ewell
195 Leigh Woods
196 Huntspill River
197 Valley of Stones
198 Sandscale Haws
199 Cribbs Meadow
200 Ebernoe Common
201 Sandwich & Pegwell Bay
202 Butter Hill
203 Richmond Park
204 Berry Head
205 New House Farm, Malham

National Nature Reserves in England (from English Nature, *Annual Report 1999–2000*).

ponies and Hebridean sheep, both breeds that have become very popular with nature reserve managers. Extravagantly horned Jacob sheep apparently did 'a fantastic job' of browsing off invading scrub on another Trust reserve. Other exotic animals currently grazing Britain's nature reserves include primitive tarpan and Przewalski's horses and water buffalo.

With its growing experience of managing large grassland nature reserves, the Wiltshire Trust has recently branched out into habitat restoration. With the help of the Heritage Lottery it acquired a 235-hectare-

National Nature Reserves in Scotland (from Scottish Natural Heritage, *Facts and Figures 1999/2000*).

wartime airfield and nearby farm at Blakehill, which, until recently, most wildlife bodies would not have looked twice at. Yet, although it has been neglected too long, and has grown tussocky and monotonous, this area on the stiff 'Minety clays' of North Wiltshire has never been ploughed and could be restored into rich pasture full of wild flowers. The Trust hopes to improve diversity by cutting hay and grazing it with cattle and ponies. This form of natural clay pasture is now considered to be an endangered habitat Europe-wide; with this one site, the Wiltshire Wildlife Trust will help the Government to reach nearly half of its ten-year target for restoring old meadows.

Nature reserves are often expensive to acquire and demanding to manage. Why, then, have the wildlife trusts devoted so much effort into building up networks of reserves, especially when so many others are doing the same? The traditional reason was to save the best of what is left of the wilder countryside. Another cogent reason is to provide a service for members, and perhaps also the local community: somewhere to go to on a sunny day. But nowadays, nature reserves serve other aspects of a wildlife trust's agenda. Like the National Nature Reserves, they can demonstrate ways of managing wildlife that may also retain land value and generate a modest profit from rents, sales and grants – in other words show that they can be a financial, as well as a spiritual, asset, a useful thing in these days of farm diversification and low-input organic produce. They also promote wildlife conservation in a more general way, by inspiring wonder and compassion for wildlife in a more involving way than television. Nature reserves are the trusts' shop-window displays. They show what trusts do, and what they stand for.

Hopes and dreams: island reserves

The recent history of the Hebridean islands known as the 'Small Isles' – Rum, Eigg and Canna, plus the even smaller isle of Muck – forms an interesting case study in nature conservation. All except Muck are owned by bodies that have conservation as a primary aim, and so are de facto nature reserves. Rum, the largest, is a National Nature Reserve, owned on behalf of the nation by Scottish Natural Heritage. Eigg belongs to the Isle of Eigg Heritage Trust, a partnership of the resident islanders, Highland Council and the Scottish Wildlife Trust. Canna is owned by the National Trust for Scotland 'for the benefit of the nation'. All three islands are noted for their natural beauty and wildlife, which includes seabirds and seals, eagles and otters. Perhaps their most distinctive species is the Manx shearwater, which nests in large numbers on Rum, with smaller colonies on Eigg and Canna. Each island has a crop of protective designations: SSSIs, National Scenic Areas, Special Protection Areas for birds, and so on. Some might see them as an island paradise, others (we hard-nosed ecologists) as overgrazed landscapes functioning at well below their biological potential: Fraser Darling's 'wet deserts'. The people that actually live there have to find ways of reconciling heritage with community income and quality of life. What do conservationists have to offer here?

The brooding hills of Rum, a mountain range in miniature. (Derek Ratcliffe)

Physically the Small Isles are surprisingly diverse. Rum is the wildest, a mainly mountainous landscape penetrated by glens, with almost no cultivated land. This is an accident of history as much as geography. Owned between 1888 and 1957 by the Bullough family, who made their money from Lancashire cottonmills, Rum was off limits to all but a privileged few, and gained a reputation as 'the forbidden isle'. There were no crofters or vested interests to worry about, as everyone who lived there worked for the Bulloughs. In 1957, the Nature Conservancy purchased Rum – at £23,000, less than a pound an acre – as an 'outdoor laboratory'. Acquired in a spirit of 'optimism for the future of nature conservation' (Boyd 1999), Rum offered opportunities for ecological investigations 'requiring complete quiet and immunity from interference'. The long-term goal was to transform the island into what it might have been before sheep and deer had laid it waste, Fraser Darling's dream of a wet Eden of mountains, woods and bubbling, trout-filled streams.

In reality, for all its hallowed status as 'a jewel in the crown' of National Nature Reserves, the experience of Rum has fallen well short of that vision. For the goals of the 1960s, the island had its uses. The sheep were all removed in 1957 after the Conservancy took over, and it now offered controlled conditions for Professor Tim Clutton-Brock FRS and his colleagues at Cambridge University to study red deer. But because it was important to preserve the deer herd, the pasture quality suffered, and natural regeneration has been limited. Instead, native trees, raised inside fenced nurseries, were methodically planted in Kinloch Glen, 15 hectares per year. In 1970, the first of two herds of Highland cattle was introduced to improve grassland diversity. In 1975 the first introduced sea eagles were raised and released on Rum.

But what must have seemed like a wonderful idea in 1957 had become something of an administrative nightmare 30 years later. Running Rum required a considerable financial outlay, and in terms of 'wildlife gain' what was happening there scarcely justified it. There was the heavy expense of maintaining a grand Edwardian folly, Kinloch Castle, which the Conservancy tried unsuccessfully to offload onto the National Trust for Scotland. While the Conservancy needed to own Rum in order to provide suitably stable and isolated conditions for long-term research, this inevitably entangled its successors in tasks that are closer to routine estate management than nature conservation: stalking, tree planting and animal husbandry. Doing all this on a remote offshore island poses severe practical and logistical problems. Today, some question whether a nature reserve, with all that that implies about exclusion and depopulation, is the right use of a significant Hebridean island like Rum (on the other hand, maybe we need places that are difficult to get to, and pretty rough when we get there).

Canna is a softer, more fertile island than Rum with a related but distinctive set of problems. Unlike Rum, Canna has enclosed farmland and several crofts. The island was given to the National Trust by its previous owner, John Lorne Campbell. Crofting has all but ceased, and Canna's single farm runs at a loss despite its sizeable headage of sheep. The island's diversity is declining. Most of the former patchwork of small hay and crop fields, cattle-grazed heaths and 'lazy beds' is now a uniform baize tablecloth of grass, much to the pleasure of the island's many rabbits, which keep it cropped short. This has been a disaster for the corncrake, for which Canna in its crofting days was a stronghold. Trust volunteers and summer wardens are hoping to tempt it back by fencing plots of tall grass and marsh, and planting 'gardens' of nettles and cow parsley, but the only long-term solution to this and other problems is the revitalisation of traditional agriculture. The island's character and the survival of its community go hand-in-hand, but the Trust finds itself struggling to prop up a failing island economy on a relatively modest budget. Unless it can solve the social problems, there is no likelihood of finding long-term solutions to the conservation issues, since the one tends to depend on the other (Johnston 2000).

The third of our island trio, Eigg, has the same sorts of things that you find on Rum, Canna and many other Hebridean islands – seals, shearwaters, puffins, various mountain flowers and Atlantic bryophytes, and, on a clear day, sensational views. Perhaps Eigg's most individual features are its singing beach of white sand, its natural scrublands of hazel and willow, and the plinth-like Sgurr, 'the highest pitchstone ridge in Britain'. Nature on Eigg is now primarily the responsibility of the Scottish Wildlife Trust. But what is a small trust with a modest income doing in a place like Eigg? Its presence is part circumstance, part opportunity. Having established a small nature reserve (the hazel scrub) there in 1979, the Trust had a representative living on the island who led the successful buyout in 1997, which resulted in the formation of a community trust to manage the island's

The Sgurr, Eigg's main landmark. (Derek Ratcliffe)

affairs. But Eigg also happened to be the right size, and with sufficiently varied natural features, for an experiment in integrating nature conservation into island life. The SWT's hopes for Eigg are not dissimilar to those of the NTS on Canna or the SNH on Rum. It wants to introduce more native trees, and give the natural scrub there a chance to regenerate and spread. It wants to introduce 'conservation-friendly' farming, which in the Hebrides tends to mean 'corncrake-friendly' farming. It wants to control bracken by cutting the expanding fronds with a tractor-and-swipe, and it wants to establish a better grazing regime, probably with more cattle and fewer sheep, with temporary fences to allow orchids to flower. It cannot afford to pay for all these things, but hopes it is in a position to act as honest broker for the island community and pull the right levers.

In terms of wildlife management – or at least aspirations of wildlife management – Rum, Eigg and Canna are on a converging course. Allowing for the islands' physical differences, the conservation policies of the various bodies look remarkably similar. However, the main issue in each case is not nature conservation in isolation, but the sustainable development of land and support for the island communities. Even on Rum, the emphasis is changing from research and experiments towards a broader policy of sustainable land use with a proper regard for the island's cultural inheritance. All conventional land uses in the Hebrides are hopelessly uneconomic on their own terms, and can be maintained only by subsidy. Sustainability, in other words, has to be paid for. Wildlife conservation is arguably as valid a land use as any other, especially now that tourism brings in more income than agriculture. But without farming, some of the wildlife may not be sustainable either. One cannot help wondering whether conservation bodies might have been tempted by the siren voices of the isles into entering projects that will strain their limited resources without necessarily much to show for it at the end. Perhaps a solution, as Laughton Johnston has suggested, lies in pooling their resources. For example, Rum's Highland cattle could be moved to Canna, which, unlike Rum, could probably sustain a beef enterprise. The animals could then be used to service both isles from a secure farm base. Or is it all just a dream, given the harsh climate, unfavourable soils and inaccessibility of the western isles? The significance of the Small Isles experiment lies in how far the idea of a nature reserve has changed from concepts of simple exclusion towards social and economic integration. Nature reserves form part of the wider land ownership debate in Scotland, and places like Eigg, and, on a larger scale, the Cairngorms, have become 'outdoor laboratories' of a different kind. If these experiments in subsidy management succeed, they may influence the future of the Western Highlands and Islands. If they don't, the experience may suggest to some that, while conservationists may preach the virtues of sustainable, wildlife-friendly land use, they are less adept at putting their fine words into practice.

6

The Farmed Environment

Between the coast and the mountain tops most British wildlife depends in some way on farming. Livestock grazing maintains grassland and heath, and arable farming creates an open, desert-like environment suitable for annual flowers that produce vast amounts of seed. One of the reasons why farming within living memory supported so much wildlife was that farming was itself such a varied activity. There was crofting – mixed smallholdings, mainly in western Scotland, and commoning – communal use of land by people with property rights. There were specialist 'habitat' farmers – fen-men, shepherds and open-range herdsmen – part-time Cornish farmers who grew bulbs and violets, and big estates that could afford to devote much of the land to country sports. Poor land was often left fallow for a season or two. Most farmers grew their own hay to feed the horses. And most fertiliser came from a horse, the river or the seashore; everything was farmed organically. The variety of farming contributed to Britain's natural variety of habitat. Some habitats for wildlife were created by farmers: meadows, hedges, arable, most ponds, some copses. It is worth emphasising, though, that others were not – no one planted the heather or the wild grass, nor dumped sand and mud around the coast, nor were the limestone pavements created by a supremely talented rock-gardener, though they may look like it (Plate 5).

The history of farming is no more serene and stable than any other industry. There have been violent lurches from corn to grass and back again, and the remembered landscape of the 1930s, with its boundless acres of 'permanent pasture', was in fact the product of empire: cheap imports and consequent lack of investment. The period between 1940 and about 1985, however, saw unprecedented changes in farming. It became reliant on technology and factory methods that boosted efficiency and, for a while, farm incomes, at the expense of the environment. It was foolish for farmers to pretend during this period that the countryside was safe with them. It was not, everybody knew it was not, and we, and they, are living with the consequences now. Here, as elsewhere in this book, we are chiefly concerned with the effect on wildlife of this agricultural boom-time, and also of the various subsequent schemes to reduce surpluses and create a more attractive countryside.

The bad old days

Agricultural policy between the 1940s and the 1980s aimed at increasing food production. The golden goal, which, until the surpluses started to pile up, was rarely questioned, was to make sure home-grown food was cheap and plentiful, and farmers prosperous. A vast increase in production

The flood plain of the River Kennet near Ramsbury in Wiltshire. It looks pleasant enough, but it represents a farming system in ruins. The sluice gates that maintained a network of drains that 'drowned' the fields in winter no longer function. Most of the nearby downland was ploughed in the 1950s and is now in set-aside. The scraps that remain are probably too small to prevent gradual loss of biodiversity. The woods overlooking the valley are no longer coppiced for hazel springs and have become too shady to support many wild flowers and flying insects. The breed that supported all three habitats – the Wiltshire sheep – is extinct.

was made possible by the postwar revolution in agricultural technology. Crop breeding produced new varieties of barley and wheat with ultra-high yields sustained by heavy dosages of chemical fertiliser and pesticides. Government grants paid for underdrains and the removal of hedges. On suitable land, the old mixed farms of prewar Britain turned into generally bigger arable units, with huge investments in machinery and grain silos. The big tractors with their spray booms, and the even bigger combine harvesters, required big fields; by 1980 the ideal wheat field was about 20 hectares, preferably in a perfect rectangle. Some fields were even larger: 'square miles of clods', as Jeremy Purseglove described them (Purseglove 1988). In the end, science and technology overtook human needs. By 1984, Britain's arable farmers produced 26 million tonnes of cereals, which was 10 million tonnes more than we could eat, and yet was unsaleable on world markets without an uneconomic price subsidy (that is, a bribe paid by taxes). Of course, this level of production came at a cost. That year, support to agriculture was around £5,000 million (Blunden & Turner 1985), so our food was not really as cheap as it seemed to be. The price we paid in environmental terms was even greater. Roughly half of the hedgerows of Britain – some 160,000 kilometres – were ripped out between 1950 and 1995, creating open prairies in place of the traditional landscape of small

No room for wildlife: farming in the Lincolnshire wolds. (Natural Image/
Bob Gibbons)

fields and copses. Less visibly our water and soil became awash with nitrates
from farm fertilisers, causing a wholesale eutrophication of the environment: murky water and lower natural diversity.

In dairy farming areas, more hedges were retained, but most of the
meadows inside them were reseeded. The dairy farmer was encouraged to
get rid of traditional breeds and invest in cows that yielded more milk and
adapted better to modern farming methods. To increase his stocking density he replaced the old, ill-drained flowery meads with shiny green crops
of rye-grass, maintained by generous doses of factory-produced fertiliser.
His hay fields were replaced by grass silage, also fertilised to replace the
flowers with more grass and allow more than one crop to be taken.
Mechanised cutting in May instead of July evicted the birds and prevented
the remaining flowers from ripening seed. To avoid parasites, the cattle
were stuffed with antibiotics and other drugs so that their dung became
effectively toxic, thus removing another wildlife habitat. Meanwhile, inflat-
ed on their diet of rye-grass and nitrogen, the cows innocently blew holes
in the ozone layer. Graham Harvey cited poor Cheshire as the ultimate
dairy farm nightmare, stinking of silage and cowpats, 'a landscape created
by the chemical giant ICI', the sound of larks replaced by that of the 'fertiliser spreader applying more nitrogen to the thick, lifeless rye-grass
sward' (Harvey 1997).

The rewards for growing cereals under high guaranteed prices encouraged farmers to drain their land. As the drainage of the Fens had shown,
even ill-drained 'levels' could be turned into fertile, profitable plains once
the water had been drawn off by deepening the drains and installing pow-

erful pumps. Drainage is expensive, but the state foots most of the bill. Decisions to drain land were taken by unaccountable local committees, chaired by a MAFF nominee and generally dominated by farmers. In some districts the main drains are the responsibility of Internal Drainage Boards, almost invariably chaired by some big, progressive farmer. At the high tide of the arable farming boom between 1975 and 1985, what Jeremy Purseglove called 'an army of engineers and machinery controlled by a drainage lobby' seemed intent on draining lowland Britain dry – dry enough to plant cereals descended from the wild grasses of semi-deserts. The results could be seen on levels and river systems throughout England, from the north Kent marshes to the Somerset Levels, from the Sussex Ouse to the Solway, as one by one former grazing marshes and other semi-natural habitats went under the plough. During my brief period as the NCC's local officer for Oxfordshire, the Thames Water Authority proposed a £1.6 million drainage scheme to lower the river Cherwell, improve the outfall of its tributary, the Ray, and drain what was left of Otmoor, that erstwhile marshy wilderness in the heart of the county (we had only recently fought off the transport department's proposal to drive the M40 through it). What saved the Cherwell was influential, articulate Oxford, enraged because punters on the river would have had to stand on tiptoe to peer over the engineered river banks. No wonder that conservationists were forever quoting Gerard Manley Hopkins' famous plea for flooded landscapes:

'What would the world be, once bereft
Of wet and wildness? Let them be left,
O let them be left, wildness and wet;
Long live the weeds and the wilderness yet.'

In the uplands, the main problem is overgrazing, which got worse in the 1990s. The EC sheep meat regime provides yet another form of subsidy, headage payments, which are supposed to guarantee the farmer a decent income irrespective of the state of the market. As a result, sheep numbers almost doubled between 1980 and 1995. In order to increase his stockage, the farmer drains and reseeds the hillside, or, if that is not possible, simply overgrazes it. Without some form of subsidy most hill farmers would go out of business, but the environmental price for continued farming has been dull, reseeded hills and grazed-to-the-knuckle fellsides. The ESA system, designed to preserve heather moors and attractive scenery, has had little overall effect, and no wonder. In the Cambrian Mountains, the farmer is paid £22 per acre for conserving heather, but £30 per head for more sheep. He earns more from the sheep. This manifestly non-sustainable use of our uplands is the result of a remote, sectoral system that addresses one interest – agriculture – but, at least until recently, disregards other claims on the land.

The system was a nonstop engine driven by ninnies and fuelled by money. A generous system of grants, subsidies and tax allowances enabled farmers to invest in land improvement and machinery that enabled them to sell more food at guaranteed prices. The system favoured those who ran their holdings efficiently as a business. Agricultural colleges taught young farmers how to work the system to squeeze ever-greater yields from their land. The agrochemical giant, ICI, set up a special department for lobbying MPs and civil servants. At the hub of the system in Whitehall were the cosy meetings between the minister and the National Farmers' Union, the free lunches between lobbyists and MPs. As John Sheail noted, the alliance of interests was so strong that conservation and recreational interests had the greatest difficulty in penetrating it, let alone influencing policy (Sheail 1998). The architects of postwar land-use policy, such as John Dower and Lord Justice Scott, had taken for granted the 'natural affinity' between farming and the protection of wild places. Technology and self-interest swept all that away, and the sheer complexity of the agricultural system created its own inertia: 'the pea brain of the dinosaur leading the industry to its own destruction', as Peter Melchett rather unkindly described it in 1980.

Government farming policy – the nearest thing we have to a national land-use strategy – is made up of the interplay of White Papers and the Common Agricultural Policy. Ever since the Agriculture Act of 1947, government has guaranteed prices for farm produce. Britain's entry into the EEC in 1973 did not change overall policy, but only reinforced it. Probably more wildlife sites went under the plough between 1940–73 than afterwards, but there was then no monitoring process to put the losses on record. The 1975 White Paper, *Food from our own resources*, the high tide of agricultural expansion, made few concessions to environmental concerns. Its successor in 1979, *Farming and the nation*, made some pretence at 'striking a balance' between production and amenity. After the test case of Amberley Wild Brooks (see Chapter 10), the protection of wildlife could, in some circumstances, claim precedence, but the onus was on the naturalists to make a convincing case, and the agriculture department would have to be convinced. Production was still the goal, and the crux of the matter was that almost all methods of increasing fertility and improving yields are harmful to wildlife. For the farmer bent wholly on agricultural production, wildlife has virtually nothing to offer. At best it is an irrelevance, at worst an obstacle.

In its 1977 paper, *Nature conservation and agriculture*, the NCC tried to suggest reasons why wildlife should matter. Some forms of wildlife might yet prove useful, it thought, and therefore it is mere prudence to conserve as many species as possible: 'This is what conservation is about – maintaining biological diversity and so keeping the options open.' Moreover, claimed the NCC, 'conservation and agriculture are interdependent. Agriculture depends upon the conservation of beneficial bacteria, soil invertebrates, pollinators, predators and parasites ... the obvious fact [is] that the conservation of the species necessary for farming is essential and provides much of the common ground between agriculture and nature conservation'. Kenneth Mellanby, writing in his New Naturalist volume, *Farming and Wildlife* (Mellanby 1981), found these arguments unconvincing. The truth is that we do not conserve wildlife because it is valuable, but because we like it. Refusing to admit that feelings have any validity denies our own humanity and threatens to downgrade wildlife to a mere resource. As Richard Mabey saw clearly in *The Common Ground*, 'a compassion for and a delight in the natural world are what turn people to act in its defence in the first place'. Just because feelings cannot be quantified it does not make them irrelevant (Mabey 1980).

In the early 1980s, more radical voices analysed the defects of modern farming, including its impacts on wildlife and scenery, and offered their solutions. Marion Shoard (1980) plumped for planning controls and more National Parks. The economist John Bowers wanted to steer tax incentives and subsidies towards more beneficial forms of production. Richard Body, a maverick Conservative MP, argued for a low input–low output system allied to free trade outside the Common Agricultural Policy (CAP). The right-wing think-tank, the Institute of Economic Affairs, proposed ditching subsidies altogether and leaving farms to sink or swim in a free market (a policy adopted, with notable success, by New Zealand in the 1990s). A few lefties from the farm worker's union got the Transport and General Workers' Union to lobby the Labour Party for full-blooded state control of land use through nationalisation. What they all had in common was the conviction that the farming system as it had existed since 1947 offered poor value, and that the urban majority who paid for it had a legitimate say in what was done. As Oliver Rackham put it, with his usual pithiness, we had 'contrived at the same time to subsidise agriculture much more than any other industry, *and* to have expensive food *and* a ravaged countryside' (Rackham 1986). Nor did the system even benefit farmers as a whole, half of whom had left the industry between 1950 and 1980.

Declining species

For the Prom concerts of 2001, the BBC commissioned a piece from the composer Sally Beamish called Knotgrass Elegy, inspired by a passage in Graham Harvey's book, *The Killing of the Countryside*. The work is a parable of the rape of the Earth by the agrochemical industry, as symbolised in the fate of the humble knotgrass weed, *Polygonum aviculare*. Hardly anyone knew or cared about the knotgrass until its loss was implicated in the decline of a species of economic importance, the grey partridge. The knotgrass happened to be the sole foodplant of a small leaf-beetle, *Gastrophysa polygoni*, whose larva featured in the diet of partridge chicks. The chicks depend on a sufficient supply of the right kind of insect as a source of protein for growing tissues and bone. So take away the knotgrass and the food chain leading up to the partridge collapses. Less knotgrass must inevitably mean fewer grey partridges. This fact became apparent only because the Game Conservancy had conducted a lot of research on partridges. The paper referred to by Harvey was written in 1982, and since then the grey partridge population has halved. Indeed, it thrives only where cereal growers have compromised by leaving unsprayed headlands or undersowing cereals with grasses. Such efforts are made mainly in places where people are interested in conserving grey partridges in order to shoot them! (Turning this on its head, the Conservancy now advises against shooting

The grey partridge survives best where unsprayed headlands are left around crop fields. (Nature Photographers Ltd)

grey partridge without habitat conservation measures.) The Game Conservancy's advocacy of six-metre field headlands saved the grey partridge in such areas, and perhaps the knotgrass too.

The cornfields of England have lost many flowers prettier and better known than the knotgrass: cornflower, corn marigold, Venus' looking-glass, corn buttercup, shepherd's needle, pheasant's eye, thorow-wax. Their pleasant names convey their one-time familiarity. Thirteen species of 'arable weeds' are now categorised as nationally scarce, seventeen are in the Red Data Book, and six have gone altogether, listed as 'extinct in the wild' (the distinction is necessary because the odd plant of unknown origin still shows up on disturbed sites). They are not the only farmland plants that are no longer everyday familiars – another missing tribe are the flowers, sedges and grasses of muddy pond margins that once flourished on mixed farms. Yet another group grew on old muck-heaps or in corners of pre-concrete farmyards. 'Arable weeds' ('arable flowers' is their 'preferred' title) used to thrive among crops by means of annual life cycles and persistent seeds that could survive for years in the ground. A few, such as the corncockle, had seeds that resembled grain, and so would be harvested and resown with the seed-corn. Agricultural advances have relegated most of them to the field margins, if not wiped them out altogether. First to go were the distinctive weeds of flax and hemp fields, which disappeared when their host crops ceased to be planted (there are no flax weeds left to colonise today's subsidised flax crops). Better ways of cleaning grain put paid to the likes of corncockle, but most of our traditional weeds are the victims of chemical herbicides developed in the 1940s and 50s. For example, the 'hormone weedkiller' MCPA wiped out the corn buttercup almost overnight. Some weeds, such as mousetail and weasel's-snout, are also vulnerable to chemical fertiliser. The last stands of cornflower and corn marigold were among root crops such as potatoes and sugar beet. I remem-

A field they forgot to spray: a crop of poppies and may-weeds. Is the time coming when weeds will be valued more than the crop? (Natural Image/ Bob Gibbons)

ber fields golden with marigolds in the early 1970s, in the days when chemists were still searching for an elixir that killed this particular weed without poisoning the crop – but by 1975 they had found it, and then it was goodbye marigolds, and good riddance. (Interesting, though, that many gardeners still plant marigolds among lettuces and other vegetables because they seem to deter insect pests.) Traditional 'weedy' fields are now rare, and mainly in areas where there are game crops and game strips for partridges, or where the spray booms miss a bank or corner. The main exception is the poppy, almost impossible to eradicate on chalky soils, thanks to its plentiful and remarkably persistent seeds. Otherwise 'weeds' are more likely to be seen as a result of roadworks or pipe-laying operations than farming. For example, a colourful ribbon of poppies, charlock and other weeds flourished for a few years by the recently opened M40 motor-way through Oxfordshire. The traditional weeds are no longer an agricul-tural problem, but chemical farming has bred a new generation of 'super-weeds', such as black-grass, cleavers and sterile brome, which thrive on fer-tiliser and are resistant to herbicides. Unfortunately they do not support the complex food webs of former weeds.

Since the 1980s, some of the rarer arable flowers have benefited from conservation schemes (meanwhile, surveys revealed that some were not quite as rare as had been feared). In a few places, rare weeds have even been conserved by abandoning any pretence at profitable farming and, in effect, harvesting the weeds – but this is only really practicable on nature reserves and experimental sites. On the National Trust's property at Boscregan near Land's End, unsprayed headlands are left for the attractive purple viper's-bugloss. In the Breckland, the Suffolk Wildlife Trust has taken on responsibility for certain sandy field margins where rare speed-wells and other rare plants grow. Unfortunately, large-scale schemes to reduce overproduction have not made much difference to our weed flora. On short-term set-aside, weeds are tackled as ruthlessly as before, while one widely used option, to sow a strip of grass around a field, actually robs them of their last refuge! Organic farming offers potentially better prospects for weeds, and there are early reports that some rare ones are reappearing where they have been spared by the hoe. Some of the more attractive ones, such as cornflower and corncockle are now a common, if usually tempo-rary, sight where 'wild flower' seeds have been sown. Wild flower sowing is a kind of anti-conservation: it replaces what is wild and natural with what we put there by choice and blurs the crucial distinction between habitat management and gardening. In the fields of the future we may even grow cornflowers as a crop. Does it matter if, in the meantime, the native corn-flower dies out, or if we can no longer tell the difference? (I only ask.)

Like the weeds they depend on, survival for many farmland birds is cur-rently on a knife edge. In essence, the problem is that they are not finding enough to eat. Like the grey partridge, many birds need a nearby supply of insects to feed their chicks. The same weeds that supported insects also produced masses of seed that fed birds in winter after the land was ploughed. Most agricultural advances mean less food for birds. Autumn sowing means fewer grain-rich stubble fields. Direct drilling does away with replenishment by the plough. Whipping out hedges deprives birds of food, shelter and nesting space. Another necessary factor is variety. Recently the MOD and RSPB carried out the first ever survey of the breeding birds of Salisbury Plain, which is normally out of bounds to the birdwatcher because of military activity. The results were astounding. Here as nowhere else the skies are still full of larks, and birds such as corn bunting, which in other parts is fast going the way of the cornflower, are still common. The reason seems to be not only the great size of the military training area – some 40,000 hectares – but also its variety of farmed habitat. There are big, hedgeless fields of corn, but they are interspersed with hillsides of well-grazed pasture where skylarks nest, knolls of rougher ground inhabited by whinchat and meadow pipit, and scrubby slopes ideal for stonechat and grasshopper warbler. The surveyors estimated that Salisbury Plain holds 48,000 pairs of breeding birds. At 14,600 'territories', the density of sky-larks there is higher than almost anywhere else in England and Wales. Thank goodness for tanks and artillery.

More often, birdwatchers will tell you doleful tales of long downland tramps in which they failed to spot a single lapwing and not once heard the familiar trill of a corn bunting. Farm birds monitored by the BTO's Common Bird Census since 1972 include some of the fastest declining

The lapwing is one of the losers: everything that could go wrong has gone wrong: field drainage, over-grazing, silage production, autumn sowing, coastal squeeze... Can it hang on in nature reserves until lapwing-friendly agriculture returns? (Natural Image/Bob Gibbons)

British species. Of Chris Mead's 'bottom ten losers' of the twentieth century, fully half nest and feed mainly on cultivated farmland. Among them is the grey partridge, whose population fell by 78 per cent between 1972 and 1996, along with bullfinch (down 62 per cent on farmland), turtle dove (85 per cent, ditto), corn bunting (74 per cent all habitats) and tree sparrow (76 per cent on farmland). The last-named has become so rare that it is reportedly being targeted by egg thieves. Familiar farm birds such as lapwing, skylark, linnet and even starling are in decline over most of Britain, and even the ubiquitous house sparrow has fallen on hard times (down 64 per cent, all habitats, since the 1970s), although no one knows why. There is, however, a ray of light ahead: Government has selected farm birds as 'good indicators of wildlife and the health of the wider environment', and so committed itself to a process that will reverse the decline. ('We value wildlife for its own sake and because it is an integral part of our surroundings and our quality of life.'.) This is a policy statement to gladden the heart, and we wait with impatience to see how they will achieve it.

It *is* possible to save declining birds, especially where they nest within limited areas and in a restricted range of habitats. Much effort over the past ten years has gone into halting the declines of three farm birds threatened with extinction: cirl bunting, stone curlew and corncrake. Before the War, the cirl bunting was a fairly widespread songbird in southern England and Wales, with perhaps as many as 10,000 breeding pairs. By the time of the first national survey in 1972, however, its numbers had fallen to just a few hundred pairs, mainly in the Southwest. By 1989 the population had all but collapsed, with just 118 pairs left, mainly in south Devon. Field study by the RSPB revealed that the bunting thrived only on traditional farmland, with thick hedges where it could nest safely, pasture rich in grasshoppers to feed

Cirl bunting, saved by countryside stewardship. (Nature Photographers Ltd)

its chicks, and weedy stubble fields to tide the adult birds over winter: a representative farm bird, in fact, but more sensitive than most because Britain lies at the cold edge of its range. The timely 'extensification' of agriculture in the 1990s probably saved the cirl bunting. English Nature and the RSPB employed a specialist to sell cirl bunting conservation to the farmers of south Devon as a special project within MAFF's Countryside Stewardship Scheme, by leaving unsprayed 'game strips' and thick, shaggy hedges. The take-up rate was good, and over the past decade Devon's cirl buntings have quadrupled from 118 to 450 pairs. At this rate the cirl bunting may reach its Biodiversity Action Plan target of 550 pairs within a few years. It will still, however, be confined mainly to south Devon; cirl buntings tend to stay on home ground, and it will be a long time before it is once again a familiar of Wealden farms or Kentish orchards. The cirl bunting had a number of things going for it: agri-environment schemes, money from its sponsors for a crash programme of research, and the fact that the birds nest in an area restricted enough to be targeted by a special scheme. The resources devoted to the cirl bunting had spin-off benefits for other farm wildlife. Birds are glamorous and capture the lion's share of whatever is going, but they are also sensitive environmental indicators. What benefits the cirl bunting will benefit what it eats, and also some of its wild neighbours.

Another rare bird that has bucked the trend, thanks almost entirely to conservation hand-outs, is the stone curlew. Twenty years ago this shy, mainly nocturnal bird seemed doomed. Stone curlews are birds of the open steppe. Britain's countryside offers only rather marginal conditions for it, mainly on the sandy heaths of the East Anglian Breckland and on the Wessex Downs, centred on Salisbury Plain. Their distribution closely follows the Chalk. They also nest at a much lower density in spring-sown crop fields, such as carrots and sugar beet. Stone curlews seek out bare, warm, sandy soil with a scatter of flints and stones. They also need a near-by supply of earthworms and insects. These conditions are harder to find than they used to be in the days when the grass was kept short and open by rabbits or range-grazed by cattle and sheep. Demand for low-quality grazing has fallen away, and since the 1950s there are seldom enough rabbits. When the vegetation grows tall and lush, the stone curlews lose interest. As a further handicap they like quiet places, well away from busy footpaths and recreation areas. In arable fields their eggs and chicks get squashed by farm rollers and harrows. It is hardly surprising, really, that the stone curlew is rare in Britain. In the 1960s it was thought that we still had as many as 300 pairs. This was a guess, and it was wrong. When detailed counts were made, in 1993, there were just 145 pairs, compared with between 1,000 and 2,000 before the War.

The RSPB has made great efforts to locate where the birds are nesting and move the eggs before the farmer rolled that part of the field. Research into the finer detail of the bird's habits revealed the importance of having enough short, open ground for foraging as well as nesting. One could tempt stone curlews to nest on rotovated strips of bare ground within short grassland – a method used with considerable success at Porton Down in Wiltshire, which has held up to 30 pairs. In 2000, assisted by favourable weather, the nationwide population had reached 250 pairs, the best for many years. The stone curlew's prospects are now rated as excellent – but only if the conservation effort is sustained. Without targeted, special treatment, the stone curlew would find survival in the British countryside much more hazardous. In a sense, we have made things better than they really are.

For the last of our trio, the corncrake, it is already too late over most of lowland Britain. From being the very evocation of still summer nights in

The rarely seen corncrake, victim of changing farming practices. (Natural Image/Mike Lane)

hay fields, haunted by the grating, monotonous call, '*crek crek*', the corn-crake now 'conjures up images of flower-rich meadows of the Hebridean machair' (Gibbons et. al 1993). Mechanised farming has pushed it to the furthest limits of the British Isles. Beyond the Hebrides and the northern isles it has nowhere left to go. There are only two choices with such a bird – let it die out or go all out to save it. In Britain we do not let birds die out without a fight. In the Western Isles, the RSPB has gone all out by, for example, buying 1,000 hectares on Coll as a corncrake reserve, and offering anyone with corncrakes on their land a subsidy of £60 per hectare for growing hay. The RSPB (and now the EC) pay out about £300,000 a year in this way, and employ up to 35 people to help the corncrakes by planting 'weed gardens' and fencing hay fields. Up to a point it works. In 1993 there were 480 calling males in Britain; in 1996 they had increased to 584, and in 2000 there were 621, mostly in the 'core areas' of the western and northern isles, but with a few others starting to nest in north-east Scotland. However, helping one rare species can sometimes hurt another. On Coll, 'corncrake friendly' management has increased the pressure to improve the pasture in places where there is another endangered species, the Irish lady's-tresses orchid (Henderson 2001). The orchid is rarer than the corn-crake – it occurs nowhere in Europe outside Britain and Ireland – but plants have fewer paid-up supporters than birds. This is likely to be an increasing problem in conservation as the authorities try to unravel the contradictory requirements of upwards of 400 rare species (see Chapter 11). The corncrake project has also attracted criticism because it turns parts of the Hebrides into bird reserves and is alien to the local culture (though surely no more so than any other kind of subsidy). Ian Mitchell (1999) points out, rightly, that the corncrake is not yet endangered world-wide (though it is in western Europe) – nor for that matter is the cirl bunting, which is common in Mediterranean maquis, nor the stone curlew. British efforts to conserve them contribute nothing much in world terms. The honest argument for preserving the full range of species in Britain is that we want to see them *here*, not somewhere else. If the nightingales desert my Wiltshire valley, it is poor consolation to know they are common in Spain. Beyond that, extinction has reverberations. The loss of a well-known farm bird would be our failure too, a funereal toll that creates a sense of guilt, reminding us of how selfish, useless and nasty we are.

Slowing the engine: surplus reduction and agri-environment schemes

The engine of destruction that removed much of the fine detail of the British landscape between 1940 and 1985, along with much of its wildlife, began to decelerate once production had gone into surplus (some might say it finally blew up and set the cornfield ablaze). It is economics, rather than environmental considerations, that has sent the machine into the garage for repeated overhauls since the 1980s. The first important com-modity to go into surplus was milk. The EC decided to apply supply con-trol by introducing quotas in the early 1980s. The milk quota system was a disaster for wildlife. While one might expect less intensive dairy farming to reduce the pressure on marginal land, it made some farmers decide to abandon dairying altogether and grow cereals instead. One result was the ploughing up of wet pastureland on the Culm measures of Devon and Cornwall, and the removal of many of the thick, tangled hedges that characterised this lovely scenery. Today, they survive mainly on nature reserves such as Dunsdon Farm in Devon, which in turn often depend on the dwindling number of farmers who refuse to move with the times.

Set-aside was the European Community's answer to cereal surpluses. In return for compensatory payments based on acreage, every farm has to set aside 15 per cent of its arable acreage every year. National governments were left some freedom of interpretation, and in Britain we botched it. There were three kinds of set-aside. The commonest was annual 'fallow', which was rotated around the farm from year to year, leaving little or no time for much wildlife interest to develop. Worse, the farmer was required to spray or plough set-aside land right in the middle of the nesting season to control weeds. This meant that any bird unwise enough to choose a weedy patch of land to build its nest was liable to lose its entire clutch or brood when the spray booms got to work. The record for what was dubbed the great set-aside massacre of 1993 seems to have been a field in the Yorkshire Dales where the farmer managed to run over two nests of snipe, three lapwing chicks, two clutches of grey partridge plus one of the parents, one skylark's nest, three meadow pipit's nests and a willow warbler! Long-term set-aside offered wider options under a Habitat Improvement Scheme, including tree-planting and grazing, but the take-up was limited, and the execution often poor. The best incentive on offer was the

Long-term set-aside near Ramsbury. It lies on the contours of former downland, but has few of its characteristic insects and plants.

Countryside Premium Scheme, introduced in 1989 as a kind of environmental top-up to set-aside plans. However, funds were limited, and the take-up rate modest. On the whole the environmental gains from set-aside were not great. The system was biased in favour of the big cereal farmer, who was paid obscene amounts of money in compensation while allowing him to nullify the scheme's intention by boosting production on the remaining 85 per cent. As the NCC put it, bluntly enough, set-aside 'provided only very limited benefits for nature conservation and we now regard it as a missed opportunity' (NCC 1991).

Another incentive to diversify farmland was the introduction of grants to manage farm woodlands (the Farm Woodland Scheme) and plant or maintain hedges (the Hedgerow Incentive Scheme). Until 1986, agricultural grants were pegged to increases in production. Surpluses forced government to change tack, and consider ways of diversifying farm activities. One possibility was putting small woods back into beneficial production, as a source of wood or shelter for game, and for wildlife. Payments were available to reinstate coppicing, a management system that favours many birds, butterflies and wild flowers, and for planting native trees. Free advice was also on hand from the local agricultural service and the Farming and Wildlife Advisory Group. The Farm Woodland Scheme brought modest benefits to nature conservation, especially by restoring some of the lost links between isolated woods and hedgerows, or preventing the overgrazing. Unfortunately, the take-up rate has been disappointing. It seems to have been most successful in pastoral Wales (where the scheme is called *Coed Cymru*) and least so in arable farming areas. Hedgerows, however, have become valued more, and, with greater incentives for their upkeep, the planting and restoration of hedges more or less counterbalances ongoing losses. A regulation introduced in 1996 protects hedges of special historic or ecological value, so long as they are registered with the local authority.

In the mid-1980s, after a damning report on *Agriculture and the Environment* by the House of Lords, MAFF came up with a scheme to bring more environment-friendly agriculture to areas still noted for their scenic beauty and wildlife. They called these Environmentally Sensitive Areas (ESA). The idea was based on an earlier scheme to preserve wet pasture in the Broads area. This had been a contentious issue since Halvergate Marshes, the largest area of spacious, uninterrupted grazing marsh in the Broads, was threatened by a pumping scheme in the early 1980s, urged by the local Independent Drainage Board. The new pumps would have lowered the water level of much of the area, enabling local farmers to plough it. The part designated as SSSI was to be spared, although the farmers expected to be compensated for loss of potential income by the NCC. One farmer, dissatisfied with the level of his compensation, started to deepen his ditches and ploughed a giant 'V' across one of the fields. Government, having stated that 'Halvergate is safe for a year', had to make it so. In March 1985 it launched a Broads Grazing Marsh Compensation Scheme, funded jointly by MAFF and the Countryside Commission, offering all

Halvergate: the largest expanse of 'grazing marsh' on the Norfolk Broads, saved from
the plough by an emergency government grant scheme, the forerunner of recent
incentives for 'environmentally sensitive' farming. (Derek Ratcliffe)

landowners a flat fee of £50 an acre to retain their livestock and the marshland that fed them. Though not munificent, the scheme was accepted as
fair, and, helped by the physical collapse of the soil on parts of the recently drained land, it saved Halvergate. (The story is told in detail in the
recently published New Naturalist *The Broads* (Moss 2001)).This was more
of a landscape issue than a wildlife one. What mattered was that MAFF was
now involved in a scheme that was based on good environmental practice,
not on increasing production. It represented a significant broadening of
MAFF's role, and a reluctant admission that the old system of blind, nonstop intensification was no longer what Britain needed or wanted.

Environmentally Sensitive Areas are administered entirely by MAFF, but
are chosen on the advice of the NCC and the Countryside Commission.
The basic idea is to offer flat-rate payments that help maintain a farmer's
income while enabling him to work the land in such a way that does not
destroy its natural character and beauty. Grant-aid would be based on an
agreed management plan. The details vary from place to place. In the
Cambrian Mountains, payments are on offer for managing heather moorland by reducing stockage. In other areas payments are aimed at retaining
permanent grassland and hedges. The system was voluntary and such
things work only when farmers see them as fair and advantageous.

The ESA scheme was gradually extended between 1986 and 1999. For
the pilot scheme, the long lists provided to MAFF by the NCC and the
Countryside Commission were whittled down to just eight: the Broads,
Pennine Dales, Somerset Levels, the eastern half of the South Downs, and

Cornwall's West Penwith area; the Cambrian Mountains in Wales, and Breadalbane and Lomondside in Scotland. Following a reasonably favourable inception, another nine ESAs were formed in 1987 at Breckland, river valleys in Suffolk and Hampshire, another part of the Peak District (North Peak), the Shropshire border, and the rest of the South Downs in England, and the Uist-Benbecula machairs, the Whitlaw and Eildon hills and the Stewartry's Loch Ken/River Dee area in Scotland. Since then the system has embraced much larger areas – the Lake District, Dartmoor and Exmoor National Parks, Anglesey, the isles of Mull, Islay and Jura, and much of Galloway and the Borders – and now covers about 10 per cent of England's agricultural land and 20 per cent of Scotland's. With devolution, new schemes are now replacing ESAs: *Tir Gofal* ('Land Care') in Wales (which combines ESA with the stewardship scheme, *Tir Cymen*) and the Rural Stewardship Scheme in Scotland.

Do ESAs help wildlife? Well, they do and they don't. Progress on ESAs is written up in terms of take-up, rather than the areas of natural habitat preserved or restored. They seldom result in the kind of finely-tuned land management that really benefits wildlife. But ESAs have become an institution, and a large number of farms have entered the scheme: some 2,500 in the first 18 months alone. It has helped farmers to maintain or introduce low intensity farming, and address broadly ecological objectives. The scheme nevertheless has or had many flaws. Its flat-rate payments ensured a fairly crude application of management prescriptions. The NCC would have preferred a scale of payments, linked to different degrees of environmental management (it would also like to have extended the ESA scheme over the whole country). At first the payments were unnecessarily niggardly at £8 million a year – a drop in the ocean compared with the often competing grants for agricultural production. As late as 2000, 15 years after the ESA scheme was introduced, only 4 per cent of the agricultural budget supported environmentally sensitive farming. Payments to conserve special habitats, such as water meadows and grass heaths, are available only on certain ESAs. And some ESA prescriptions, for example, for 'restoring' chalk grassland, fell well short of that in practice. While over 1,000 hectares of the South Wessex Downs has been 'restored' in this way, the result is not the naturally rich ancient turf of chalklands, but sown grassland intermixed with a few herbs. The fact of the matter is that most wildlife habitats cannot be recreated by agricultural techniques, such as sowing. They have to develop naturally, and that takes a long time. Another problem with any overall assessment is lack of data, for MAFF treated ESA farm agreements as confidential. What the scheme did do was to match incentives with local landscape character and make MAFF take direct responsibility for funding wildlife-friendly farming. In its latter stages, the scheme has helped certain declining species, such as marsh fritillary butterflies, by encouraging the right sort of extensive grazing. On the other hand, the ongoing decline of widespread species proves that the scheme is inadequate; to make a real difference production and conservation will need to be balanced more evenly.

The climate of the upper Yorkshire Dales is too harsh for growing corn. Instead the farmers grew natural grass in small walled fields and stored the hay in stone barns. However, chemical fertiliser has made it possible to produce silage, with the consequent disappearance of most of the wild flowers. They survive mainly on verges, river banks and odd corners. (Derek Ratcliffe)

Things are now moving in that direction. In 1999, a new set of CAP reforms known as Agenda 2000 were agreed. In essence, they are another step towards an integrated rural policy, with more funding for extended agri-environment schemes. By 2006, expenditure on environmentally sensitive farming is set to double to over £200 million per year – about 10 per cent of agricultural support – while subsidies for production will fall by 5 per cent. This sounds wonderful, but we should be prepared for disappointment. In his analysis of the reform proposals, the naturalist and farmer Eric Bignell warns that wildlife gain may be limited. While they contain much unfathomable financial tinkering to reduce overproduction and bring EU prices closer to world prices, real opportunities for wildlife are hard to spot. As before, the language may raise expectations only to dash them in practice. Clever exploitation of the new rules should help to 'extensify' farming, and maintain beneficial regimes in difficult and remote areas, but it will require a new kind of business canniness, combining husbandry with a keen eye for harvesting grants (Bignell 1999).

Countryside Stewardship is close in conception and practice to ESAs, but on a smaller scale and it is confined to England. Stewardship was introduced in 1990 by the environment White Paper, *Our common inheritance.* Administered by the Countryside Commission, it seeks to combine wildlife and amenity interests with farming and land management through the usual system of perks and agreements. However, unlike the ESA system, the

The Dorset coast where
Countryside Stewardship
has had a beneficial effect
on the scenery, if not the
wildlife.

latter are linked to habitats rather than areas. The pilot Stewardship
scheme chose chalk and limestone grassland, lowland heaths, 'waterside
landscapes', coastal land and 'uplands'. Later, 'historic landscapes', 'old
meadow and pasture' and 'hedgerow landscapes' were added. It has
proved popular, and is, indeed, massively oversubscribed; some 4,000 con-
tracts were signed during the first three years of operation, covering some
800 square kilometres and 1,470 kilometres of hedgerow. In 1996, the
scheme was taken over by MAFF, perhaps a sign of success. From my
impression as a walker, Countryside Stewardship has made a visible contribution to improving the scenery of downs and coastlands by restoring
areas of grazed permanent pasture from surplus arable. Parts of the Dorset
coastal footpath, for example, have changed from a strip of green between
corn and cliff to a more spacious band of grass, a distinct coastal landscape,
and I have seen blue butterflies fluttering in fields that until recently were
wildlife deserts. Stewardship's main limitation is that it is a thing of shreds
and patches, limited mainly to the agricultural margins. In Wales, this fault
has been overcome in the equivalent *Tir Cymen* (now *Tir Gofal*) scheme,

which is eligible only for whole farms, and thus prevents the farmer from doing what he often did in England, intensifying production on the rest of the holding. Under *Tir Cymen*, the farmer agrees to observe a code of practice, keep rights of way open and avoid polluting the water, as well as bringing woods, heaths and other unimproved land under positive, sustainable management. Support is available for capital works of all kinds, from fencing and bracken control to nestboxes and hedges. There is even an arable option, in which the farmer is paid to leave broad headlands between fields, something England has yet to introduce, despite much urging to do so by RSPB. *Tir Cymen* is administered by the Countryside Council for Wales; again an indication of modest success may be the evident eagerness of the Welsh agricultural department to get their hands on it.

Organic farming

Organic farmers, as Graham Harvey remarked, 'plough a lonely furrow'. In 2000 only a single per cent of agricultural land (420,000 hectares) was farmed organically – although even this represents a doubling in the past five years. An encouragement scheme called Organic Aid is co-funded by the UK government and the European Union. The sort of people who want to farm without the aid of chemicals and pesticides tend to be gentle idealists, concerned about health and the environment, and kindness to animals. The successful ones also need a hard-nosed business sense to find markets for their produce. Organic farms should, and probably do, have more wildlife than comparable units that rely on chemical warfare. The nature of organic farming tends to produce mixed farms; cows may be fed on home-grown hay and silage, which means grass and clover fields, and perhaps even wet meadows enriched by river silt. A large organic farm, or an area of continuous organic units might, in time, have better soil and cleaner water. It will have more weeds and insects, and so more of the animals and birds that depend on them. Wildlife bodies often turn to experienced organic farmers for advice and help on running their holdings in a similar way. For as long as consumers share the grower's values, the future for organic farming looks fairly rosy, though it will probably never exceed a few per cent of agricultural production in Britain (and may one day be routed by cheaper organic food from outside Europe). Organic farming will not by itself sustain the majority of wildlife species, since they are tied to natural habitats. But it offers a *vision* of a healthier, happier countryside, with bees and grasshoppers, and swallows nesting in the barn. Unfortunately there will never be enough organic farms to make a real difference to the fundamental problems of nutrient enrichment from nitrogenous fertilisers or the pesticide-based sterility of arable Britain.

Straws in the wind

Elmley Marshes on the southern side of the Isle of Sheppey is one of the finest remaining wildernesses on the North Kent marshes. A century ago, a broad wet ribbon of marshland stretched along the Thames estuary from Whitstable almost to the gates of London. It has since been extensively reclaimed for agriculture and industry; about half of it was lost between 1935 and 1982 (Williams & Bowers 1987). In 1951, the NCC had designated as an SSSI some six square kilometres of grazing marsh and 'fleets' (tidal channels) at Halstow and Cooling Marshes, the setting for the early chapters of *Great Expectations*. But SSSI designation did not stop farmers from deepening their ditches and ploughing the dried-out pasture. By 1984, when the NCC resurveyed the site, the once unbroken vista of marsh had become like a war-torn flag, 'reduced to ribbons and patches, sandwiched among wide new fields of wheat' (Purseglove 1988). Only high compensation payments saved the remainder from a similar fate. On nearby Sheppey, one farmer, Philip Merricks, was unsatisfied with the negative aspect of SSSI compensation. He felt that if he could not intensify his agricultural holding, he might as well go the whole hog and manage the land properly as a nature reserve ('If I can't grow the best wheat, I'll grow the best birds'). In 1987, Merricks established a trust to look after 729 hectares of Elmley Marshes. Under the terms of an agreement with NCC, water levels were raised, and a low-input regime of sheep grazing established there. It has, by all accounts, been a resounding success. The area of bird-rich grazing marsh now stretches along the southern shore of Sheppey covering some 2,000 hectares. Merricks enlisted experts in wetland management from home and abroad, and, being a farmer himself, has been better able than outsiders to persuade neighbours to follow suit. Populations of waders, such as lapwing and avocet, and wildfowl such as wigeon, have built up over the years, and English Nature's hides on The Swale at the eastern end of the island now offer some spectacular birdwatching (here the expression 'birds enough to darken the sky' is not just a metaphor). Elmley Marshes has attracted much attention, not only as a way of integrating grazing and wildlife management, but as a means of improving wildlife *and* retaining commercial viability. Philip Merricks is currently deputy chair of the Farming and Wildlife Advisory Group.

Increasingly, wetlands are also being created on former ploughland. The RSPB provided a template for such projects on its 298-hectare holding at Lakenheath Fen in the Suffolk Breckland, purchased with the help of the Heritage Lottery Fund in 1995. The site, on former floodland between a railway line and the Little Ouse river, had been under intensive agriculture, mainly carrots (the peaty soil is top-grade), while the last wet bit had been planted with poplars destined for the matchstick trade. The RSPB are returning it, by stages, into wetland by excavating new channels and ponds, and piping river water into bunded shallows. Like Elmley Marshes, the site will contribute to the national target of 1,200 hectares of newly created reed bed in the UK Biodiversity Action Plan (the poplars have been left for the sake of the golden oriole). Some birds have already spotted the opportunity: reed warblers, for example, increased over four years from 4 to over 110 pairs; gadwall and shoveler have bred, and in 1999 a group of bearded tits were seen exploring its potential. In 1997, they found a bug new to Britain. As a sop to the botanists, fen ragwort has been planted there, as part of its 'recovery programme'. Lakenheath Fen is an oasis of wet in a mainly dry, arable area, but it has demonstrated some of the possibilities of habitat restoration. Other places where wetland habitat, including reed beds, has been created include the National Trust's Wicken Fen and English Nature's Stodmarsh reserve in Kent. The same idea is being tried out on at least one commercial farm, Lower Farm near Oxford, which is being systematically restored to the condition it was in 50 years ago, that is, a very wet farm with winter-flooded fields, visited by geese and lapwings.

A small number of farms are run as experiments, with the object of demonstrating ways of reconciling wildlife and profit. On the Hill of White Hammers on Hoy, Roy Harris runs a sheep farm that maintains habitat variety, including coastal heathland, the habitat of the tiny Scottish primrose, and projects a 'way forward' for extensive hill farming through conservation grants and local markets. The primrose has, in effect, become the object of a new kind of subsidy, not a crop, but an attractive and symbolically potent 'wilding'. At Loddington in Leicestershire, a project (the Allerton Project) run by the Game Conservancy and an educational trust, demonstrates that thoughtful use of odd corners and headlands can increase wildlife significantly with little loss of income. High numbers of hare and grey partridges are maintained by diversifying crops, minimising pesticides (or at least by restricting spray times) and leaving field headlands unsprayed. Although this 333-hectare farm has only 43 hectares of permanent grass, barn owls hunt along the field borders, larks sing over most of the farm, and frogs and newts breed in the ponds. With the help of the Countryside Stewardship scheme, all this wildlife costs £5.77 per hectare per year, and even this modest figure is set to fall as agri-environmental schemes come on board (Boatman & Stoate 1999).

Again, the RSPB has taken a lead by purchasing Grange Farm in Cambridgeshire, to 'trial, demonstrate and advocate new farmland management techniques that favour farmland birds'. It took on the 180-hectare arable farm without knowing what birds were there (they are, in fact, surprisingly diverse), and with the intention of carrying on farming, if possible to normal production levels. By experimenting with cost-effective techniques, capable of emulation by any farmer, the RSPB hopes to demonstrate ways of providing more food for finches, sparrows and buntings in winter, and encouraging birds such as skylark to breed there. Over the next ten years, about a quarter of the farm will be given over to cropping trials, with the portion in set-aside being used to provide extra cover, feeding ground and nest sites. In recognition thereof, they have renamed it Hope Farm.

Since 1986, the farm environment has been in a state of flux, with an increasing amount of agricultural support being switched from production to environmental schemes. This is set to increase over at least the first two decades of the twenty-first century, probably accompanied by diverging systems in England, Scotland and Wales. It is likely to create a yet more extreme polarity between countryside farmed for production and areas maintained mainly for tourism and amenity. It has the potential to make at least parts of the farmed countryside more wildlife-friendly than it is at pre-

Freeman's Marsh, near Hungerford, owes much of its natural diversity to flooding and clean water. The beds of stream water-crowfoot, a key species in the chalk-stream ecosystem, became much reduced in the 1990s by low flow-rates and increased silt loads.

sent. There are fresh problems ahead, with genetically modified crops, invasive species, eutrophication and the shaky state of stock farming in Britain (see Chapter 13). But there is room for measured optimism too. At least we are living in interesting times.

7

Woods and Forests

Foresters and afforestation

In some ways conservationists and foresters are much alike. Both are often keen on natural history, both enjoy working outdoors and believe their work will be of lasting value. Above all they share the same single-mindedness. Just as a local conservation officer will tend to weigh up all the land in sight in terms of its 'conservation value', so a forester will view it in terms of its potential for planting trees. The forester plants trees in the belief that their benefits are self-evident. In his eyes, conservation designations threatened to sterilise land that was perfectly good for trees. As an NCC local officer, I well remember the aggressive questioning – almost the sense of outrage – from visiting foresters as they pointed out that this or that nature reserve or SSSI was perfectly capable of producing timber. Where was my willingness to compromise? (Answer: compromise invariably meant we lost.) The farming community was often half-apologetic about damaging SSSIs; foresters, in my experience, never. Instead they laid great store on concepts such as integration and multipurpose use, which generally allowed room for more planting. Local plans in Scotland and Wales tended to be very generous to forestry interests (new forests meant new jobs, thought the planners, and weren't trees good for wildlife?). This shared single-mindedness meant that when forestry and nature conservation came into conflict, as they increasingly did in the 1970s and 1980s, the resulting struggle was bitter and protracted.

To begin with, though, the Nature Conservancy saw forestry almost as an ally. Some of the first generation of senior Conservancy staff came from a background in forestry, often in the colonial service. Fraser Darling considered that 'the more forest there is the better', and was inclined to see even the Forestry Commission's 'square green rugs' of Sitka spruce as a benefit, since 'after it is thinned out ... other stuff could be used which are pleasanter trees altogether' (Mackay 1995). Dudley Stamp (1969) thought that planting policies had improved a great deal since the 1930s, and that 'introducing a range of conifers' improved a dull old oak wood. As late as 1978, the NCC was publicly congratulating the FC on its recognition of 'the needs of conservation and amenity' and looked forward to 'fruitful collaboration in the future' (NCC 1978).

The problem with forestry in Britain lies in its scale, in the kinds of trees that are planted, the method of planting them and, often, in the places they are planted. By the late 1940s, most of our larger woods, as well as many smaller ones, were in a bad state, having been plundered of their timber during two World Wars. At the same time, the timber market had shifted from slow-growing hardwoods such as oak and beech towards fast-growing softwood conifers grown mainly for pulp and pit-props. In the cold, wet climate of the hills, where most plantations are situated, the most successful tree is Sitka spruce, a native of oceanic north-west America. Great things were claimed for the universal spruce: it would create a strategic reserve of timber, save imports and create jobs in deprived areas, as well as diversify the habitat. Other favourites are larch from Japan and lodgepole pine, also from North America. In the 1950s, the Forestry Commission was still planting oak and other native hardwoods on suitable lowland soils, but by 1984, 98 per cent of its plantings were non-native conifers, above all Sitka spruce. Conifers grown in commercial plantations are harvested, generally by clear-felling, at around 40 to 60 years old, and rarely achieve the graceful maturity of their native counterparts. Since the Commission's main job is to grow marketable timber as cheaply and efficiently as possible, the crop trees are planted densely in straight rows, which, from the air, make the plantations look like carpet rugs. They replace the natural irregularity and variety of vegetation on the open hillside with, as nearly as possible, total uniformity. Plantations lack the structural complexity of native woods, with their developed layers of shrubs and herbage. In some there is practically no natural plant growth at all, apart from the odd wisp of fern and hummock of moss. Of course, even monocultures of foreign trees support wildlife, in the tree canopy, along the broad surfaced rides created for timber lorries, and in the clearings left after fellings. But modern tree-planting has closer affinities with arable agriculture than traditional woodmanship: the ground is ploughed, fertiliser, and sometimes pesticides, is applied, the nursery-grown seedlings are planted in rows, and later the crop is harvested in extensive clear-fells. Modern techniques, with their reliance on machines and the agro-chemical industry, change the physical and chemical nature of the soil and alter the natural drainage. Dense conifer plantations acidify the soil and drainage water. Forest hygiene demands the removal of the dead wood habitat. Broadly speaking, and despite all the propaganda to the contrary, commercially managed conifer plantations are not good for wildlife. Moreover, they block access to the open hill. Walking in them can be a funereal experience, despite the FC's former chairman, Sir Robert Robinson's evidently sincere belief that his plantations were much nicer to walk in than open country (Mackay 1995).

Even so, it was possible to persuade yourself, as Darling and Stamp did, that the austere new forests would eventually mellow into a mixed-species, pseudo-alpine environment with more concessions to amenity and wildlife. Back in the 1960s, afforestation was not widely regarded as a major environmental problem in Britain, mainly because there was still so much open hill land. The importance of ancient and natural woodland was not then widely understood, and the value of open moorland and blanket bog less appreciated. Even the NCC 'improved' some of its woodland nature reserves by planting trees, albeit native ones. It was assumed that pretty well all the great trees of the British landscape had been planted at some time;

Near natural woodland at Bramshaw Wood in the New Forest – mature oak, an understorey of holly, open glades and plenty of dead wood. (Derek Ratcliffe)

the rest was dismissively regarded as 'scrub'. That was why there was so little resistance to what happened to our native woods between 1945 and 1985, when nearly half were clear-felled and replaced by crop trees (30 per cent) or farmland (10 per cent). By the 1970s, however, blanket afforestation was transforming entire landscapes in out-of-the-way places such as Knapdale and Kintyre, Galloway and the Borders, and parts of central Wales. There any requirement for 'an acceptable balance with agriculture, the environment and other interests', which the Forestry Commission was bound by statute to respect, seems to have been overlooked. The underlying rationale for the onward advance of Sitka spruce was that Britain should save on imports by becoming essentially self-sustaining in timber. But this was a pipe dream. By 1985, 2 million hectares – nearly 10 per cent of Britain's land surface – lay under crop trees, but still met only 12 per cent of domestic needs (Sheail 1998). The FC's prognosis was for a further 1.8 million hectares of new forest (other forecasts went as high as 2 million). If achieved, this would cover nearly two-thirds of the remaining afforestable land (NCC 1986) and have a drastic effect on the wildlife of

No more ferns at Ferny Knowe. This picnic site from hell was hastily dismantled after its appearance in Steve Tompkins' book, *Forestry in Crisis: the Battle for the Hills* (1989). (Derek Ratcliffe)

the open hill. Astonishingly this misguided policy was accepted without much demur at Westminster, thanks to the influence of the forestry lobby in both Houses of Parliament. As protesters quickly discover, politicians, particularly in Scotland, tend to be forestry-friendly, seeing it as a source of 'jobs' while regarding open land as inherently inexhaustible and, in any case, as 'barren wilderness'. The same is true, and for the same reason, of local authorities. Before 1980, opposition to afforestation had been led by amenity groups, such as the Friends of the Lake District, and individual naturalists. It was only in the 1980s that the NCC and the RSPB began to question the entire basis of forestry in Britain, and the public alerted to its lack of accountability in works such as Steve Tompkins' *Forestry in Crisis: the Battle for the Hills* (1989). Unfortunately the NCC was unpopular in Scotland, while the Forestry Commission, whose headquarters were in Edinburgh, was regarded as part of the Scottish establishment. Before recounting these hill battles, let us briefly review the story of this body, whose decisions have been so important for wildlife in Britain.

The Forestry Commission

The Forestry Commission was set up in 1919 for the purpose of buying up cheap land for planting trees and ensuring that Britain had sufficient strategic reserves of timber to withstand another war (at that time the Royal Navy still depended on coal, and coal depended on an adequate supply of wooden pit-props). Although technically part of the agriculture departments, the FC has a unique status as an undevolved department of state, while having some of the characteristics of a quango. For many years it has been Britain's largest landowner, owning 1,165,000 hectares (6 per cent of the land surface) in 1987. It has powers to buy and sell land, and provides grants and loans for others to plant trees. Until recently it also operated a dedication scheme whereby a private owner received tax exemption in exchange for devoting part of his land to timber production. In such cases, the owner in effect became the FC's factotum under a management agreement. One way or another, the FC decides where, how and how many trees are grown in the UK.

In 1943, the Forestry Commission pressed successfully for a large-scale programme of reafforestation in order to achieve adequate timber reserves to fight another war, and also to help the balance of payments. It calculated that the nation needed some 2 million hectares of further planting over the next 50 years. Existing woods and forests could supply less than half of this. The rest would have to come from planting bare land. Some of the FC's early plantings were an embarrassment – crude green rectangles and lozenges that made no concessions to the lie of the land, and so, quite apart from looking ghastly, proved vulnerable to storms and floods. There were many complaints, and, as a result of them, the Commission agreed to limit its activities in the Lake District and Snowdonia. With the appointment of Sylvia Crowe as an adviser on landscape in 1964, forest design improved in England and Wales, but she had little influence in Scotland where large-scale afforestation was being increasingly directed. Already planting in Scotland exceeded that of England and Wales put together, and was set to expand. From the 1970s, much of the planting target of 35,000 hectares per year would be met by private, tax-break forestry (see below). The FC turned itself into 'a kind of sponsor for the private sector ... not of promoting standards within that sector but of standing behind grant applications and assisting them in minimising the impact of agricultural and amenity objections' (Mackay 1995). It forcefully opposed planning regulation, sought vainly by the Countryside Commission for Scotland for plantings of more than 50 hectares or within designated National Scenic Areas. It often gave planting permission to SSSI owners, thereby forcing up the land value, and making compensation more difficult and expensive. It was not until very late in the day, in 1985, that the Commission officially accepted a duty to achieve a 'reasonable balance' between forestry and conservation – and even then it interpreted this as 'tarting up' the forest, rather than refraining from planting up areas of nature conservation importance. On its own estate, very few natural woods had escaped partial or total felling and replanting with conifers. That many of the original trees did in fact survive was in spite of the FC's best efforts, not because of them.

The FC began to reconsider the value of native hardwood trees and woods in the 1980s, prodded first by a parliamentary committee (which was, in turn, much influenced by the submissions of Oliver Rackham and George Peterken, incorporated in the NCC's evidence), and then, in 1982, by an academic conference in Loughborough that brought leading

Fleet Forest advances like a dark green tide breaking against the Galloway Hills. The picture chosen to front the NCC's critique of modern forestry, *Nature conservation and afforestation in Britain* (1986). (Derek Ratcliffe)

foresters and woodland ecologists together. In July 1985, the Scottish Secretary, George Younger, announced a new policy aimed at 'maintaining and enhancing' broad-leaved woods. The amount of timber that could be removed without a licence would be reduced. The FC was not allowed to license the clearance of more woods for farmland 'without very strong reasons'. Grants and tax incentives for planting slow-growing broad-leaved trees were increased, thus implicitly recognising that there were reasons other than productivity and profit for planting trees (for Britain's oaks and ashes can never hope to compete on the timber market with cheap tropical hardwoods). The policy of 'coniferising' native woods was, in effect, abandoned. And the Commission now had the duty to balance the needs of both timber production and wildlife. All in all, it amounted to a formidable enforced U-turn. Good things followed. In 1986 the FC signed an agreement with the NCC over 344 SSSIs in its care, covering 70,000 hectares. By the early 1990s it had accepted the NCC's Ancient Woodland Inventory (see below), which identified such woods, and provided guidelines for their care and maintenance. In 1988, an embargo was placed on 'predominantly coniferous' afforestation in England, and, from 1992, the Forestry Commission offered special management grants for woods of high conservation value. Today all SSSIs on the Commission's estate are managed by agreement with the country agencies under formal agreements or statements of intent. Moreover, a substantial part of forestry research has been switched towards biodiversity and conservation.

In 1992 the Forestry Commission was reorganised into two branches, a

Forestry Authority, which continues its regulatory function, and Forest Enterprise to manage its estate. Under pressure from the Royal Forestry Societies, and from the public, concerned about restrictions on access, the Government decided against the wholesale privatisation of FC land. Conservation bodies found themselves defending the FC, presumably on the principle of 'always keep ahold of nurse, for fear of finding something worse'. Personally, as the above remarks may betray, I think I would have risked it.

The planting of the uplands

In the 1970s a new phenomenon began to affect the upland scene – the private-sector forestry companies. A sharp-eyed accountant had spotted an opportunity to make money out of planting trees by exploiting a tax loophole. What was more, technology now enabled timber growers to grow trees on peat bogs. Experiments using lodgepole pine as a nurse crop for the ubiquitous Sitka spruce suggested how it could be done: the growing pine dries out the peat, and, having done its job, it dies, leaving a good root bed for the spruce. In its new role as facilitator for private forestry, the Forestry Commission was only too pleased to license afforestation that helped to achieve its timber targets. The land was cheap and the planting grants generous. By 1987, the Perth-based Fountain Forestry company had bought up 40,000 hectares of land to sell on to its wealthy investors, earning them some £12 million by way of grants and tax exemption. The profits lay in public subsidies and early sale of the young forests to corporate investors, not in the crop; in effect the forests were financed by the taxpayer. Alexander Mather (1987) compared the new spruce forests of Scotland with banana and rubber plantations in the days of the Empire: decisions were taken far away, and local involvement and benefits were minimal. Even the contract labour was often imported. The forest companies failed to deliver their promises of jobs, an article of faith with credulous local councillors. For example, the much-publicised figure of 2,000 jobs in Caithness and Sutherland turned out to be a projection 50 years hence when trees were due for felling, assuming they survived that long. Until then, the true figure was closer to 60. Once planted, the trees needed minimal attention until harvesting. And to maintain even a modest labour force, there needed to be a continual supply of new land for planting. According to the National Audit Office, each 'job' created cost the taxpayer about £60,000.

Environment and amenity bodies were caught unawares by the speed in which aggressive forestry companies were buying up land. The only mechanism available to oppose damaging planting schemes was the SSSI, and even there the Forestry Commission was not bound to follow the NCC's advice; in its response to the NCC's report on afforestation, it explicitly rejected any idea that the FC should not promote afforestation on SSSIs. The 1970s saw some very contentious plantings, including the lower half of Abernethy Forest in Speyside and Llanbrynmair Moors in north Wales, both of SSSI quality. Things came to a head in the early 1980s at Creag

Meagaidh (pronounced 'meggy'), a mountain fastness north of Loch Laggan in the central Highlands. Creag Meagaidh was an SSSI, partly for the arctic-alpine flora of its montane grounds, including the magnificent north-east corrie, walled by 300-metre cliffs, but also for the natural mosaic of moor, marsh and birch wood on the open slopes below. Walkers and climbers were drawn there not only by its wild beauty, but because this was the one area in that part of Scotland which had not been rendered inaccessible by recent planting. In the early 1980s, the estate was sold to Fountain Forestry, who promptly slapped in an application to afforest practically all the slopes capable of supporting timber with Sitka spruce – some 1,100 hectares. The Forestry Commission dutifully offered a grant, the SSSI notwithstanding, and Fountain Forestry rejected the NCC's subsequent offer of compensation for 'forgoing profits'. The case went to the Secretary of State, George Younger, whose Solomonic judgement was to half the area covered by planting permission, and to urge the NCC to negotiate with Fountain Forestry over the remainder. Totting up their sums, the developers decided to abandon the project, refusing the NCC's offer of a management agreement ('voluntary agreements are doubly difficult in such circumstances because the tax advantages to an owner who wishes to afforest his land may outweigh any other consideration', noted the NCC in its annual report). The NCC was left with no alternative but to assume responsibility for the whole 3,940-hectare estate. Creag Meagaidh was 'declared' a National Nature Reserve in May 1986. Having bought the land for £300,000, the company sold it to the NCC for £431,000, thanks to the added value brought to it by planting permission. Lucky investors.

Creag Meagaidh was only the opening skirmish in the struggle that culminated in the so-called Battle of the Bogs over the peat-flow country of Caithness and Sutherland, a series of flat, patterned bogs covering a vast, virtually uninhabited area in the far north of Scotland. This was a true wilderness, and made a great impression on any naturalist lucky enough to have seen it in its original state (see Plate 4). I vividly remember my own visit in the late 1970s, before afforestation had got underway. You could see for miles across countless lochans and colourful cushions of bog-moss to a still unobscured far horizon notched by the hills of Morven and Scaraben. At any step you might put up a greenshank, dunlin or golden plover; there were divers and scoter on some of the lochs, and the odd arctic skua or hen harrier hawking over the moss. There was a dreamlike sense about the place: the birds were different, the scale was Siberian; even the sky looked twice as big. Whatever small uses generations had made of it had left the flow country pretty much as they found it, wild, wobbly and weird.

Unfortunately, forestry technicians had discovered a way of growing trees on it through a combination of deep drains and the use of nurse crops on raised cultivation ridges, together with heavy doses of fertiliser and insecticides sprayed from the air. As at Creag Meagaidh, the tax and grant system allowed all concerned to make money out of the venture, including the flow-country laird who had at last found a buyer for previously almost worthless land. The Forestry Commission and Fountain Forestry targeted the area, and, by 1987, had bought up 65,000 hectares within it. The aim was a full 100,000 hectares, in a single vast estate that would achieve the hoped-for economies of scale. This would, of course, mean goodbye to the flow country of the bog-moss, greenshanks and the wide open skies, particularly as the physical and chemical effect of planting extended much further than the plantations themselves. The foresters offered to leave a few clearings for the birds, but, as Fountain Forestry's director memorably observed, if the greenshank 'cannot survive on 650 acres, it doesn't bloody well deserve to survive'.

Since there had seemed no urgent need to do so, the area's wildlife had never been surveyed in detail. It was only when, in response to gathering events, the NCC sent teams of peatland and bird experts to the Caithness flows that its full importance was recognised. To the bemusement of foresters and politicians alike, the NCC now claimed that the flow country represented 'possibly the largest single expanse of blanket bog in the world' (NCC 1987), containing significant proportions of the European Community's nesting dunlin and golden plover and fully two-thirds of its greenshank (which was not all that many since the EC did not at that time include Sweden and Finland). The minister was reminded of the international treaties he had signed up to protect birds and bogs, including the EC's directive on wild birds. The problem was that by the time all this was realised, afforestation was already under way.

The NCC 'went public' on the issue by publishing a glossy, full-colour report, *Birds, Bogs and Forestry*, summing up its scientific findings and plead-

Financed by tax breaks. Planting trees in wet blanket bog requires deep drainage, plenty of fertiliser and sometimes a blitz of insecticides to combat pests, such as the pine beauty moth. The crop is barely economic. (Richard Lindsay)

ing for a moratorium on further planting in the area pending a land-use strategy for the region. The Scottish establishment and media chose to regard this as an unwelcome intrusion in Scottish affairs, 'without regard for the delicate economic and social fabric' of the area. The NCC's scientific claims were 'preposterous', said the local MP, Robert McClennan. The forests were a godsend and the NCC seemed bent on 'sterilising' the land, just like the Highland clearances of evil memory. In retrospect, the NCC might have done better to leave the high-profile campaigning to the RSPB, and concentrate on the assessment of the vegetation and bog structure it eventually produced, two years too late, in 1988. But the NCC's chairman, William Wilkinson, had recently visited Galloway and the Borders, and had been appalled at the scale and impact of recent afforestation there. He now regarded upland afforestation as the most serious nature conservation issue of the past 30 years. Pressed by the NCC on the one hand and the formidable Scottish forestry lobby on the other, the Scottish minister made another Solomonic decision and invited the NCC to designate up to half the plantable area as SSSIs. Eventually this became the largest agglomeration of SSSIs in Britain. The immoderate publicity that followed *Birds, Bogs and Forestry* damaged the NCC's reputation in Scotland. Moreover, it lacked the resources for designations on this scale, especially as they would involve potentially endless wrangles over compensation. The subsequent working party, convened 'to examine land use options' in Caithness and Sutherland, was chaired by Highland Regional Council, to whom jobs were generally the foremost issue. The noisy 'Battle of the Bogs' had at least exposed taxation-driven forestry investment for the disreputable racket it was. In the budget of 1988, the Chancellor (Nigel Lawson) stopped the gravy train dead in its tracks by removing the tax incentives for commercial woodland in one swoop, so that the expenses of planting and maintaining the new forests were no longer deductible. That effectively solved the problem at a stroke, but not before a lot of lasting damage had been done.

In its place came the Woodland Grant Scheme (see below) and the Scottish system of 'indicative forestry strategies', regional plans for forestry which gave the local community more of a say. They came just too late to prevent more outrageous blanket plantations at Glen Dye in Deeside, at Lethem in the Borders, and at Strath Cuileannach in Sutherland, all grant-aided from the public purse, but, significantly, these attracted far more censure in Scotland than had much bigger schemes a decade earlier. In future, forestry in Scotland will have to take greater account of public opinion, and, as in England, will take on roles and guises other than timber production, such as public amenity and urban renewal.

Ancient woods and nature conservation

Most of our oldest woods are small woods. Except in the Fens, nearly every rural parish has at least one, their historic purpose being to supply timber to the owner and firewood to the parishioners. Some woods also supplied local trades with small-bore wood for making hurdles and fences, thatching spars and rustic tools. In iron-smelting districts, 'colliers' cut wood for

Plate 1

Beinn Eighe in Wester Ross, Britain's first National Nature Reserve. His orders had been to bid for the small native pine wood, but the Nature Conservancy agent got carried away. (Derek Ratcliffe)

Plate 2

Remote country. The high tops of Ben Avon in the eastern Cairngorms, an arctic desert of granite and three-leaved rush. (PRM)

Plate 3

The hills of Glencoe as seen from a bog pool on Rannoch Moor. A seemingly primeval scene, but hulks of oak trees lie entombed in the peat, and the bare hills once wore a scrubby apron of birch and willow. (Derek Ratcliffe)

Plate 4

The Sutherland 'flow country' of peat pools and mosses: one of the world's largest blanket bogs. Ben Griam Beg in the distance. (Natural Image/Bob Gibbons)

Plate 5

Contrasting fates for limestone pavement: in the foreground, a nature reserve (Clawthorpe Fell NNR), in the background, a quarry. (English Nature/Peter Wakely)

Plate 6

Natural woodland: the interior of a native pine wood at Abernethy, with its lush understorey of blaeberry, cowberry and juniper. Each tree is a distinct individual, suggesting wide genetic diversity and consequent resistance to disease and insect attack. The stumps are good for rare flies. (Derek Ratcliffe)

Plate 7

Natural woodland: valley woodland of oak and birch at Ceunant Llenyrch
National Nature Reserve in Snowdonia. (Derek Ratcliffe)

In the woodland's humid interior, a blanket of moss covers boulders,
stumps and stream sides. (Derek Ratcliffe)

Plate 8

A quintessential English habitat – a bluebell wood in late April. By great good fortune, bluebell grows better here than anywhere in the world. (English Nature/Peter Wakely)

Plate 9

Wetland: the Insh Marshes, an 850-hectare RSPB reserve of pools, sedge-bed and river shingle in the Spey valley. (English Nature/Peter Wakely)

Plate 10

Minsmere, the RSPB reserve on the Suffolk coast, famous for its marsh harriers, bitterns and other rare birds. In the distance, contending uses of wilderness – Sizewell nuclear power station and the Forestry Commission's Dulwich Forest. (Natural Image/Peter Wilson)

Plate 11

Dungeness: a unique natural wonder with specialised plants and insects able to live in a desert of shingle, and even a dwarf wood of holly. It was also a handy source of gravel for Britain's motorway network and a suitably remote place to park another nuclear power station. (Derek Ratcliffe)

Plate 12

North Norfolk dunes and salt marsh at Holkham, part of Britain's most extensive natural soft shore. (English Nature/Peter Wakely)

Newborough Warren: 2,000 hectares of sand dune and salt marsh facing the Irish Sea on Anglesey. The hills of Snowdonia lie beyond, separated by the Menai Strait. (CCW/Linda Williams)

Plate 13

Upper Teesdale has been called 'a botanical Mecca' and 'England's arctic outpost'. Here its special 'sugar limestone' forms a low scar near the summit of Cronkley Fell. The grass is cropped as close as downland turf by sheep and rabbits. (Derek Ratcliffe)

Another use for a botanical Mecca: flood it with a reservoir to tame the River Tees and help the chemical industries at Teesmouth. (Derek Ratcliffe)

Plate 14

The Muir of Dinnet National Nature Reserve in Royal Deeside: a grouse moor turning into a birch wood. In this part of north-east Scotland the moor is a rich and unusual mixture of heather, bearberry and cowberry, with numerous associated herbs, including wood anemone. The birch woodland, on the other hand, could be anywhere. The process seems inexorable. (PRM)

Plate 15

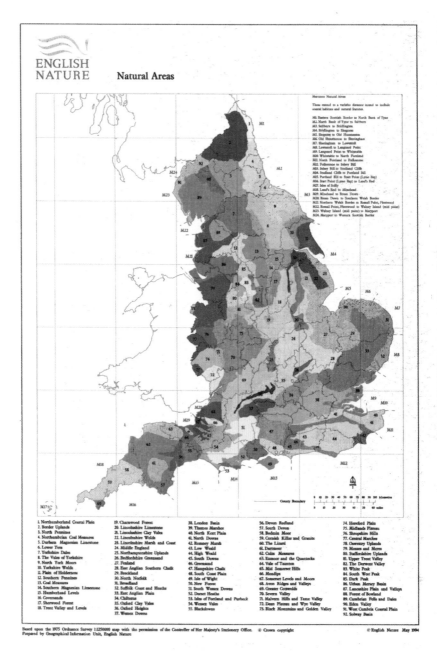

'Natural Areas' in England: areas of similar landscape and other natural features offered as a framework for conservation policy by English Nature and the Countryside Commission. (© English Nature)

Plate 16

Derek Ratcliffe, the NCC's chief scientist, about to inspect a merlin's nest.
(Des Thompson)

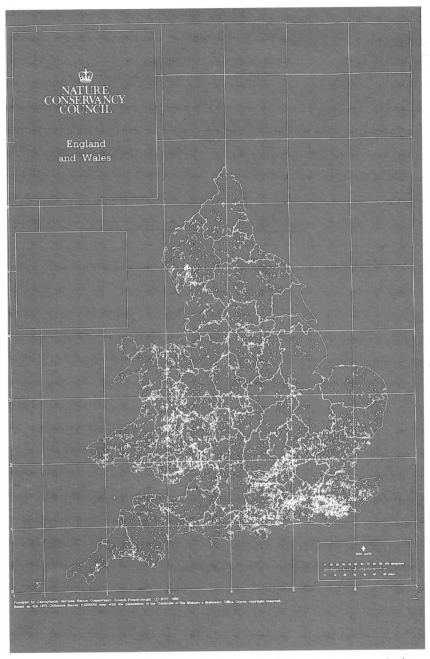

NATURE
CONSERVANCY
COUNCIL

England
and Wales

The pattern of ancient woodland in England and Wales is broadly the same as in the time of the Domesday Book, over 900 years ago – though the woods are smaller, fewer and often converted to timber crops. (English Nature)

baking into charcoal. Tanners relied on the bark of oak and alder to cure hides into leather, bakers on bundles of faggots to heat the ovens. Well into the twentieth century, a whole rural industry was kept supplied with locally grown wood through a traditional form of management known as coppicing, which capitalises on the fact that most British trees regenerate well from recently cut stumps (so long as deer and livestock are prevented from browsing them). Where timber was also wanted, generally for sale outside the parish, some trees, especially oak, were grown on as 'standards', and felled when about 70 to 100 years old. Judicious management preserved a layer of coppice shrubs beneath an open canopy of standard trees, linked by a system of open rides, with glades where the underwood had been cut recently. Such woods, shaped (but generally not planted) for purely utilitarian ends, formed our richest wild habitat. But, like most wild habitats that have been influenced by centuries of use, they demand maintenance. If too many deer get in, the system fails, as it also does when the wood is neglected, and allowed to grow too shady.

It was not until the 1970s that, largely through the work of the NCC's woodland scientist, George Peterken, the astonishing variety of natural woodland vegetation became understood. Peterken identified a dozen broad types of woodland dominated by different kinds of tree: oak, beech, lime, hornbeam, ash mixed with elm, ash mixed with lime, ash mixed with hazel and field maple. In a category of their own are the ancient pine woods of Scotland (see below). These are all semi-natural woods, in which

Coppiced ancient woodland at Brasenose Wood near Oxford with scattered oak and ash 'standards' over a layer of ash, hazel, birch and other trees cut on a regular cycle. Traditional management preserves ideal conditions for woodland flowers and sun-loving insects.

most of the trees (beech and pedunculate oak are frequent exceptions), and the vegetation beneath them, are self-sown and wild. What many commentators failed to appreciate about our woods is that they can be exploited for all kinds of sustainable uses, and yet retain much of their natural character. Those which have existed for at least 400 years are known as ancient woods. Not all natural woods are ancient, nor are ancient woods necessarily natural. For example, much of the naturally regenerated woodland of Highland glens is of twentieth-century origin, while many ancient woods have been restocked with conifers and other planted trees. But a great many woods – generally the richest ones – are at least partly both ancient and natural. England and Wales have 397,700 hectares of ancient woodland, but most of the individual woods are small. The median size is 7 hectares in England and 5 hectares in Wales. Only 2 per cent of our ancient woods measure over 100 hectares (Peterken 1996). Ancient woods can be compared with parish churches, each one an individual, often with an ancient core on which later generations have added bits or removed them. You can often distinguish old woodland by 'indicator species'. Tapestries of wood anemone, primrose and wood-violets are characteristic of ancient wood banks, and older woods have generally acquired a greater biodiversity, including scarce species such as wild service tree or wild daffodil. Woods rich in insects and other invertebrates generally contain a good deal of rotten wood, either in old standing trees or as stumps and fallen logs. Lichens and mosses also like old trees, along with moisture and clean air. Fungi, it seems, like undisturbed soil.

The twentieth century was not kind to our ancient woods. Two World Wars wrecked their timber value, but this was often rather marginal in any case. Worse, as rural economies ceased to need much wood, many woods were used as livestock shelters ('barns with leaves') or for rearing pheasants. Neglect changed their character from relatively open, flower-filled places to either a more shaded, damper environment less friendly to flowers and butterflies or an overgrazed, moribund scatter of trees. Between 1940 and 1980, about 10 per cent of all ancient woods were cleared away for farming, especially in the east of England. Many more were replanted. Until 1985, the Forestry Commission's broad advice to woodland owners was to start again by cutting down most of the existing 'scrub' and planting crop trees, generally conifers with a leavening of oak and other hardwoods around the edge. It set an example on its own small wood properties. Oddly enough, ancient woods were often better protected in the suburbs, where they have been preserved for amenity, than on farms where they were just in the way.

Between 1984 and 1998, the Forestry Commission sold off 144,000 hectares of woodland, including many of its smaller properties. Many of them were restocked ancient woods of great character and importance to wildlife, and some were acquired by the Woodland Trust or by one of the county wildlife trusts. For example, the Worcestershire Wildlife Trust dug deep into its pockets to purchase Shrawley Wood and Tiddesley Wood, both replanted ancient woods in which some of the native vegetation had survived. Of Shrawley Wood, 'half of which is supposed to be a conifer plantation', Oliver Rackham (1990) could gleefully report that it was now once again 'a magnificent lime coppice in which, in places, careful search reveals the remains of a conifer'. Happy as the restoration of such woods is, this was a classic instance of digging a hole and then filling it in again. We ought to preserve a few of those half-wrecked woods as a monument to the time when forestry meant getting rid of native trees.

In 1985, the Forestry Commission introduced its Woodland Grant Scheme, marking a dramatic shift from 'coniferising' to preserving native deciduous woods, especially ancient and natural woodland. For this reason it was important to identify exactly which woods were ancient. This became another of the NCC's grand-scale surveys of the 1980s, undertaken mainly by contract workers under the supervision of Drs George Peterken and Keith Kirby of the NCC's scientific team. Fortunately parts of Britain were well surveyed already, especially by Oliver Rackham of Cambridge University, the master of woodland 'eco-history'. The NCC's 'county inventories' of ancient woodland, with their tables and computer-generated maps, recorded relative size as well as distribution. They also provided a mass of information on the history and character of our woodland cleaves and dingles, hangers and copses, dumbles and ghylls, and the intricacy of their relationships with community and landscape. More than any other branch of ecology, woodland study has helped provide a temporal dimension to natural history, demonstrating the slow evolution of natural habitats over time. In ancient woodland you can feel the past, and sense the perpetual thread that runs through time, especially when you learn to read the clues.

Since the 1980s, schemes that help to conserve and regenerate small woods have proliferated. Perhaps the most successful is the Woodland Trust (see Chapter 3), which has acquired hundreds of interesting woods and opened them to the public. Another welcome boost has been the resurgence of charcoal burning to provide a home-grown product for garden barbecues in a happy marriage of nature conservation and commercial harvest. Making quality charcoal requires the maintenance of a constant supply of coppice wood. At Bradbury Woods, Suffolk, one of the most celebrated ancient woods, a hard-working 'collier' doubles as the reserve's de facto warden, while selling sacks of premium quality charcoal cut from the nature reserve. Contrary to popular belief, it is perfectly all right to chop down and chop up woodland trees, so long as it is done sustainably and so long as that has been the historic use of that particular wood. Most woods contain too many trees. Paradoxical as it may be for lovers of trees, most wildlife lives in the gaps inbetween.

Other locally based schemes aim to bring small woods into beneficial management by helping owners find markets for traditional woodland products, offering advice and organising training schemes. The basic idea is to promote small woods as things of value rather than 'barns with leaves'. Coed Cymru, the Welsh Wildwood Campaign set up in 1985, a partnership of statutory and voluntary bodies, saves small woods from sheep by pro-

Making charcoal the traditional way by slow-cooking wood under a layer of turf. This demonstration was laid on to celebrate the 40th birthday of Yarner Wood NNR in 1992.

moting regeneration and a more vigorous market for home-grown, low-grade hardwoods. For example, it has persuaded parts of the tourist industry to use Welsh oak instead of chemically treated softwoods for fence posts. Similarly, the Silvanus Trust in Devon 'works to develop the viable and sustainable management of woods' in south-west England through a crafts training scheme combined with market promotion of native woodland produce. Although the emphasis in both cases is on economic development, this can also a force for nature conservation by reinstating coppicing and related activities that diversify woods and benefit wildlife.

The Scottish pine woods

The native pine woods of Scotland stand apart from other natural woods in Britain. They used to be regarded as our least disturbed woods, their reputation as the last fragments of a vast ancient forest enhanced by their majestic mountain settings and the grandeur of the older trees. Their interiors, too, are quite unlike lowland broad-leaved woods (see Plate 6). In the place of familiar primroses and bluebells are heather, cowberry and blaeberry (the English bilberry) rooted in thick moss, or, in places, with a shaggy understorey of juniper bushes, some resembling clouds, others towering like cypress trees. Seedling pines need sunlight and mineral soils, and so these woods regenerate mainly in clearings created by felling or fire (which burns off the surface skin of peat), or around the edge. In their unenclosed moorland setting, natural pine woods are dynamic, changing their shape over the years, regenerating in one place and dying in another, expanding or contracting depending on the amount of grazing they receive. They have also been shaped by the interaction of natural events, namely storms and fire (being full of inflammable resin, pines are our least fireproof tree), and human usage such as timber management and stock grazing. On the lower slopes most of the pines are of planted origin, from seed collected from selected nearby trees. On the upper slopes, or the remotest part of the glen, decades of sheep and deer grazing has often brought regeneration to a halt, with the result that former woodland is turning into an open park landscape of moribund trees – beautiful to behold, but on the way out. The pine wood of Glen Falloch, which has seen no regeneration since 1820, now consists of only a scatter of elderly trees. Everywhere grazing has suppressed the natural tree line, except in a few special places such as Creag Fhiachlach in the Cairngorms.

Some 35 native pine woods were identified by the Aberdeen foresters Steven and Carlisle in their well-known book, *The Native Pinewoods of Scotland*, published in 1959. These were all Highland woods dominated by Scots pine, but native pine also survives as a component of lowland mixed woods and as fragments of original vegetation in plantations such as Glen More. A Forestry Commission survey in the early 1990s found in all some 16,000 hectares of native pine wood, including 3,500 hectares scattered in mixed woods and so overlooked by Steven and Carlisle. The largest area of some 4,700 hectares lies in Speyside, mainly in the great forests of Abernethy, Rothiemurchus and Glen Feshie, followed by 2,287 hectares in Deeside, concentrated in Ballochbuie, Glen Tanar and the southern Cairngorm glens. Other large pine woods lie in the east-running glens of Affric, Cannich and Strathfarrar, with smaller ones further west at Rannoch, Tyndrum, Loch Maree and elsewhere. We know better now than to regard them as fragments of a vast 'Forest of Caledon' supposedly destroyed by Vikings or Sassenachs. Pollen evidence shows that while many sites have been under pines for the past 10,000 years, the idea of a single vast forest is a myth (though that did not stop the Green Party from promising to recreate it in its Manifesto for the Highlands in 1989). In fact, Scotland has been poorly wooded for 3,000 years, and today's native pine woods probably reached something like their present extent in the distant past. They are, in effect, ancient, semi-natural woods, like many of those in the lowlands. The difference is that the Scottish pine woods have long been recognised as special by foresters and conservationists alike. One of their first conservators was Queen Victoria, who saved Ballochbuie from the timberman's axe by purchasing it in the 1850s. All but two of them are SSSIs – ironically, Ballochbuie is one of the exceptions; being personally owned by the Queen

A hen capercaillie on the nest. A catastrophic decline in this magnificent pine-wood bird may be due to a range of factors, including wet Junes and overgrazing of its favourite blaeberry (bilberry) plant. There is a voluntary ban on shooting, but the outlook is still grim. (John Young)

it is exempt from scheduling. Recently many have been Euro-designated as Natura 2000 sites, and native pine wood is also listed as a priority habitat in the Biodiversity Action Plan. Moreover they are home to many species listed in the BAP from red squirrels and capercaillies to obscure insects, spiders and lower plants known only by scientific names. Rothiemurchus alone has 16 BAP species (Smout & Lambert 1999).

Even so, until the 1990s most of the larger native pine woods were essentially 'working woods', with timber production as their primary purpose. Back in the 1970s, the Forestry Commission and the NCC agreed on a scheme whereby the larger pine woods would be divided into zones for planting, using seed of local provenance, and for natural regeneration, with a further area set aside as natural non-interference forest. In practice, the plan was weighted towards forestry. In the 'extraction zones', management was close to normal commercial timber practice, while natural regeneration was speeded up by felling to create open clearings, in which the ground was often 'screefed' by bulldozers to expose the mineral soil, thus destroying the natural vegetation. Unless fenced, the non-interference zones, generally at the upper limits of the forest, became shelters for red deer and sheep, which put paid to any possibility of regeneration there. The scheme had the effect of eroding the natural character of the woods while doing little or nothing to prevent their gradual decay through overgrazing. It was thought more important to conserve the different 'races' of Scots pine, which are based on the chemistry of their terpene oils. Any seed used to replant native pine woods was supposed to be 'of local provenance', that is, from nearby natural trees.

The pendulum has now swung from timber production towards sustainable use and conservation. Since 1990, the FC's 'Caledonian Forest Reserve' has been managed mainly for amenity and its environmental value rather than commercial production alone. Planting is still the policy, but the trees planted are Scots pine and other natives rather than the more commercially profitable Sitka spruce. The new policy was not without its opponents. *Forestry and British Timber* magazine, the mouthpiece of the industry, argued that 'commercial conifer forestry has as much right to a place in the Cairngorms as elsewhere'. And as late as 1996, the FC granted Lord Strathnaver a felling licence to chop down some of his native pines. Today's priority is to increase the area of native pine. Scotland's forestry strategy aims at a 35 per cent expansion of the native pine wood resource by 2005, and 35 per cent more in the following 20 years. Along with that, there is a new emphasis on diversity. We now acknowledge there is more to pine woods than pine trees. Monocultures of Scots pine are an artefact of past management. Natural pine woods should contain boggy hollows and lochans, an admixture of pine, oak, ash and hazel along the lower slopes, and natural glades on rocky ground, or where flood, wind and fire has opened up the canopy. There is considerable current interest in restoring 'wet woodland' or muskeg, a habitat of scattered dwarf trees in a boggy glade that was scarcely recognised in the past – and so was often drained and planted. In 1998, the European Union funded a three-year project to restore some 300 hectares of wet woodland in waterlogged hollows and river valleys by blocking drains and removing exotic trees. If successful the project will help to enhance the natural biodiversity of pine woods, and also create a habitat of rare beauty and wilderness appeal.

Pine wood conservation is taking place amid the usual impenetrable blizzard of partnerships and acronyms. Behind the bland words, there still seems to be a rooted conviction that Scotland should look like Norway, and that absence of woods is tantamount to land degradation. In the Cairngorms the overall strategy is primarily the responsibility of the Cairngorms Partnership, destined to be replaced in a few years by a National Park. Here the situation has changed for the better since the bad old days not so long ago when the forest bogs of Loch Morlich were drained and planted, when half of Abernethy was chopped down, and visitors were about as welcome as harriers and stoats. Today the Cairngorms area is almost entirely in benign ownership, either by conservation bodies such as Scottish Natural Heritage, the RSPB and the National Trust for Scotland, or by enlightened estates such as Rothiemurchus and Glen Tanar. The Forestry Commission has joined in not only by grant-aiding woodland habitat restoration, but by restoring parts of its own estate at Glen More to a more natural condition, removing imported crop trees and planting local Scots pine. Elsewhere, the partnership is working towards a defined 'desired future condition' for each native pine wood, based not on some theoretical forest of the antediluvian past, but on the historic and natural character of that particular wood. The woods still provide timber, but it is to be harvested on a sustainable basis, and for a range of social pur-

Rum's new woodlands – an attempt to restore long-lost woodland by planting native trees. (Derek Ratcliffe)

poses (it is still grants, not timber, that generate jobs). The Caledonian Partnership of conservation, amenity and forestry bodies has so far brought some 8,000 hectares of native woodland into 'restorative management', with the help of nearly £4 million of Euro-funding.

Compared with the 1970s, the future of our native pine woods now looks fairly rosy, cherished as they are as a kind of symbol of Scottish wilderness and culture. Even so, expanding the resource does not necessarily guarantee the survival of its component pine-wood animals and plants. As a recent report by Plantlife has pointed out, the distinctive pine-wood flora, which includes rare and delicate species such as twinflower and one-flowered wintergreen, has been entirely neglected (Coulthard & Scott 2001). Numbers of black grouse and capercaillie are dangerously low, thanks to the overgrazing of their favourite food plants by deer. The planting may help to fulfil targets, but it is creating plantations of Scots pine, not natural forest. A better nature conservation strategy would be to massacre the deer, fence the woods and then leave things to nature. But in a less than ideal world the attainment of a consensus that native pine woods are worth conserving is something.

Forest management for nature conservation

The idea of multi-purpose use came late to state forests in Britain. From 1949, the FC was allowed to consider the possibility of using forests for pleasure as well as profit. The following decade saw the establishment of National Forest Parks, the foresters' contribution to access to the countryside, in which car parks, trails, picnic sites, holiday chalets and other attractions were designed. At first, however, there were few concessions to nature conservation. The Parks provided some long, shady walks with a fine view at the end, but as far as the trees were concerned, it was hard to spot any difference between, say, the Queen Elizabeth Forest (1953) or the Border Forest (1955) and any other large plantation. The past decade has seen real improvements. The advantage of public access is that it obliges forest owners to provide things the public likes, notably broad-leaved trees, watersides and glades, and they are good for wildlife too. There is, of course, a price to pay, in terms of badly sited car parks, trampling and disturbance, and the lopping of overhanging boughs and even whole trees as a safety measure. When the FC bought Hamsterley estate in Durham in 1927, the object was to plant as many conifers as possible. As its leaflet reminds us, 'when foresters began planting at Hamsterley in the 1920s they were instructed to plant forests as strategic reserves in time of war. *They did as they were told* [my italics] and had to plant new forests with very little informed help, after all, people had been used to cutting down forests not planting new ones ... How different things are today!'. Well, up to a point. To turn Hamsterley Forest into an amenity, Forest Enterprise planted more broad-leaves, removed conifers from certain beauty spots, and have started to create a more varied structure of fellings, mature stands and mixed woodlands. Meadows and unimproved pastures within the Forest (which the FC was not allowed to plant) have been designated SSSIs and the lower part of the Forest is now designated a Forest Nature Reserve. Some state-owned forests, such as Grizedale in Cumbria, have more elaborate facilities, including a forest centre, play areas and sculptures. The biggest of all, Kielder Forest, was beautified by amenity planting of birch and other trees after Kielder Reservoir was created, and has become popular as a kind of 'English Scandinavia'.

Amenity use does not by itself serve nature conservation aims, but it usually implies some concessions in that direction. In large southern broad-leaved forests, such as Dean, Savernake and Alice Holt, the main problem is one of too many trees. Historically these areas were much more varied than today, with wood pastures grazed by cattle and ponies, and compartments cut over as coppice, as well as tall trees managed for timber. It was their structural complexity and, above all, their openness that made them so rich in wildlife. Forestry policy has either turned them all into similar kinds of high forest, or done away with native trees altogether, and replaced them with conifers. The foresters also felled most of the over-mature trees on the grounds that their timber value was deteriorating. This is rather like demolishing most of the chapels and ornaments of a church, while enlarging the nave into a single, plain, whitewashed hall. Wildlife thrives on detail, and detail thrives on variety. A single-minded policy of timber production produces a dull wood. The most notorious example of this triumph of profit over pleasure is the Crown inclosures of the New Forest, which were once famous for their butterflies. The replacement of the oaks and beeches with conifers in the 1960s reduced their interest considerably, but the *coup de grâce* came in 1969 when the FC let livestock in, thus grazing the rides flat and eliminating nectar sources for the butterflies and the food plants of

Only those with long memories or collections of old postcards remember how much Dutch elm disease changed the landscape. In the 1980s the woods of the east Midlands were full of dead elms, creaking and groaning spookily, like ghosts of the lost trees. This is Overhall Grove in Cambridgeshire.

- DANGER -

Due to Dutch Elm Disease Many trees are falling. Visitors are advised to use the diversion until felling work is completed.

their caterpillars. It had banned butterfly collecting there in 1962, but there are now few butterflies in any case. The unenclosed 'ancient and ornamental woods' of the Forest were supposed to be managed as an amenity, not for timber. Unfortunately, foresters were convinced that these woods were not regenerating properly and needed a helping hand. This meant chopping down and removing many of the mature trees in operations that sometimes created awful mud baths out of the forest soil (as Colin Tubbs shows all too clearly in his book on the New Forest in this series). In 1970, the minister put a stop to it, but as late as 1996 there were still reports of old trees important for wildlife being felled in the Forest (the great winds of 1989 and 1991 blew down many more). By the 1990s, previous assumptions about ancient woodlands had been stood on their head. The 'regeneration fellings' are yesteryear's policy and minimal intervention is now the fashion. The ancient and ornamental woods are assuming a more natural, most would say more attractive, character, with probably more dead wood than at any time in the Forest's 1,000-year history. Coniferisation has been halted, but overgrazing remains a serious problem. Since 1982, the importance of

Britain's largest beetle, the stag beetle, is a 'flagship' species for the many insects that need dead and decaying wood. Yet there are more stag beetles in the hospital grounds and suburbs of South London than anywhere in the countryside – perhaps a reflection of the scarcity of suitable old trees in today's woodlands. (Natural Image/ Bob Gibbons)

the New Forest has been recognised by local authorities and by the Forest's own administration. It is now, in effect, a National Park, and some of its woods are de facto nature reserves. With great care, and a lot of money, the New Forest may one day be nearly as good for wildlife as it was in 1950. 'Old growth' forest of the kind we were busy chopping down only a generation ago is now rare throughout Europe, even in Scandinavia, and the last remnants are cherished accordingly.

Today, woodland is arguably the best preserved natural habitat. Many woods are in a better state now than they were in the recent past, though some of the species they once supported have departed. Some are looked after by conservation bodies who now know more or less what to do – and, just as importantly, what not to do. In recent years the National Trust and some of the county wildlife trusts have made significant contributions to woodland management for conservation, especially in East Anglia, Nottinghamshire, Hereford and Worcester, Devon and North Wales. Public perception of what is good for woodland wildlife lags some way behind. At one extreme, people think no trees should be felled in nature reserves – but this would be disastrous. Our neglected woods are crying out for thinning and more openness. Regeneration, and much of woodland wildlife, thrives on sunshine, as every peasant and forester once knew. On the other hand, old trees are precious and more middle-aged trees should be left to grow old. Each veteran oak or table-sized ash stool took several hundred years to grow. Once gone, we cannot bring them back.

Crutches for an ancient tree: the Major Oak in Sherwood Forest.

Forestry policy in England has become much more 'people-friendly'. The FC's new strategy, subtitled 'a new focus for England's woodlands', swarms with colour pictures of children having the time of their lives, skipping along forest trails, snacking on straw bales and being enthralled by some puppet squirrels. By comparison, there is only one picture of a man sawing up a tree into planks. The minister, Elliot Morley, said he had 'firmly closed the door on single-purpose plantations'. It is now conceded that not all places of value for conservation would be improved by planting trees, however well designed. Henceforth, forestry in England will have to pay court to national policies: rural regeneration, recreation (especially near cities), restoration of industrial wasteland and even nature conservation. This is described as 'a flexible menu-based system allowing greater targeting to deliver public benefits from public money'. Behind the strategy lies an implied recognition that the glory days of coniferisation, in England at least, are over: the land is too well protected by National Parks, SSSIs and registered commons, and the go-it-alone conifer lobby has run out of friends. If commercial forestry has a future in Britain, it will be in Scotland.

It all looks like a remarkable turnabout. But is it? The propaganda is reassuring, but there is a suspicion that conservation projects are being used as a figleaf behind which, if not quite 'business as usual', a subtler form of afforestation continues. Large-scale plantings such as the National Forest in the English Midlands and the Millennium Forest in Scotland (see also p. 200) may have multi-purpose goals and respect open land of scientific interest, but they also assist forestry in its aim of growing as much timber as possible. Back in 1990, the Commission's new director might have let the cat out of the bag in his remarks in *Forestry and British Timber* magazine: 'The British public does not have a good feeling about Sitka plantations. It wants to see hardwoods so we have got to support these aspirations. I am not talking about a dramatic change in direction ... We must support [the new forest in the Midlands] to enhance the public's view of forestry: *then we can get on with the job of planting timber*'. And 'planting timber', Steve Tompkins has pointed out, still means draining moorland and planting Sitka spruce. At the end of 2001, the Forestry Commission completed a census of England's trees. It seems there are now some 1,300 million trees, or 25 trees for every man, woman and child, covering 8.4 per cent of the land area. Though lower than the European average, this is still twice as many trees as there were a century ago, and probably more than at any time since the Middle Ages. The resurgence is entirely due to planting. The results were hailed uncritically by the press as a good thing for the environment. If so, it is a man-made environment, often produced at the expence of the natural one. The good news is rather that modern foresters are more ready to see trees as objects of delight, and not merely of profit. Some think the bad old days of commercial forestry are over. Time will tell. Perhaps the last word on this subject should be left to the man whose vision and patient persistence made much of this possible, George Peterken:

> 'It is tempting to conclude that woodland nature conservation can now be safely left to the forestry profession, thereby releasing conservationists to devote their limited resources to less tractable problems elsewhere. In the short-term, this may seem reasonable enough, but in the longer term we have to recognise that pressures to cut back on nature conservation within forestry will periodically increase, and that Government policies can be altered to reinforce them. Many conservation organisations appear to have placed woodland conservation on the back burner, but this is unwise. In the long term, it would be prudent to maintain an active role for conservation organisations within forestry, for a strong, independent voice will periodically be needed to maintain environmental standards against other pressures on forestry.' (Peterken 1996).

8

Bricks and Water

Until the 1970s, nature conservation was mainly a rural activity. The only National Nature Reserve within a town was Wren's Nest, at Dudley, and that one was designated for fossils; its wildlife – trilobites and sea-lilies – had died out about 300 million years ago. It was also a land activity. For a maritime nation, the wildlife of the sea had been strangely neglected. But the sea has its own institutions, resistant to the bureaucratic ideas of land-lubbing conservationists. Conservation in the sea was about fish and fish-eries, and that was the province of the agriculture departments. Otherwise, the sea was useful as a junkyard. You could dump whatever you liked in the sea, and it would sink to the bottom, never to be seen again.

Popular interest in the marine environment was growing in the 1970s, encouraged by sport-diving and events such as Underwater Conservation Year in 1977 (which led, as such things do, to the formation of an Underwater Conservation Society, now the Marine Conservation Society). The NCC was by then dipping its toe into marine matters, through surveys and its contribution to oil-spill contingency plans. It took a lead by setting up a working party on marine wildlife, resulting in its 1979 report, *Nature Conservation in the Marine Environment* (NCC 1979) describing Britain's marine habitats and the impacts on them of human activity, such as pollu-tion, barrages, reclamation and fishing. A key recommendation was the establishment of Marine Nature Reserves. The point of these was not only to protect special bits of coast, but also to give the NCC a statutory role in marine conservation, and so make its voice heard.

At the same time, the conservation world was taking a greater interest in towns. Here there was potential common ground. The Government was interested in revitalising the inner cities, and an element of this was the provision of more green spaces, with community involvement in their care. The NCC chipped in with a report on Birmingham's wildlife, written by W G Teagle and called *The Endless Village* (1978). At the bidding of its direc-tor, Bob Boote, thousands of copies were printed and distributed, and as a result some large spiders living under Spaghetti Junction became almost as famous as the kestrels nesting at the Law Courts of The Strand. The hun-dred-odd copies mouldering in the basement at the NCC's office in Aberdeen (what did we care about Birmingham?) inspired me to write a similar book about the wildlife of my then home city, although I am not sure our star roof-nesting oystercatchers ever acquired quite the same notoriety as Bunny Teagle's spiders. As the NCC recognised, nature con-servation in towns and cities is all about 'getting people involved and tak-ing imaginative advantage of the many opportunities available' (*NCC 5th Report*). For example, while permanent nature reserves might be unrealistic, there is a lot you could do with wasteland and temporary open spaces. George Barker, the NCC's former officer in the West Midlands, made it his mission to propel urban conservation into the nature conservation main-stream, with its emphasis on communities and education. Today urban nature conservation is a multimillion pound business, involving a labyrinth of partnerships between local authorities, businesses and local people. It has already gone well beyond the mainstream. It is positively fashionable.

In this chapter, I have tried to unravel some of the key threads of nature conservation in towns and the coast, beginning with the great survey of our shallow seas undertaken between 1987 and 1997.

Marine habitats and species

Britain's shores and shallow seas contain more variety, it is said, than any European country facing the Atlantic – our combination of geology and warm currents ensures a large diversity of marine habitats. Yet although the life of some individual sites (like Lundy and certain estuaries) had been well surveyed, there were, before 1987, few detailed habitat maps, and no assessment of the coast as a whole. Such an assessment was needed to form a factual basis for marine nature conservation, for example, for choosing candidate Marine Nature Reserves. Hitherto, only estuaries had been part of the conservation agenda, mainly because their intertidal mud was used by birds as feeding grounds (their other wildlife was therefore of interest mainly as bird food). Otherwise conservation stopped at the high water mark. In 1987, an NCC team of eight under a leading marine biologist,

Sunset star coral, *Leptopsammia pruvoti*, part of a rich community of soft corals, sponges and sea squirts in underwater caves on the Knoll Pins, part of Lundy Marine Nature Reserve. (JNCC)

Keith Hiscock, began to survey shallow-water marine habitats. Their remit was to describe and map the variety of marine habitats, and identify sites and species of nature conservation importance. The review was based initially on existing 'field surveys' by marine institutions and diving groups, but for much of the coast, the team had to start from scratch. Like 'rescue-dig' archaeologists, they concentrated on habitats under threat such as natural harbours, rias (flooded valleys) and chalk shores (which have a peculiar algal flora of their own). Their fieldwork even took them to Britain's most isolated islet, Rockall. The team developed a computer database, incorporating habitat data and bibliographic details, and state-of-the-art analytical tools to organise and classify marine ecosystems. Their pioneering work, analogous to the National Vegetation Classification on land, has made a significant contribution to marine conservation Europe-wide, and helped to raise the subject's profile in the 1990s. Arguably, the review was the last significant, fundamental research commissioned and carried out by a British nature conservation agency. Although the original hopes for a network of Marine Nature Reserves never materialised, the Marine Nature Conservation Review (MNCR), as it is called, has been used to identify informal 'marine consultation areas', and candidate SACs under the EC Habitats and Species Directive. In 1998 it was transmuted into the JNCC's Marine Information Project, an information service on marine conservation, that also promotes common standards for surveying and monitoring sea-bed habitats. It has also influenced the selection of habitats and species for the Biodiversity Action Plan [see Box]. Alas, under current pricing

---

Marine habitats and species listed in the UK Biodiversity Action Plan

Maritime habitats BAPs:
- Maritime cliff and slopes
- Coastal sand dunes
- Coastal vegetated shingle
- Coastal salt marsh
- Machair
- Littoral and sublittoral chalk
- *Sabellaria alveolata* (honeycomb worm) reefs
- *Sabellaria spinulosa* (tubeworm) reefs
- *Serpula vermicularis* (tubeworm) beds
- Sheltered muddy gravels
- Maerl beds
- Horse mussel beds
- Seagrass (*Zostera*) beds
- Sublittoral sands and gravels
- Mudflats
- Tidal rapids
- Saline lagoons
- Deep-water coral reefs
- Mud habitats in deep water

Maritime species BAPs:
- Baleen whales
- Small dolphins
- Toothed whales (other than small dolphins)
- Harbour porpoise
- Marine turtles
- Basking shark
- Common skate
- Deep-water fish
- Commercial marine fish
- Fan shell
- Northern hatchet shell
- Native oyster
- Sea-fan anemone
- Ivell's sea anemone
- Starlet sea anemone
- Pink sea-fan
- Sunset cup coral
- *Ascophyllum nodosum Ecad mackii* (a brown alga)
- *Anotrichium barbatum* (a red alga)

rules for government publications, buying the dozen volumes of the MNCR published by 2001 requires a deep purse. As with similar government surveys of Britain's wildlife, the public who paid for it all deserve at least an affordable summary volume in readable, non-technical language, which explains what is special about Britain's shores and what this work is doing to protect them. Unfortunately this does not seem to be a priority. The choice is between dumbed-down leaflets or technical tomes.

Since 1981, government has given protection to a steadily growing number of rare and threatened species on the advice of the NCC and its successor bodies. These include several marine animals and plants. Fisheries apart, marine wildlife has been far less well surveyed than on land, though knowledge has nonetheless increased enormously in the past 15 years. Most of our listed species are cetaceans and sea turtles, which are protected internationally, or ones belonging to a restricted and threatened habitat, such as saltwater lagoons or our cold-water equivalents of coral reefs. One such is the pink sea-fan, *Eunicella verrucosa*, which grows on deep water 'reefs' in Lyme Bay and elsewhere along the south coast. Divers report that it is not as common as it was. The species is slow growing, and probably vulnerable to over-collecting, but very little is known about its ecology. What statutory protection brings to it is not underwater policemen but attention and, with luck, research funding. Divers from the Marine Conservation Society and Devon Wildlife Trust are now videoing and photographing marked sea-fans to see how fast they grow, and studying its population dynamics. Early results suggest that it is being damaged not so much by collecting but by scallop-dredgers and weighted fishing nets (Munro & Munro 2000).

Vulnerable to trawler nets. Pink sea-fan in deep water off Pencra Head, Cornwall. (JNCC)

There has been much recent interest in the little-known deep-sea corals off northern Scotland and Ireland, a world only now being revealed by submersibles equipped with powerful lights. An example is the Darwin Mounds, a series of small reefs between the Outer Hebrides and the Faeroes, dominated by a large cauliflower-like coral, *Lophelia pertusa*, along with soft corals, sea-fans and sponges. The presence of these deep-sea gardens was first revealed when fragments of them were brought to the surface by trawl nets. They are brittle, and easily smashed to pieces by the weights used to keep the huge modern fishing nets open. The WWF has called for an Oceans Act to protect such places. Unfortunately many of the trawlers in these waters are from other European countries, especially Spain, and without international co-operation such safeguards would be worthless. The Prime Minister has promised 'measures to improve marine conservation including a series of marine stewardship reports' (Sullivan 2001).

What can happen when would-be protection is not accompanied by an adequate knowledge of the natural history of a species is well illustrated by the story of the lagoon sandworm, *Armandia cirrhosa*. This animal was protected mainly because it is confined to a habitat that itself is under threat – saline pools sheltered behind a bar or sea wall. The seaworm was found in just one such pool, Eight Acre Pond near Lymington, Hampshire, which had been excavated on the site of a saltworks about a hundred years ago. The pond is a Local Nature Reserve, and used to be drained each winter to prevent wave damage and to provide mud and shallow water for feeding birds. When the worm was discovered in 1985, its population was estimated at 15 million individuals, buried in sediment at the bottom of the pond. In the interests of the worm, so it was thought, they stopped draining the pond. In 1991, when the species was due to be surveyed again as part of the five-year monitoring process for protected species, search as they might, investigators could not find a single worm. The worm, it seemed, had relied on regular periods of drought, which was no doubt why it had chosen that particular pond in the first place. Unfortunately the nearest known population is in Portugal.

One species providing good evidence of a decline through overfishing is the basking shark. The legal protection at last given in 1998 crowned a ten-year campaign led by the Marine Conservation Society to save this harmless but enormous shark, the world's second largest fish. The basking shark, whose liver is rich in oil, has long been hunted and harpooned by Norwegian whalers and small-scale shark fisheries operating from home waters, mainly from western Scotland (the writer Gavin Maxwell set up one of them in the 1950s, and wrote about its tribulations and eventual failure in *Harpoon at a Venture*). What may be the world's largest concentration of basking sharks congregates off the west coast of the Isle of Man each summer. By the 1970s, only one man, based on the Clyde, was still harpooning them, at the rate of ten to 100 a year, though there was nothing to stop others, like Maxwell, from having a go. More seriously, the Norwegian whalers were allowed a quota of 400 tonnes of shark liver – perhaps 800 to 1,000 individual sharks – from the Irish Sea, as part of a quid pro quo arrange-

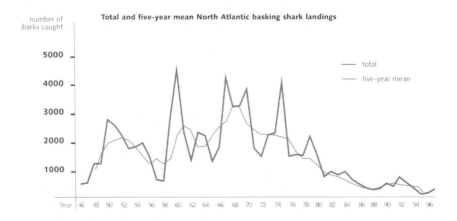

The fall and fall of the basking shark measured by fishery catches. It ceased to be commercial around 1975: that is, it cost more to find one than the fish was worth. (JNCC)

ment, allowing British fishermen to take some of their cod stocks. There was no stock assessment to find out whether this was sustainable; indeed at that time we knew almost nothing about basking sharks. Since the Government would not do it, the Marine Conservation Society organised a national shark watch campaign to find out where the sharks were congregating inshore and try to monitor their numbers. Seawatch, the conservation arm of the Sea Life centres, joined in to support a project using satellite tracking to follow the movements of the shark (fortunately the basking shark feeds on plankton near the surface, and so can be spotted from land). Combined with fishery catch data, the results of the survey suggested a collapse in the North Sea population since the late 1970s. Eventually the whalers, finding fewer and fewer sharks, departed. Since the closure of the last British basking shark fishery had removed any strenuous objections to protection, the basking shark was duly listed. A proposal by the Countryside Minister, Elliot Morley, to add it to the CITES Convention (q.v.) in 2000 unfortunately failed. In effect, this shark has become an honorary whale, enjoying legal protection from 'intentional or reckless harassment' as well as harpooning. But sharks of all kinds are still being slaughtered unsustainably by Spanish fishermen to supply Orientals with the basic ingredient for shark's-fin soup. Meaningful protection for such far-ranging species can only come from better knowledge and international co-operation.

Another species that has attracted much interest is the bottle-nosed dolphin. This is the one most often exhibited in dolphinariums, big, active and intelligent, and with a pleasant expression on its bottle-nosed face. It also makes attractive whistling sounds. It is not rare, but, like all dolphins it tends to get caught in fishing nets. Since dolphins are, of course, air-breathing mammals, when they get tangled up in fishing nets they drown. By 1998, large fast trawlers, working in pairs to catch bass and mackerel, were towing nets the size of a football pitch. Also hunting bass are the dolphins, which are unable to escape in time and are swept up in the nets as a 'bycatch'. After suffering a prolonged and agonising death, they are thrown overboard, sometimes with a rope around their tail, and it was these bodies, washed on shore by the tide, that first alerted authorities to the problem. In 2001, over 800 dolphin 'strandings' were found on the Cornish, Devon and Breton coasts, and the full scale of the slaughter might have been as high as 2,400 (Deere-Jones 2001). This is manifestly unsustainable, and if continued unchecked, will result in the extinction of dolphins and porpoises in UK waters. Fishery scientists accompanying one boat in 2000 were shocked to find 12 dead common dolphins entangled in the net. Another trawler caught two dozen dolphins that day, and these two ships were only part of a fishing fleet that stayed in these waters for six weeks. The minister, Elliot Morley, claimed this was the first time he had been given any 'real evidence' on the scale of the problem, but avoided promising immediate action. It is the old chestnut: fishery problems lie 'outside the competence of individual states', although the Maastricht treaty does in fact allow any state to take action to protect the environment. One way or another, it is no joke being a British dolphin. A recent study found that, for whatever reason, Moray Firth dolphins have the highest level of skin lesions known. Will the EC's protection of the bottle-nosed dolphin help to improve their fortunes? Two large SACs, in the Moray Firth and Cardigan Bay, are being established to protect our two resident populations of this species, and in this case are extended well beyond the three-mile limit. But marine 'sites' are difficult to police, and dolphins can come and go as they please.

Marine nature reserves and Euro-sites

Marine nature reserves were the dream of a few divers and marine biologists who, unlike the rest of us, had seen what was going on below the waterline. Traditional nature conservation in Britain always stopped at the low tide mark except in enclosed tidal waters and estuaries. Anything beyond that was under the jurisdiction of nautical authorities. Although underwater habitats were relatively poorly explored, it was known that, as on land, some places have more wildlife than others, and that disturbance to the sea bed through trawling, dredging, fish farming or sheer boat traffic was reducing habitat quality and biodiversity. The NCC wanted to conserve at least the best examples of shallow-sea habitats from damage. By the 1970s, it had reached agreements in a few places, such as the Isles of Scilly, based on voluntary self-regulation on sport-diving and power boats, but they had no force in law. The NCC pointed out that the UK was lagging behind other countries in this matter, and failing to honour commitments it had signed up to under international conventions (NCC 1979). This was a pity, because our marine life, warmed by the Gulf Stream and sheltered by hundreds of inlets, islands and estuaries, was as rich and remarkable as anywhere in Europe.

The NCC wanted to create protected areas below the watermark. After due debate, the Government gave it the power to do so under sections 36 and 37 of the Wildlife and Countryside Act, which took effect towards the end of 1981. The NCC could, with the co-operation of the responsible authorities, now make bylaws within a three-mile limit to protect an area designated as a formal Marine Nature Reserve (MNR). There was, however, a catch, and a big one. The NCC had no power to override existing uses of the site. Many bodies have vested interests and regulatory powers. Fishing from shore or in tidal waters has been a common right since the Magna Carta. Since 1966, the main bylaw-making authorities are the Sea Fisheries Committees, and without their co-operation the NCC could do nothing to control fishing in a sensitive area. The NCC had to reach full agreement with each and every authority and user before offering its recommendation to the Secretary of State for his signature. Did anyone expect it to succeed? Even if it had, Marine Nature Reserves offered rather limited protection. The NCC could, for example, control the navigation of pleasure boats, but not trawlers (unless the fisheries committee agreed). It could prosecute divers for removing sea urchins, but not prevent fishing lines from wrecking the sea urchins' habitat. Nothing, of course, could prevent a reserve becoming degraded through pollution or the effects of overfishing from outside.

The NCC drew up a list of seven potential sites covering a wide geographical and ecological range: the waters around the islands of Lundy, Skomer and Bardsey, the Isles of Scilly, the Menai Strait, Loch Sween in Argyll and Strangford Lough in Northern Ireland. The NCC envisaged some explanatory noticeboards and lectures, and better opportunities for research, but essentially the Marine Nature Reserves would be areas where everyone agreed not to disturb the wildlife. A voluntary code of conduct would operate for divers.

Unfortunately not everyone agreed this was a good idea. Fisheries committees and other authorities were new to this sort of thing, and tended to see it as an unnecessary piece of bureaucracy from a small group that had temporarily captured the minister's ear. There were also strong objections from local fishermen and others, especially on restrictions on the use of dredging gear. While it was realised that people would need a lot of persuading, the NCC hoped to be able to set up two or three 'MNRs' within a couple of years. In practice, it hit a brick wall. At Lundy, the main obstacle was that bylaws were needed to prevent trawling and dredging. These could be made only by the local Sea Fisheries Committee. After protracted discussion, in which the original proposals were much watered down, it agreed to do so, thereby enabling the NCC to draw up a public consultation paper. Britain's first Marine Nature Reserve at Lundy was duly declared on 16 October 1986. But it opened no doors, and it was not until 1990 that a second MNR was made at Skomer and the nearby Marloes Peninsula. To obtain restrictions on scallop dredging, the NCC had had to prove to everyone's satisfaction, by means of a lengthy experiment, that the huge iron rakes were harming wildlife of the sea floor. (Today no one should be in any doubt. A recent study by the University of Liverpool showed exactly how tow nets and dredges degrade the sea bed, replacing soft corals, sea urchins and

A bed of horse mussel. The shells are tied to the sea bed with strong threads, which bind together the sediment and form a kind of reef in which other life, like soft corals, can colonise. (JNCC)

sponges with sediment-living communities, and with knock-on effects for flat fish, such as turbot and plaice.) A third contender, at Loch Sween in Argyll, was withdrawn because SNH did not want to upset anybody. Perhaps the most successful MNR was at Strangford Lough in Northern Ireland, said to be the most outstanding sea lough in all Europe, which was designated in 1995. There the conservation authority took a different tack, with the emphasis not on exclusion, but on sustainable use. Thus it drew on the threads that combined the interests of users and conservationists in wanting to preserve a clean marine environment. Equally, there was by now plenty of evidence that unregulated dredging for scallops was diminishing the resource, not only here but throughout western Europe, and that conservation was in everyone's long-term interests. Crucially for success, Government became actively involved, rather than taking refuge behind a smoke screen of bland phrases, as before.

So, more than a decade's effort produced just three Marine Nature Reserves in the UK; one in England, one in Wales and one in Northern Ireland, plus a non-statutory 'marine park' around the Isles of Scilly. Attempts to make further MNRs were quietly abandoned in the 1990s, partly because they were fruitless, partly because European legislation offered a potentially better route through the EC Habitats Directive (see below). Interestingly, some European countries have had much greater success with Marine Nature Reserves than Britain. Perhaps their vested interests are less entrenched, or maybe their politicians are more interested in fish.

In a recent review of marine environmental regulation, the DETR had
this to say about Marine Nature Reserves:

> 'The ambition to create a comprehensive network of MNRs is
> widely perceived to have failed, as evidenced by the very small
> number of areas declared in the past twenty years, through the
> extreme difficulties of securing agreement to take these forward.
> Failure may relate not so much to the inadequacy of the legisla-
> tion, as to the policy position that total agreement between all rele-
> vant interests is needed before any MNR can be put in place.
> Where established, however, MNRs have proved useful test beds
> for marine conservation management.' (DETR report on marine
> conservation, 2001).

Since the mid-1990s, nature conservation in Britain has taken a
European turn. Under the EU Habitats and Species Directive, which every
EU country has signed up to, the UK must create a network of conservation sites called Special Areas for Conservation or SACs (q.v.). A key difference between SACs and the British system of SSSIs is that the former
include marine habitats. They are chosen to protect the best examples of
scarce and vulnerable marine communities, such as maerl beds, saline
lagoons and beds of horse mussel. While legislation merely *allowed* the
establishment of MNRs, the Habitat Directive *requires* governments to
establish SACs.

A maerl bed off Gairloch in the Scottish Highlands. Formed by purple nodules of
dead coralline seaweeds, maerl beds form a reef habitat rich in marine life.
Unfortunately maerl is in industrial demand and also vulnerable to trawling and
harbour improvements. (JNCC)

The Directive's main requirement is that SACs should be protected from developments that are likely to harm them, except in cases of 'overriding public interest, including those of a social or economic nature'. Drawing up candidate SACs dominated the work of the conservation agencies during the second half of the 1990s. For selecting marine SACs, Britain had a flying start in the Marine Nature Conservation Review, which might have been created for just such a task. However, as experience with Marine Nature Reserves has shown, it is one thing to know where the best sites are but quite another to reach agreement on their conservation. The advantage of European schemes is that Europe pays for them. During the past few years, the EU has funded a project to 'set up partnerships' and produce management plans for 12 key areas covering a wide range of UK shores and marine habitats. These are Papa Stour, Loch Maddy, the Sound of Arisaig, the coast near Lindisfarne, the Solway Firth, Morecambe Bay, The Wash and North Norfolk Coast, the Lleyn Peninsula, Cardigan Bay, Plymouth Sound and Estuaries, Chesil Bank and The Fleet, and Strangford Lough. However, there are as yet no proposals to extend the SSSI system to cover marine habitats; the minister merely has to 'have regard to the requirements of the Directive' in making decisions, and ministers generally prefer voluntary means wherever possible.

Choking in the black gold

Oil has several qualities that make it bad for birds. First, it floats on water, which means that a relatively small amount of spilt oil can form a slick that spreads for miles. Second, the volatile content of the oil soon evaporates, leaving a sticky, tar-like residue. This matts the feathers of seabirds, especially those that dive or swim on the surface, reducing their buoyancy and resistance to cold. Not unnaturally the poor birds try to preen the oil from their feathers, and end up swallowing some of it. This does them no good either. Lastly, there is a lot of it about, both in the sea and onshore. The birds most vulnerable to oiling are guillemots and razorbills, which do most of their feeding in inshore waters, close to shipping routes. Seabirds that congregate in sheltered waters in winter, such as long-tailed duck and common scoter, have also been badly hit by recent oil spills. Every time a large slick is reported, television journalists are eager to talk up the number of oiled birds 'rescued' and released back to the wild after a clean-up, as though they were now safe. In fact most of them will die. Cleaning up oiled birds is more a matter for our consciences than a contribution to the survival of the species; and, horrible though it is, the bird populations can well afford the losses.

Minor oil spills from ships or pipelines go on all the time, either as the result of accidents or when a skipper decides to flout international law by cleaning out his tanks at sea (this once serious problem has been overcome in large tankers by changes in ship design). Perhaps 60,000 tonnes of oil are discharged into the North Sea in an average year, most of it illegally (Evans 1997). For example, autumn 1989 saw separate spillages in the Mersey (from a broken pipe), the Humber (when two tankers collided) and the Solent (from a tanker at berth). 1996 was another busy year for coastguards: apart from the wreck of the *Sea Empress* (see below), there were 46 'casualty spills' from boats and 300 leakages from oil and gas installations amongst a total of 678 reported incidents from the Straits of Dover to the Moray Firth.

Of course the cases one remembers are the really big spills, which seem to happen about twice a decade in British waters. None made a greater impact than the *Torrey Canyon*, which, while travelling at full speed in clear weather, managed to impale itself on a rock in the Seven Stones reef off Lands End on 18 March 1967. Over the next week, most of her cargo of 119,000 tonnes of Kuwaiti crude oil flooded out of the ruptured hold, polluting 230 kilometres of Cornish coast, including 40 popular beaches, before the wreck was set on fire with the help of RAF Buccaneer bombers. But what caused most of the damage to shore life was not the oil so much as the three million gallons of toxic detergent sprayed over the beaches and rocks in an attempt to emulsify and disperse it. In a panic over the threat to the tourist trade, the authorities ignored the Nature Conservancy's warning that emulsifiers kill shellfish larvae and thus destabilise whole food webs, and that, once the oil had come ashore, it was better to let the wind and waves do the dispersing. In Brittany, which received more of the oil than Cornwall, the French, more alert to their shellfish stocks, used sawdust and bales of hay to mop up the worst of it. Some 8,000 guillemots and razorbills were cleaned up by the RSPCA and others at makeshift 'rehabilitation centres', but only a few lived long enough to release. An estimated 25,000 birds were killed by *Torrey Canyon* oil, which sounds a lot, though it was probably fewer than those perishing in routine, unreported spills every year, or what pollution by PCBs (polychlorinated biphenols) would do to birds in the Irish Sea the following year. For the public, the *Torrey Canyon* was a disaster, but it had no known long-term ecological consequences; nothing irreplaceable had been lost, unlike the grubbing up of half of Bradfield Woods in Suffolk that year, or the flooding of Cow Green in Upper Teesdale at about that time. The main point of the *Torrey Canyon* wreck was as an object lesson in how not to deal with oil slicks. Fortunately the experience led to the use of less harmful chemical dispersants, and not on land.

In the 1970s, emergency procedures for dealing with slicks were much improved. The NCC appointed a full-time marine expert, Robin Donally, to advise on wildlife impacts, and published an *Oil Pollution Manual*. The worst oil spill of that decade was probably the *Esso Bernicia* incident in December 1978, when the tanker collided with a jetty at Sullom Voe in Orkney. Although only 1,174 tonnes of heavy crude oil leaked from the ship – far less than from the *Amoco Cadiz* in the same year – it hung around in the enclosed waters of Yell Sound at a time when many arctic birds were there. The result was an unusually high toll of uncommon seabirds, including 176 great northern divers, 306 long-tailed ducks and 632 black guillemots, as well as sundry otters and seals. To those that picked up the corpses and knocked the dying birds on the head, it was a scene of horror and desolation that they will probably remember for the rest of their lives, but, like

Cleaning up after the
wreck of the *Sea Empress*.
(CCW)

the Torrey Canyon, it probably did no more than temporarily depress the
numbers of some species.

In the 1990s we had two well-remembered wrecks, the *MV Braer*, which
foundered on rocks off the coast of Shetland in January 1993, and the *Sea
Empress*, which managed to hole itself in well-charted waters as it neared
the port of Milford Haven in February 1996. Fortunately the bulk of the
Braer's cargo of 84,000 tonnes of light oil was dispersed by gales or sank to
the sea bottom where it does not seem to have caused lasting damage,
except to the local shellfish industry (the one incident I remember from
the *Braer* is the television crew who, looking for an oiled otter to film, succeeded in running over a healthy one). The 72,000-tonne spill of crude oil
from *Sea Empress* was a greater embarrassment. A special 'environmental
evaluation committee' was set up to assess the damage and monitor the
recovery, and make recommendations. Most of the severely polluted coastline, it seemed, suffered only short-term damage, although the temporary
loss of limpets to graze the intertidal rocks led to a dramatic increase in
seaweeds, changing the appearance of many shores. There was concern

about a rare starfish, *Asterina phylactica*, and the impact of oiling on the common scoter, which congregates off the coast of South Wales in winter. But, as before, the conclusion was that far more seabirds probably die every year from accumulated minor incidents than in the occasional major spill.

The slick of the millennium came from the *Erika*, which foundered and broke up off Brittany in December 1999, releasing its crude oil cargo into bays and inlets important for wintering seabirds, especially auks. Five times as many birds were oiled by the *Erika* as by the *Torrey Canyon*, many of them individuals that nest in Britain (we know this because the *Erika* disaster led to the largest haul of ringed recoveries in the 90-year history of the bird Ringing Scheme). Chris Mead estimated that 10–20 per cent of young auks from the Irish Sea may have perished, together with perhaps 5–10 per cent of those that nest in Western Scotland. The *Erika* probably killed some 50,000 guillemots.

Even tragedies on this scale did not seem to cause overall demographic change to a bird whose prospects Mead rates as excellent, despite the large number killed annually by oil or drift nets. Over a million guillemots – a third of Europe's total – nest in Britain. It seems that the harm we dole out to them with one hand is more than compensated by the way fishing activity has changed the population structure of commercial fish, reducing numbers of large, predatory fish and producing an abundance of small fish of just the right size for a busy guillemot with a hungry chick to support. Whether oil spills add significantly to the irregular breeding success of some seabirds, such as the kittiwake, or to the natural losses caused by winter storms, is uncertain. What they do do is to kill individual birds in a horrible way, and spoil some of our most beautiful coastlines and beaches for at least a season. A major wreck makes a good media drama with pictures of people rescuing suffering animals. But the environment recovers from that sort of disaster.

Fish farms and other marine conservation issues

Fish farming or aquaculture is an important enterprise in Scotland. The industry has grown from about 800 tonnes of fish per year in the early 1980s to 125,000 tonnes in 2000. Practically every sea loch on the Scottish west coast has at least one fish farm. The industry currently contributes £180 million to Scotland's economy and is an important provider of jobs in the Highlands and Islands.

On the other hand, fish farming has almost certainly helped to reduce the numbers of wild salmon and sea trout. Farmed salmon often escape. In 2000 alone, some 440,000 salmon escaped from Scottish fish farms (Scott 2001). In the previous year, about a quarter of a million more salmon escaped from a single fish farm, at Shapinsay, Orkney, when storm-force winds broke open their cages. There is nothing to stop these fish interbreeding with wild salmon. The fear is that interbreeding will produce genetically weakened fish that lack the instincts needed to survive in the wild. There is also evidence that interbreeding reduces their resistance to disease and to parasites, such as sea lice.

Another problem is pollution. Like farms on land, fish farms depend heavily on chemicals: the feed is factory-produced. So are the pesticides that control sea lice and the highly toxic paints that prevent the nets from fouling. The excreta of caged salmon is rich in nitrogen and phosphorus. In 2000, the estimated 7,500 tonnes of nitrogen and over 1,000 tonnes of phosphorus that fell on the sea bed were equivalent to the sewage of a large city. An average-size fish farm will release nitrogen equivalent to the sewage of 11,000 people while the phosphorus output is comparable to 33,000 people. On Scotland's remote sea lochs, caged salmon produce more sewage than people. Not surprisingly, plankton blooms are increasing. In the sea, this often takes the form of alarming-looking 'red tides' made up of dense masses of diatoms and dinoflagellates. Some of these microscopic beasties produce toxins that accumulate in filter-feeding animals such as mussels and scallops, and render them inedible. Hence fish farms are not popular with shellfish farmers. Red tides, formerly rare, are increasing. The Scottish Executive denied that they had anything to do with fish farms, but their denials failed to carry much conviction, especially when a television documentary, *Warnings from the wild*, revealed a culture of secrecy in which scientists were warned to keep quiet about the environmental risks.

Regulations on marine fish farms have been tightened up. Until 1998, all a would-be fish farmer needed to do was to apply to the Crown Estate Commissioners, responsible for the sea bed, for a lease. No planning permission was necessary, and therefore there was no democratic control or environmental assessment. Local authorities were eventually given a say in the process in 1998 under an interim arrangement, while consultations on the transfer of responsibility from the Commissioners to planning authorities continued. In 2001, the Scottish Executive turned down a petition for a public inquiry into fish farming. Instead, it proposed to set up an 'aquaculture strategy group' to develop 'a strategic framework that balances the environmental impacts against the socioeconomic benefits'. Presumably it is the Executive who will decide exactly where this balance should lie. In the meantime, WWF published a damning report, *Bitter Harvest*, calling for a moratorium on new fish farm developments until a strategy for aquaculture in Scotland is produced.

The reckless over-exploitation of cod, herring and skate has presumably had knock-on effects on seabirds and other marine life. In the Irish Sea, the, alas, now misnamed common skate seems to be on the verge of extinction, reduced, as far as anyone knows, to two small populations. It has earned the skate a place in the Biodiversity Action Plan. Elasmobranchs such as sharks and skates are particularly vulnerable to overfishing because of their reproductive biology. They lay fewer eggs than other fish, and take longer to mature. Even common species, such as the thornback ray, are in decline. With fast modern fishing vessels armed with sonar and huge nets, fishing, if never exactly easy, is just too efficient.

The overfishing of one key species, the sand eel, was almost certainly responsible for a series of poor breeding seasons for arctic terns and shags

The kittiwake is more sensitive than other gulls to fluctuations in fish stocks. Its decline in Shetland in the late 1980s has been linked to overfishing of sand eels.

during the 1980s. Sand eels are trawled mainly by vessels from Denmark, where the tiny eels are used for all kinds of things – cattle food, fertiliser, even as fuel in power stations. The Shetland sand eel fishery was closed in 1990 – but to protect the fish stocks, not the birds. The terns revived immediately, but when the fishing boats turned to other grounds in the Firth of Forth, there were similar consequences for the birds of the Isle of May. The island's large shag population failed to raise a single chick in 1993, when 90,000 tonnes of sand eels were hoovered up from their shoaling grounds on 'the Wee Bankie'. The current EU quota of 950,000 tonnes of sand eels from the North Sea exceeds the highest catches of sand eels in recent years. It is impossible to prove scientifically that overfishing caused the fall of seabird numbers, since the evidence is circumstantial. The conservation argument is rather that, with the current quotas, sand eel fishing in the North Sea is non-sustainable, and that the situation merits a precautionary approach to further fishing. The whole debate really highlights how little is known about many of the fish on which seabirds feed (Avery 1993).

A last issue concerns beaches. In response to Britain's reputation as 'the dirty man of Europe', local authorities have made great efforts to clean their popular beaches and so earn the EC's coveted 'blue flag' award. Blue flags mean more tourists, on whose spending seaside towns depend. Machines have once again come to their aid. One man driving up and down the beach can keep a beach more or less clean of seaweed and other debris by hoovering up the surface sand and filtering out anything that isn't sand. The weed is dumped in piles and left to rot. The problem with these machines is that they wipe out the top few inches of the sand where the wildlife is, and so leave a beach virtually sterile. Shore birds such as sanderling and turnstone have to look elsewhere, as do beachcombers and naturalists. While no one wants litter, tar-boils or sewage on their beach, many do like to see *some* signs of life, such as gulls picking over the weed and small birds running along the strand-line. Moreover, machines alter the natural profile of the beach, preventing sand from accumulating and forming dunes. Dunes are desirable things, for they are not only rich in wildlife but natural coastal defences. For these reasons local authorities in South Wales have started employing teams of hand-pickers instead of using beach-cleaning machines, especially in places designated as SSSIs. They have found that, where there is weed, there are usually birds, but that invertebrates such as sandhoppers are slow to return to mechanically cleansed beaches. Perhaps the solution is to encourage the EC blue flag people to regard clean weed as a blessing: perhaps a green, brown and red flag, incorporating the main seaweed colours, would acknowledge that ecological health is better than blue sterility.

Nature conservation in towns and cities

In a crowded island such as Britain, nature conservation in towns and cities is important partly because towns and cities cover so much of the land (including wildlife 'hot spots' such as the Solent and the lower Thames valley). It is also important because most of us live in towns and cities, and want to have wild places close to home, not a congested road journey away. Urban nature conservation merges with the wider demand for a clean, sustainable and attractive environment. It is not so much about saving natural habitats and species (though it does involve that too) as about living with wildlife. It is a community-based, grassroots activity, and urban wildlife projects tend to reflect local desires and priorities. The players are many, far too many even to mention them all (a good source is English Nature's Urban Wildlife newsletter). An important one is the network of Groundwork Trusts, that began in St Helens and Knowsley in 1981 and has since extended to include many of Britain's towns and cities, co-ordinated by a national Groundwork Foundation. The Groundwork Trusts work with local communities and businesses to create wildlife habitat on derelict land and restore abandoned habitats, such as mill dams or disused railway banks. Another is Landlife, which produces wild flower seed for sowing on derelict urban land. Its National Wildflower Centre, among the tower blocks of Liverpool, is 'a focus for creative conservation', working to revitalise civic confidence through wildlife projects. London has its own institution, the London Ecology Centre, which, with the help of the London Wildlife Trust, has surveyed the city's wildlife in great detail and drawn up conservation plans for many individual sites. Britain's first 'ecology park' was established in an abandoned London lorry park in 1978 as an outdoor educational centre, and although the site was disgracefully redeveloped in the 1980s, the idea has spread to other cities (London meanwhile has a new one in Camden). Most local authorities today have evolved 'greening' policies, including local Biodiversity Action Plans, and although some of this is no doubt lip service to a fashionable cause (when it comes to the crunch, development and 'jobs' still generally win), it also reflects a strong, perhaps irreversible, sea change in public opinion towards environmental policies. Moreover, under EC law, major developments now require environmental impact assessment – a useful job opportunity for freelance ecologists. Sensible businesses talk to environmental bodies at the planning stage; it can save them a fortune in time and legal expenses later on. Local opinion also counts more than it did; it is probably the reason why Rainham Marshes (see pp. 226–9) is now a nature reserve and not a business park.

Nature reserves are now part of the everyday amenities of any city, as well as of the countryside. When I wrote a book about the wildlife of Aberdeen in the early 1980s, that city had no nature reserves of any kind. It did have a country park, but that had to double as a rubbish dump, and Government refused to recognise it. Elsewhere, the Granite City seemed to have an obsession with daffodils, planting them in every wild corner and river bank it could find. Things have certainly changed since then. Today Aberdeen has no fewer than five Local Nature Reserves, more than any Scottish city except Glasgow, as well as a local Biodiversity Action Plan and a policy of open access to all the countryside under its control. Civic pride has broadened from beds of roses and daffodils to embrace bogs and estuaries. Urban nature reserves may vary from acre-sized plots barely dignified by a label to elaborately planned public amenities, often on restored land. The Wetland Centre at Barnes in London is among the most ambitious: a truly man-made wilderness on the site of a sterile reservoir, restructured into lakes and islands, planted with native vegetation, and with perhaps Britain's most sophisticated birdwatching facilities. The problem with this sort of thing, apart from the expense, is vandalism. In some areas, nature reserves, like any public amenity, demand a high level of maintenance. The search is still on for a vandal-proof sign.

The largest exercise in habitat creation is being done not in towns but around them. The public has been sold the idea of Community Forests as multipurpose woodlands that will beautify the landscape, benefit wildlife, act as 'carbon banks' and, not just incidentally, contribute to the nation's timber targets. At the turn of the twenty-first century, forests were being planned for many conurbations, including a National Forest in the Midlands and another in the Scottish lowlands (the 'Central Scotland Forest Initiative'). The scheme is being promoted in England by the Countryside Agency (formerly the Countryside Commission) with the support of local boroughs and other authorities. By 2000, about two million trees had been planted on over 100 sites. Any advantages for wildlife will be a long time coming, and will depend on how the forests are managed, whether the objectives are right, and whether the planners have the patience and ability to see them through. To judge from their literature, some of those concerned think that if you plant enough trees you end up with a forest. In practice, it is not as simple as that. If successful, you get a plantation, and if not you get a waste of time. Some of the derelict opencast mines and old railway banks destined for planted trees would be better left as they are. Planted forests have the significant demerit of being

Community Forest planting at Great Wyrely, Staffordshire. (English Nature/Peter Wakely)

much the same, whether you live in Watford, Sunderland or Falkirk. Newly planted areas resemble nothing so much as military cemeteries, with their close rows of tree shelters in a setting maintained by mowers and herbicides. The result might be a planner's version of an ideal countryside, with something for everyone, so long as we are not all that fussy. But in practice, many of the trees will probably never reach maturity, and by the time some of them do, we may well have changed our minds about them. Trees are awkward things, and best left to grow where they choose. They are not the ready made answer to all 'green' situations.

Urban wildlife habitats

There are two main kinds of wildlife habitat in towns. One is based on man-made structures: roofs and streets, derelict land, railway sidings and canal banks, urban parks and churchyards, all of which can support interesting wild plants and animals. The other consists of islands of former countryside that were swallowed up by the expanding city and are now maintained as public open spaces. Some cities contain wonderful wild

Churchyards can be a won-
derful oasis for wild flow-
ers, such as this one at
Hadlow Down, Sussex.
(English Nature/Peter
Wakely)

spaces, true *rus in urbe* ('country in towns'), such as Plymouth with its headlands and coves, or Bristol with the fabulous Avon Gorge running through
it, or Edinburgh, with its great wild core at Holyrood Park. Some of these
places are important enough to be designated as SSSIs or even as National
Nature Reserves, for example, Sutton Park in Birmingham. Some of
Britain's oldest effective nature reserves lie in London, such as Ken Wood
in Hampstead, left to the nation in 1925, or Perivale Wood in Brent, preserved by the Selborne Society since the 1880s. Today, some 2,500 hectares
of ancient woodland lies within Greater London – a greater density than
in some rural areas.

A classic urban nature reserve is Possil Marsh in Glasgow, Scotland's oldest formal nature reserve, designated in 1931. It lies in a small, waterlogged
depression close to the Forth and Clyde Canal, and was purchased by the
then Scottish Society for the Protection of Birds as a bird sanctuary. Some
50 species are known to breed there, and many more pass through. Possil
Marsh is rich in wildlife largely because the canal forms a surprisingly clean
arterial route through the town. Hence the marsh forms a stopping place,

a kind of service station for water birds and other wildlife. Today its role has expanded. It has its place in the city strategy for nature conservation and urban regeneration, which includes a multimillion pound scheme to restore the canal for tourism. Such places survive best when they are valued, and where local residents are involved in their conservation. I had reason to reflect on this when a few of us once tried to save an equivalent place in Aberdeen called Scotstown Muir. This small corner of wetness and wild in an expanding city was theoretically protected as an SSSI and by the local structure plan, but it seemed that nothing could prevent it from being surrounded by housing estates and being treated as dumping ground. In the end, what was left of it did become a nature reserve of sorts, but only after a seedy arrangement in which part of it had to be surrendered for housing (it still leaves a nasty taste behind 20 years later).

The plant life of industrial cities and ports has a fascination of its own. Plants from around the world establish themselves in unusual circumstances: heated canal water, or baked brick or cinders, or spoil-heaps high in salt or toxic minerals. A city's flora can mirror the city's history. In Nottingham, for example, a yellow, cress-like flower called false London rocket, common on waste ground, is believed to have originated in consignments of liquorice root from Iran, imported by The Boots Company (Shepherd 1998). One can also find there a kind of ironic floral comment in the tussocks of Japanese millet growing on the ruins of the Raleigh bicycle factory. Such plants come and go in response to trade and manufacture. Conservation is scarcely the point in such a dynamic situation, but

Kennet and Avon canal. Its tranquillity and wildlife owed much to relatively low usage. With the opening of canals to more boat traffic, the waterweeds and their associated wildlife disappear.

even so, Nottingham botanists have persuaded developers to spare the
trailing snapdragon, which colonised the walls of a yard at the back of the
main shopping centre; it is thought to be a relic of a herbal pharmacy that
once stood there. I hear a certain civic pride is also taken in Sheffield's
wild fig trees, which grew up by canals warmed by the city's Bessemer steel
furnaces.

Part of the fascination of post-industrial derelict land is that it can
mature into diverse combinations of wild plants and animals rarely if ever
to be found in nature, or, at least, not in so small a space. The potential
variety is great – lowland lakes created by gravel diggings, marshes or
'carrs' formed by mining subsidence, 'bings' of oil shale in central
Scotland and flooded brick-pits, home to crested newts and other pond
life. Former opencast coalmines form miniature hill ranges with marshy
bottomland, combining wild grassland, heath, scrub and fresh water. The
former colliery at Grimesthorpe in South Yorkshire now attracts snipe and
Jack snipe in winter, and breeding ringed and little ringed plover. Finches
feed on the teasels, thistles and other weeds, and the slurry lagoons are car-
peted in June with purple and pink marsh orchids. The flora that develops

Part of the Cotswold Water
Park, former gravel dig-
gings now an attractive
wildlife refuge of gravelly
banks and secluded pools.

naturally can surprise even experienced botanists. Who would have expected to find sea campions at inland Grimesthorpe, or rare helleborine orchids on coal bings in Lanarkshire, or clubmosses on pottery spoil at Telford? With just a little encouragement, spoil heaps can take on the appearance of a wild garden, dense with colourful vetches and melilots that would be regarded as a horticultural triumph in a fashionable wild garden of the 1990s. A few, such as Nob End in Manchester, have even been designated SSSIs.

Another industrial wasteland that has unexpectedly produced a botanical bonanza is pulverised fly ash (PFA) from coal-fired power stations. Despite their high level of salt, and toxic elements such as boron and arsenic, which make colonisation a slow process, tips of PFA can develop in interesting ways, producing slightly surreal landscapes, such as sloping purple and yellow lawns of vetch, or of wild orchids in young willow scrub, or floating islands of flocculent ash, held together by plant roots. The dynamic of vegetation can produce temporary salt marshes (complete with seaside plants) before the salt leaches out, and the presence of complete suites of habitats from open water to woodland makes them potentially ideal places to teach the principles of ecology.

Derelict industrial land has had its champions, most notably Professor A.D. Bradshaw at Liverpool University, and since 1991 government guidelines have included nature conservation as a legitimate use for it, for which land grants can be paid. But even so, there is a wedded belief that such places are eyesores, and that turning them into lifeless mounds of grass or plantations of trees represents an environmental improvement. For example, of the 1,400 hectares of fuel ash mounds existing in the mid-1990s, two-thirds has been, or will be, restored to farmland (Shaw 1994) (and one can't help wondering where all the boron and arsenic goes; surely not into the crop?). Because of tougher waste-disposal legislation, such habitats may not be seen again, and, unless some are deliberately preserved as nature reserves, future generations may be denied the sight of willow glades purpled with orchids, or neat round islands of willow, fern and moss, slowly drifting in the breeze.

Creation and translocation

Natural processes are often slow and random. Urban managers often try to speed things up by creating a custom-designed habitat. Habitat creation is currently popular – one might argue, too popular – and one can see why. Developers regularly use the tactic of 'planning gain' – offering 'exchange land' created by planting trees or sowing wild flowers as compensation for destroying part of an SSSI – as though a fake by some jobbing artist is just as good as the original masterpiece. Certain kinds of wildlife-rich habitats, such as reed beds or salt marsh, can indeed be created quickly, especially if there are wild seed sources nearby. Other kinds, such as mature ancient woodland or chalk grassland, cannot be created at all; they need to evolve at nature's pace and the most we can do is give them a flying start. Habitat creation is valid so long as it is not seen as a substitute for the real thing, or the facts twisted in some developer's glossy leaflet. But, more often than not, it serves a landscape and amenity function rather than contributing much towards nature conservation. And it is not as easy as it looks.

Oliver Gilbert cited the example of Tinsley Park, on the edge of Sheffield, as a habitat creation scheme where 'the best laid schemes gang aft a-gley'. On the face of things the local authority did not make too bad a job of it. The area looks quite attractive, with satisfying contours and some eye-catching ponds, and is popular with local residents. But in fact, nearly everything went wrong. The designers, four ecologists working for different agencies, planned to produce a natural-looking mosaic of grassland and scrub, with heath on the higher ground and a chain of ponds in the valley. The planning was impeccable, but the contractual work fell far short of its aim. A grass sward was sown in autumn to prevent erosion, but had to be sprayed out the following year and resown with what was supposed to be a native wild flower and grass mix. Like most such mixes, it was nothing of the sort, and the main survivor is a tall, fodder variety of bird's-foot trefoil. The designers specified a late summer cut for the flower meadows, but they never received it. Meanwhile broad-leaved trees were planted in fenced plots, but many of them died due to lack of maintenance. The main survivor, alder, was planted in the wrong place. None of the woodland flowers specified for the woodland appeared – perhaps they failed to take, perhaps someone forgot to plant them. The hand-sown gorse and broom were much more successful, and now threaten to take over the whole site. Creating heathland using turf transplants was a complete failure, but even if it had succeeded, airport regulations later required the levelling of that part of the site. Even so, the local authority regards Tinsley Park as a success. Only the ecologists were disappointed (Gilbert & Anderson 1998).

Perhaps this is the place to mention that other popular conservation technique of the 1990s, translocation. The term means transporting an animal or plant from one place to another, supposedly from danger to safety. For example, the showy marsh orchids that appear on fly ash tips have sometimes been translocated, presumably because orchids have an exotic glamour not possessed by other wild plants. The Department of Transport has publicly patted itself on the back after removing wild flowers from the path of roadworks to the verge. But wild flowers prefer to grow where they choose, and there is no point in trying to transplant them unless the receptor site is suitable for them, and managed appropriately. Translocation has obvious attractions to a developer or local politician, and to the uninitiated it looks like conservation. European legislation may encourage its use, given the Habitats Directive's requirement to provide 'mitigation' for the loss of important wildlife habitat. The only report to date (Gault 1997) found that although translocation was quite a widely used technique in the 1990s, remarkably little evidence was available on whether it has succeeded or failed. The precise aims were rarely set out, and 'without an aim or criteria, success is an entirely nebulous concept'. Hardly anybody had a policy on translocations, and it was all being done on an ad hoc basis. There was, however, evidence that translocation was being trumpeted by developers for PR purposes. At Selar Farm in South Wales some marsh fritillary butterflies were moved to make way for an opencast coal mine. As required under EU legislation, they were dumped on a 'receptor site' nearby. The developers, Celtic Energy, told the local press that 'as far as we are concerned it is a successful project'. The butterflies might have disagreed, since they did not take to their new home and died out – but, as the investigator found, success in such matters is a 'conveniently flexible' term.

The best evidence comes from that much translocated creature, the great crested newt. It is a protected species that often lives in flooded mineral pits earmarked for development, and so poses a dilemma to conservation agencies. Moving the newts out of harm's way has been a popular solution. The most celebrated (or notorious) example was at Orton Brick Pits in Peterborough, where 15,000 newts were caught and conveyed in buckets to a prepared site nearby, while their original home was drained and filled in. At the crux of the issue was English Nature's reluctance to designate as an SSSI a site that had planning permission for housing. WWF took the view that it was EN's duty to designate on scientific grounds, and the minister's to decide whether houses or newts should come first. There were other reported cases of inconvenient newts, such as Peter's Pit in Kent, where once again English Nature found itself piggy-in-the-middle, this time between the developers and the county wildlife trust. The chances are that these exercises in 'mobile newting' will fail. Of the 86 occasions between 1970 and 1990 when crested newts were caught and moved from one place to another, only 34 were subsequently monitored, and of these only 13 were judged successful, in that they resulted in stable, self-sustaining colonies (Latham 1994). Most of the successes were ponds in nature reserves or gardens where there was someone to keep an eye on them. Arguably translocation is less a means of conserving wildlife than as a process by which conservers and developers can work as 'partners'. Today there is a broad consensus among conservation bodies and local authorities, if not developers, that it should be seen as a last, not a first, resort.

People and wildlife

The effect of large numbers of people on wildlife has been surprisingly little studied. There has been no calls from the conservation agencies to reduce visitor numbers in places such as Richmond Park or Hampstead Heath, and even in heavily visited rural beauty spots, such as Box Hill or Ivinghoe Beacon, there is little observable impact on wildlife other than erosion here and there. Entomologists actually like paths, since the compacted soil is used by basking insects or for burrowing into by solitary bees and wasps. Moderately trampled short grass can be richer in plants than untrampled long grass. In woodland, birds, especially songbirds, seem tolerant of people so long as they keep to a regular route. On the other hand, urban woods are apt to be vandalised, or mismanaged by local authorities supposedly in the interests of health and safety (one of the blessings of temperate broad-leaved woodland is that it does not burn well). One interesting difference between urban and rural woods is that the former tend to be invaded by introduced plants from nearby gardens and waste ground. Oliver Gilbert found that 39 per cent of the vegetation of Ecclesall Wood, an ancient and natural wood in Sheffield, consisted of non-native species, including montbretia from Africa, snowberry and pick-a-back plant from America, balsam from central Asia and knotweed from Japan. Some of the plants found in Ecclesall Wood are cultivars, such as 'Highclere holly', *Ilex x altaclerensis,* of purely garden origin. Hence Ecclesall Wood, and presumably many others like it, are becoming a special kind of habitat, not found outside towns. A large proportion of these exotic species have edible berries, and were probably sown by birds, roosting in the wood, but feeding in the surrounding gardens. Hence urban woods are adapting to their circumstances, and seem to be gradually developing a cosmopolitan, rather than a native, flora. Does this represent undesirable 'contamination'? On the contrary, Gilbert regards it as 'a beautiful example of ecology in action' (Gilbert & Bevan 1997). Urban wild spaces are the melting pot for perhaps the most dynamic phase in the development of our wild vegetation since the melting of the ice caps. To

1980s housing on former heathland at Parley Common, Dorset. Almost half of the 10,000 hectares of Dorset heath present in 1960 has gone, much of it to housing estates. (English Nature/Peter Wakely)

some scientists, it is places such as Ecclesall Wood, rather than pristine but dynamically dead-end habitats, that are 'scientifically interesting'.

Other habitats sit much less easily with urban development. Sand dunes erode easily, and can then be blown away by the wind. Peat bogs turn into quagmires. Lowland heath tends to catch fire. It was a misfortune for our wildlife when Bournemouth was built (and London, too, for that matter). When chunks of Canford Heath and Parley Common in Dorset were reclaimed to build housing estates ten years ago, it was not just the loss of a precious and declining habitat that concerned people. It was also the certain knowledge that the rest of the heath would regularly go up in flames. Sure enough, the 412 hectares of Canford Heath suffered 179 fires between 1990 and 2000. The similarly sized but more rural Hartland Moor experienced only two during that time (Haskins 2000). As a result of the fires, bracken, gorse, birch and grass are spreading at the expense of heather (the grass is also encouraged by dogs, whose accumulating droppings have turned the paths bright green). Ground-nesting birds are vulnerable to disturbance. At Canford Heath, fires have seen off the woodlark, and probably the nightjar. Cats and rats are making life harder for sand lizards.

The subject of disturbance will loom large when large areas of unenclosed countryside are opened to the public, as government has promised. There has been virtually no research, and so people will be able to take up strong positions for or against with few facts to worry about. The closure of public footpaths in the spring and early summer of 2001 offered an unparalleled opportunity to investigate the matter, but unfortunately foot-and-mouth restrictions included ecologists. There were anecdotal reports that lapwings nested closer to footpaths than usual, and that deer became less nervous, and were more often to be seen away from cover in daylight. This is not very surprising. Some birds are shy of human contact especially during the nesting season, but, in a country where the hills are overgrazed from top to bottom, and there is nothing left in the fields for birds to eat, disturbance is unlikely to be the main reason for their plight. One argument, voiced on a radio nature programme recently by the RSPB's Roy Dennis, is that access to remote, wild country should remain difficult for quite different reasons. If the small island of Britain is to become an important player in world conservation terms, we need more wild places where nature truly does come first. Perhaps we might then find room for some of the bigger animals and birds that deserted us long ago: cranes, beavers, wolves, maybe even moose!

Gardens

Wildlife gardening is enormously popular, combining as it does the felicities of tending one's garden with the virtues of doing something for the environment. A garden full of birds, butterflies and pretty, non-invasive wild flowers is good for us. But do gardens make a significant contribution to nature conservation? If they do, it must be by default. Gardens are important because the wildlife value of the surrounding, mostly agricul-

The very image of garden wildlife: tortoiseshell butterflies on buddleia – simple things make conservationists of us all. (Natural Image/Bob Gibbons)

tural, land has deteriorated so much. However, this is a murky subject, because private gardens are, after all, private. Gardens are a considerable resource, covering at least 485,000 hectares or 3 per cent of the land area of England and Wales, far more than some semi-natural habitats. The BTO's 'Garden Birdwatch' scheme has accumulated a lot of data on their use by birds, showing that gardens support unusually high densities of some species, such as blackbirds and hedge sparrows. How much wildlife an average garden will support, however, depends on what is done there. The typical small, high-maintenance suburban garden of lawn, privet hedge, borders and a vegetable patch, maintained with the help of pesticides and fertiliser, will be relatively sterile. A bird table is a start, and a pond in partial shade with a marshy fringe a big improvement. Those with enough space can attract wildlife to feed and stay with flowers rich in nectar such as buddleia and ice-plant, nettle corners, compost heaps (popular with grass snakes and slowworms) and log piles. A garden with bushes and trees has more structure, and potentially more 'niches' for wildlife, than most comparable areas of non-wooded habitat in the wild. Those on the outskirts of town are usually richer than more isolated ones nearer the centre. Best of all are gardens deep in the country or at least linked to larger green spaces, such as railway lines, river banks, parks and allotments. Next to birds, the group that likes gardens best is insects. The potential diversity of quite ordinary gardens is astounding. Dr Jennifer Owen identified some 2,204 invertebrate species, mainly insects, in her modest suburban garden in Leicestershire between 1972 and 1986, including nearly half of

British harvestmen (ten species), 35 per cent of hoverflies (91 species) and 29 per cent (263 species) of moths, as well as several ichneumon flies new to science. Of course not all of them would be residents: perhaps most were just flying through. However, gardens support more species of some insects than most wild habitats. For example, most of our species of lady-bird live in gardens, as do at least a third of our larger moths.

Some insects probably depend on gardens, such as the sizeable group associated with currant bushes (though changes in garden practice – who grows gooseberries these days? – now confine them mainly to allotments). Garden ponds have helped to compensate for the loss of farm ponds, and form an important refuge for frogs and smooth newts. Gardens help some birds to survive the winter, and probably influence the behaviour of species such as the siskin, and allow blackcaps to overwinter in Britain. The RSPB even recommends gardeners to leave out grain in autumn and winter, as a lifeline to beleaguered farm birds. But, of course, gardens can be a trap as well as a refuge. Dirty bird tables can infect birds with salmonella. Stale peanuts can develop harmful toxins. Cats take far more birds than any wild predator, and everyone knows the propensity of birds to collide fatally with cars and windows (the impressive casualty list from the reflective glass windows of English Nature's national office in Peterborough was a standing joke in the conservation world). Hedgehogs love gardens, but they also tend to get run over by the gardener's car. Following its recovery from persecution, the sparrowhawk is often blamed for taking songbirds in gardens: successful wildlife gardens naturally attract predators, and the hawk would have more trouble catching birds on open farmland. It is best to regard its visits as a compliment.

Most species that occur frequently in gardens are widespread, and not under serious threat – but not all. The broad-leaved helleborine orchid, much declined in wild woods, is colonising the rougher corners of many Glasgow gardens. Unfertilised lawns are an important habitat for grassland fungi. Sickle-leaved hare's-ear, one of our rarest wild flowers, can be almost a weed in gardens, self-seeding and spreading from its original plantings. Part of the charm of wild gardens is not knowing what will turn up next. But gardens, however well designed, and, however sparingly pesticides and other chemicals are used, are no substitute for an adequate conservation policy in the countryside. The fact is that most species of wildlife do not live in gardens. Those that do tend to be mobile species, good at spotting opportunities to feed or breed. And while individual gardens offer plenty of variety, small gardens as a whole tend to support much the same wildlife wherever they are. Wildlife gardens are fun and worthwhile, but in the last analysis it is perhaps we that need them most.

Golf courses

The game of golf contributes to nature in two ways. First, because of its popularity. There are over 2,500 golf courses in Britain, covering about 1,500 square kilometres or 0.5 per cent of the land surface (Tobin & Taylor 1996). And because golf courses are often built on good wildlife habitats, especially sand dunes and heaths, their wilder corners can preserve some of what was there before. Of course dunes and heaths are better for wildlife than golf courses, but golf courses are better than concrete. While a golf course's greens, tees and fairways have to be close-mown and managed to as lawn-like a state as possible, about a third of the course is given over to 'rough': tall grass, bushes or marshland. Enough is left of the natural habitat for at least 81 golf courses to be designated SSSIs. A few are absolute gems, such as Royal St George's near Deal, which contains most of Britain's lizard orchids, or the Royal Birkdale near Southport, which has breeding sand lizards and natterjack toads. In 1990, the NCC even produced a pamphlet in praise of golf courses, advocating the creation of more golf courses as a nature conservation measure (its author was a keen golfer). One hopes that this advice was not taken too literally by Horsham District Council, which built a golf course on its own nature reserves and claimed it was an improvement. Golf courses have, in fact, tamed or wrecked some wonderful wild places. On the other hand, the game has undoubtedly saved places from less benign urban development.

Like gardens, whether golf courses make good or bad habitats for wildlife depends on their design. Generally the older ones have the most wildlife because they have had longer to mellow and mature. Planting trees and bushes in the rough (or, better still, leaving natural ones alone), digging ponds and ditches, sowing wild flowers can make a difference, but those that are already on wildlife sites, such as sand dunes, should be left as they are. In 1995, the Golfers' Association published environmental guidelines for course maintenance and planning, which include ideas of this sort. Against that, however, is the fact that golf clubs tend to be conservative and take great pride in long vistas of short, neat grass.

PART III

Living With Wildlife

9

Development: Causes Célèbres

Wondering how to summarise the thousands of cases where development has come into conflict with nature conservation since 1970, I decided to choose just six, three in England, two in Wales and one in Scotland. You can learn more about conservation in practice from the detail of a long-running dispute than from any number of generalisations about procedures. My six examples raise basic questions about how we treat wild places. For example, the long-running soap story of the Newbury bypass focused attention on the nation's road-building policies, and made people think about whether it is really in our interests to put road transport ahead of all other considerations. The case of Amberley Wild Brooks was similar to many others at the high tide of arable agriculture in the 1970s, but it was the one that settled an important issue: was agriculture *always* to have primacy over nature conservation? (answer: no). The periodic attempts to expand the skiing resort at Cairngorm raise wider questions about amenity and wilderness. In every case, the issues go beyond nature conservation in the narrow sense. But then, nature conservation is no longer defined narrowly, but as one of the defining principles by which we try to live. Cases such as these helped to make it so.

Amberley Wild Brooks

Jeremy Purseglove (Purseglove 1988) has written about the peculiar English attachment to landscape, which often finds its way into the stories that we loved as children. The river bank and the Wild Wood of *The Wind in the Willows* were based on real places (the Thames, near Pangbourne). So was the playground of Winnie-the-Pooh and Christopher Robin (Cotchford Farm in the Ashdown Forest), the chalklands of *Watership Down* (near Kingsclere, Hampshire), and the ancient mines and sandstone edges of Alan Garner's novel, *The Weirdstone of Brisingamen* (Alderley Edge in Cheshire). Even the chessboard fields in *Through the Looking-Glass* are said to have been inspired by Otmoor, near Lewis Carroll's Oxford. As far as I know no such tales have been woven around Amberley Wild Brooks, a maze of dykes and wet pastureland on the flood plain of the River Arun in West Sussex, but all the same it is the sort of place that has bone-deep appeal – unexpected, secluded, a corner of wild cradled by downs and woods, overlooked by a ruined castle. Although covering only 360 hectares, Amberley Wild Brooks contains 400 species of flowering plants, including all the duckweeds, all the water-milfoils, 35 kinds of grass and practically every pondweed known from south-east England. Birdwatchers came there to see the flocks of Bewick's swans that arrive in winter from Siberia, bringing, as Purseglove describes it, 'an improbable flavour of the

Amberley Wild Brooks. The grassland is rather ordinary – which is why the extraordinary quality of the place was overlooked until the 1970s. Its secrets lie beneath the water. (Natural Image/Bob Gibbons)

Russian steppe to the domesticated landscape of the Home Counties'.

Most of these plants and animals survive there for one reason above all: Amberley Wild Brooks is wet, and the water is clean and clear. In the rainy winter of 1974–75 it was very wet. Since the War, and particularly since Britain's entry into the Common Market in 1973, agricultural policy had attempted, as far as possible, to 'tame the flood' in order to grow more food. Improved drainage technology enabled a water authority to pump practically any wetland dry enough to grow barley, so long as the taxpayer footed the bill. And so they did, on a broad scale, heedless of the environmental cost. In the case of Amberley Wild Brooks, local farmers wanted to drain the flood plain to build up their livestock grazing, which would in turn free them to plough the nearby downs for growing more cereals. Acting on local complaints about the wet fields, Southern Water Authority, keen to oblige, prepared a suitable plan and applied to the Ministry of Agriculture for £340,000 under the recent 1976 Land Drainage Act. Amberley Wild Brooks was not at that time a protected site, and that would normally have been that: more cows and no pondweeds. However, this was

Sussex, full of well-heeled, middle-class, weekend countrymen, steeped in the culture of Beatrix Potter and Kenneth Grahame. They complained, as did conservation bodies such as the CPRE and the RSPB. For this exceptional case, the minister called in the application and a public inquiry was held in March 1978.

The conservationists won the argument. Unusually, they had been given access to the figures – a serious mistake as it turned out – which enabled the economist, John Bowers, to show that, even leaving the environment issue aside, Southern Water Authority had overstated the benefits. Since society was paying the bill, he argued, there should be some return to society. Here, who gained but the farmers? What do people want – more cows or natural beauty? As Norman Moore, the NCC's Chief Advisory Officer and main witness, cogently argued, Amberley Wild Brooks was a test case: 'if conservation is not paramount in this case, no site can be considered safe'. If one of the richest wetland sites in the country, home to numerous rare plants and animals, could be drained with public funds, then conservationists might as well give up. The inspector upheld such arguments, and grant-aid was not forthcoming. Instead, the NCC held 'valuable and constructive' meetings with all parties for an alternative scheme that would leave room for some agricultural improvements without spoiling its special character (for at that time even conflicts won produced compromise agreements).

Amberley Wild Brooks was a test case that established a precedent: henceforth agricultural improvements would not have an automatic priority over wildlife. Others, not just the local land users, would have a say in what happened there. The sense of a turning point being reached was reinforced a few years later when the Inspector ruled against the reclamation of Gedney Drove End, an SSSI on The Wash, for much the same reasons. The agricultural lobby was a sore loser. In future, the water authorities and the Ministry of Agriculture would be more guarded about revealing their cost-benefit figures. At least one Amberley farmer said he would stop naturalists visiting his land. And even here, the drainage had been improved before the rumpus of 1978, and there were fewer and fewer wild duck and swans roosting in the winter-wet fields below the ruined castle. But such was the power of the agricultural lobby in those heady days that Amberley Wild Brooks was regarded as a victory for the little green David against the Goliath of the CAP. There was still hope for wildness and wet.

'The Third Battle of Newbury'

Between the county library and the local museum in Newbury is a gallery and coffee shop called Desmoulins, named in honour of a tiny but now famous snail called *Vertigo moulinsiana*, the Desmoulin's whorl snail. Everybody in Newbury knows about 'the snail that nearly stopped the bypass'. It has become a symbol of the environmental price we pay to get from A to B that bit faster. Judging from recent cases in the media, small, rare animals seem to make a habit of living in the path of bypasses. In the case of the Dersingham bypass in Norfolk it was a moth, *Choristoneura lafauryana,* about the size of a fingernail and ironically christened the 'mighty moth' by the tabloids ('Bypass Mothballed'). For the Winchester bypass at Twyford Down the best they could come up with was a colony of the chalkhill blue butterfly, although one that was said to be the largest in Britain. But with Newbury, obscurity returned in triumph with a snail the size of a breadcrumb that the European Union in its wisdom was about to list, along with its habitat, as endangered. It meant that the British government could in theory be taken to court if it heedlessly squashed the snail in its haste to complete a fast highway from the port of Southampton to the industrial Midlands.

The Newbury bypass, built amid scenes of considerable turbulence in the mid-1990s, follows a route west of the town through some of the loveliest countryside in the Thames valley. This was the route the Department of Transport had wanted, and the one the Inspector recommended after a public inquiry held in 1988. Because Newbury sits in a valley surrounded by woods, heaths and marshes, any road bypassing the town was bound to conflict with wildlife and amenity interests. At the 1988 inquiry, the NCC gave evidence on the impact on wildlife of alternative western and eastern routes. Both would cause damage, but the NCC was more opposed to the grottier route east of the town since it would cross an established SSSI and Local Nature Reserve at Thatcham Reedbeds. The western route, though passing through beautiful countryside, threatened fewer protected sites: a corner off Snelsmore Common SSSI, a local trust reserve at Rack Marsh and a couple of non-statutory ancient woods. This route was defended mainly by local residents (SPEWBY – the Society for the Prevention of the Western Bypass) – although the NCC did have reservations about the way the Department intended to cross the Rivers Kennet and Lambourn, by embankment rather than a less environmentally damaging bridge. No full impact assessment was ever made, and the presence of otters, dormice, badgers and other protected animals along the western bypass route was overlooked or underplayed. Had the Countryside Commission joined forces with the NCC, it might have been different: the Commission was more concerned about scenic beauty than the NCC. Unfortunately, for reasons of its own it decided not to become involved. At a subsequent public inquiry, held four years later in 1992, the Inspector refused to reopen matters, such as wildlife, that he said had been covered four years earlier. Two years after that, the Secretary of State for Transport, Bruce Mawhinney, deferred his final decision on the bypass pending a review, then changed his mind and gave the go-ahead in June 1994.

Shortly before the road engineers moved in, someone found colonies of the Euro-listed Desmoulin's whorl-snail living in the river marshes directly in the path of the road. This meant that the marshes might qualify as a Special Area of Conservation (SAC) under European legislation, which gave those opposed to the bypass the basis for a legal challenge. And many *were* opposed (although the residents of Newbury, sick of traffic congestion, were mostly for it). Shortly before roadworks began in the winter of 1995–96, a petition signed by 10,000 people was handed in. About half that number walked in quiet protest along part of the route. Professional protesters moved in, mostly young people who camped in the woods and, to make it harder to evict them, built tree houses, tunnels and aerial walkways. Visiting the camps, there was a strong sense that the environment had become a new religion. Cabalistic totems – feathers, stones and bones – were tied to trees[1]. Many of the protesters were vegetarians. 'Look over there,' said my guide, as I left my goodwill gifts of whisky and cake. A small group crouched around a fire were seemingly segregated from the others. 'Do you know who they are?' she asked. No, I didn't, but suspected they might be some sort of secret weapon. 'They are ... the *meat-eaters*'. She made them sound like cannibals.

Thanks to the protesters, the Newbury bypass became the most expensive non-tunnel 14 kilometres of road ever built in Britain. Some 500 security guards were hired to protect the workforce and evict the protesters from their camps. Television reported every stage, and an archetypal protester called Swampy became a national hero. As a tactic, conservationists led by Friends of the Earth decided to champion the snail. The protesters at the aptly named 'Ricketty Camp' on Speen Moor tried to avoid eviction by claiming that they were protecting a species against illegal persecution. Their case was taken to the High Court in March 1996, but failed since, in the judge's view, claiming to protect the snail was no excuse for illegal occupation. He did, however, urge the Government to try to minimise the damage to Desmoulin's Snail and its watery home. The Environment Secretary, John Gummer, responded by proposing eight isolated colonies of the snail as a kind of polka dot candidate SAC that left plenty of room for a dual carriageway. After 15 years of argument, the Government was determined to have it built.

A coalition of conservationists fronted by Friends of the Earth and WWF-UK returned to the High Court to seek leave for a judicial review of the Highway Agency's treatment of Gummer's proposed SAC. The judge's refusal this time seemed rather technical. The snail sites were not a candidate SAC, he ruled, but only a *possible* SAC. However, as before, Mr Justice Sedley tempered his judgement with remarks that were helpful in the long run. By pressing ahead with the road at the same time as it was going out to consultation over its environmental impact, 'the Government is apparently foreclosing the possible answers to its own question'. He went on, 'One can appreciate the force of the view that if the protection of the natural environment keeps coming second, we shall end by destroying our own habitat'.

The snail story now moves into realms of farce. The Environment Agency in consultation with English Nature came up with a plan to move the snail out of harm's way, along with chunks of its habitat, by digging up giant turves and transporting them to a prepared site further upstream. It did not seem to matter that no one knew anything about the snail's life cycle

---

[1] The protesters' reverence for nature did not necessarily mean they were keen on natural history. In the immortal words of one of them, 'Yeah, we're here because we love nature, and all that s***'.

or ecology, nor that the silly little snail was obviously being used as a means to an end, not as the end in itself. Besides, even on its own terms, digging up turves in the hope of saving an obscure snail was probably a waste of time. As the snail expert at the Natural History Museum warned, 'to think you can sustainably recreate a complex ecosystem that has taken centuries to develop in the space of a month does not make biological sense'.

To cut a long story short, the Newbury bypass was built. It opened without ceremony in November 1998. In the meantime, the junior Transport Minister responsible, Steven Norris, appeared on television to admit he had since changed his mind and agreed that always 'pandering to motorists' at the expense of the environment was wrong. The fuss over the Newbury bypass had undoubtedly convinced many in Government that its roads policy was unpopular and becoming a serious burden on public expenses. The fact of the matter was that while protests could not physically prevent a road from being built, they could delay the process until Government was tearing its collective hair in frustration and dismay at the growing cost. At nearly £6 billion in 1994–95, Transport had become Britain's third-largest spending department, with a budget twice as large as Agriculture and three times that of Environment. The proposed link road through Oxleas Wood in Greater London, which was, in the end, not built, involved eight years of public wrangling; the Winchester bypass at Twyford Down, which was, took another eight. Potentially the next set of damaging headlines would be the new bypass at Salisbury, involving damage to Harnham Meadows, an SSSI and another prospective SAC. In the Government's subsequent U-turn on roads, this road, and dozens of others, was put on hold or scrapped. [By 2001 there were signs that Government was changing its mind again, approving a long list of new roads, such as the previously cancelled Birmingham Northern Relief Road.]

Today one of the prettiest corners of Berkshire has been irredeemably spoiled. On the other hand, Desmoulin's little snail is not doing too badly; indeed it is now known to be widespread by chalk streams in the south of England. It even has its own leaflet, published by English Nature, flying the flag for threatened snails everywhere. It would be interesting, though, to learn how it managed to find its way into Annex II of the mighty EU Habitats Directive, and why no one bothered to investigate its status in Britain first. I was less than surprised – and not terribly pleased either – when a conchologist friend found what he suspected to be *Vertigo moulinsiana* in a sedge bed a few yards from my home in the Kennet valley. But I suppose it may come in useful one day, should government decide to build a bypass through the back garden.

The Berwyn saga

Y Berwyn, the Berwyn mountains, lie in an underpopulated corner of Wales on the borders of Denbighshire and Powys. They form a rolling plateau swelling to 827 metres (2,713 feet) at Moel Sych, with heather on the lower slopes grading upwards into natural grassland with blanket bogs along the watersheds. In times gone by, the area was managed as grouse

A view of the Berwyn plateau, now part of a National Nature Reserve. (CCW)

moor, but over the past half-century grouse numbers have plummeted. Today the Welsh red grouse is not so much a sport species as an endangered one. In 1957, the Nature Conservancy designated 3,900 hectares around Moel Sych a Site of Special Scientific Interest, as a good example of moorland and bog, noted for its bird life, especially breeding raptors, such as merlin and hen harrier, and waders, such as golden plover. But most of this 62,000-hectare range of hills was protected only by its inaccessibility and naturally infertile soil.

By the 1970s, agricultural subsidies were making the Welsh hills more profitable by increasing the sheep stockage with the help of drainage schemes and heavy doses of fertiliser. Forestry companies were also interested in this part of Wales, which had largely escaped the blanket afforestation of so many former sheep-walks and grouse moors elsewhere. Knowing what was in store, in 1977 the RSPB surveyed the bird life of the Berwyns and found it to be even more important than had been thought. Some 60,000 hectares, in several large blocks, qualified as SSSI, said the RSPB, and the NCC should fight hard to preserve them. However, the NCC felt that notifying so large an area would injure relations with the local farming community. Its director for Wales, Dr Tom Pritchard, decided that a softer approach was necessary. The NCC had a statutory duty 'to take account of other land uses', but as he wrote in the NCC's Annual Report, 'success in nature conservation is more often achieved by understanding and goodwill in the field than by legislation and compulsion'.

Normally the decision to notify an SSSI would be taken purely on scientific grounds. In this case, because of the sensitivities involved, the NCC decided first to consult the Welsh Office, the Agricultural Development and Advisory Service (ADAS) and the Forestry Commission. Their firm and unanimous view was that the economic development of the area was more important than wildlife. The NCC looked for a way out of the dilemma. It decided to commission Reggie Lofthouse, formerly Chief Surveyor at ADAS, as an independent arbiter. Lofthouse came up with an alternative scheme that effectively shared out the cake among the interested parties. In addition to the SSSI there would be informal 'consultation zones' where any publicly funded developments likely to damage the scientific interest would be 'scrutinised'. The NCC appointed a full-time official for the Berwyns, whose job was to build bridges and negotiate agreements with the dozens of owners and occupiers in the area. Disgusted with this performance, the RSPB threatened to take the NCC to court on the grounds that it had not fulfilled its statutory duty to notify the entire area of scientific interest (it eventually desisted, having noted NCC's more robust stance on West Sedgemoor in the Somerset Levels). The Welsh Office, on the other hand, applauded this 'most valuable fresh approach'. The undersecretary, Michael Roberts, hoped that the solution hammered out in the Berwyns would spawn similar initiatives to reconcile nature conservation with agricultural and forestry interests. In 1980, some 70 local owners and occupiers formed a Berwyn Society to represent their interests and oppose further SSSIs in the area.

Meanwhile the waters were muddied further by an afforestation plan that left very little room indeed for compromise. In 1979, the Economic Forestry Group (EFG) acquired another hill area in the Berwyn, also of SSSI quality, called Llanbrynmair Moors. This contained a particularly high density of breeding birds, as well as one of the best peat bogs in Wales. The EFG wanted to plant up all 1,800 hectares of it. After some hesitation, the NCC decided not to oppose the plan, apparently as part of an arrangement with the Forestry Commission. It also turned down an offer to sell the site due to 'the pressure of other priority cases on [its] limited finances'. All the NCC could win by way of mitigation was EFG's agreement to leave certain small areas unplanted. The rest was drained and planted, with the help of a £400,000 grant. Commenting in *Birds* magazine, the RSPB's director Ian Prestt contrasted 'the vigour and determination with which forestry interests pursue their objectives', with 'the relative weakness and uncertainty of the NCC's attempts to defend important ornithological sites'. Plainly, he saw the NCC's actions as evidence of a worrying lack of confidence.

The problem with the conciliatory approach exemplified at the Berwyns is that nature conservation has to play second fiddle to more powerful interests. The Forestry Commission's views about the extent of plantable land in the Berwyns were never questioned, nor did the NCC take issue with the agricultural assessment. The NCC's proposals, on the other hand, were put under the microscope, and whittled down bit by bit. 'Whatever the outcome,' wrote John Sheail commenting on the case, 'it was a sign of weakness to be always responding to the demands and criticisms made by others' (Sheail 1998). Bill Adams considered that for all the fine talk about 'co-operation and integration', the Berwyn experience was one of retreat and defeat' (Adams 1986). In defence of the Welsh NCC's position, its powers were weak and the local hostility to SSSIs was great. All that its statutory role won was a place at the negotiating table, where the NCC was heavily outgunned. Tom Pritchard was convinced that his conciliatory tactics lay within the spirit of the law, and that he had achieved all that could realistically have been expected.

Today part of the Berwyns is a National Nature Reserve, the NCC and its successor body having bought out farms and tenancies as they came on the market to save them from afforestation, while doing their best to offer grazing and estate work to those still doggedly farming on. The end of tax benefits for forestry in 1989 brought a respite, and ironically some of the trees planted in the 1970s are now being removed (also with public money, needless to say). With the benefit of hindsight, the future of hill areas such as the Berwyns might be better served by inclusive schemes that recognise the primacy of sustainable use. The Berwyns saga is a monument to how things were done 30 years ago, and why nature was the constant loser.

Tidal barrages and Cardiff Bay

Judging from the statistics of 'loss and damage' trotted out annually by the conservation agencies, while it is easy to damage a Site of Special Scientific Interest (for example, by simply neglecting it), it takes a great effort to destroy one. Even after the land has been drained, fertilised and reseeded, there is usually *something* left, stubborn remnants of wild habitat that still justify an SSSI label. The instances where an SSSI has been completely obliterated, such as Herald Way near Coventry (drained and then built on) or Selar Farm in South Wales (it became an opencast coal mine), are few enough to be individually notorious. The best-known example of all is the Taff-Ely Estuary SSSI, also in South Wales, whose once teeming mudflats and tidal marshes are now permanently under water. It is better known as Cardiff Bay.

Tidal barrages have three main purposes. In the 1970s feasibility studies were carried out for barrages at the Morecambe Bay and The Wash that would act as impoundment reservoirs of fresh water. In the event, Government opted to build inland reservoirs instead, at Kielder in the Borders and Rutland Water in the Midlands. The second purpose of a barrage is to create tidal energy. Alternative sources of energy began to be explored seriously in the 1980s, when oil and gas prices were high, and when enthusiasm for nuclear power was falling, especially after the Chernobyl disaster and embarrassing rumours of leaks at Sellafield. Two major barrage schemes, on the Mersey and on the Severn, were investigated. Plans for a barrage on the Usk were eventually turned down after a public inquiry in 1995, largely on environmental grounds, but there are current proposals for barrages on the Loughor and Neath estuaries, as well as the giant, though presently mothballed, scheme for a Severn barrage. Which brings us to the third reason why authorities like barrage schemes

'Unsightly mud'. Cardiff Bay before the barrage. (Niall Burton/BTO)

– amenity and urban regeneration. This is most evident in South Wales with its decayed towns and narrow estuaries. A barrage on the Tawe was constructed between 1989 and 1992 as part of a plan to restore the Lower Swansea valley. And the local authorities had long wanted to build a barrage at Cardiff.

The main environmental problem of tidal barrages is that they reduce valuable natural habitats by drastically constricting the tidal range (that on the Tawe fell from 10 metres to 1.5 metres). They also change the character of an estuary by removing the daily scour of the tide and replacing intertidal habitats with a lagoon vulnerable to pollution from sewage and toxic metals brought in by the river. Unfortunately there is no national policy for estuaries, and so they are subject to piecemeal development that slowly diminishes their value for wildlife. At present, the economics for tidal barrages are less favourable than they once seemed, but all it may need to revive them is a rise in fuel prices, while summer droughts might also revive proposals for impoundment reservoirs.

The barrage proposed for Cardiff Bay generates no energy at all. Its role is entirely one of amenity. The City of Cardiff is formed around a bay into which two rivers, the Taff and the Ely, converge, depositing their loads of silt from the valleys inland as mud banks. These are exposed at low tide when, particularly in winter, they become one of the densest feeding grounds for waders and wildfowl in the Severn estuary. Enough dunlin, shelduck, curlew and redshank used the bay for it to have been designated as an SSSI and proposed as a Special Protection Area for wild birds. Unfortunately, mudflats are rarely seen by urban planners as an especially attractive asset. Cardiff had a down-at-heel appearance not in keeping with its dignity as the capital of Wales. In 1987, the Welsh Office and the local council came up with plans for a major redevelopment programme of the dockland area, which included covering up the unsightly mud with water to provide an attractive inland bay. This would be done by building a massive 800-metre steel and concrete embankment, paid for largely by the taxpayer. Delighted with the shimmering yacht-filled lake in front of them, developers would, it was hoped, flock to Cardiff.

Knowing that conservation bodies would object strongly to the destruction of an SSSI, the then Welsh Secretary, Peter Walker, met the NCC's chairman in 1986 to explain his plans and seek the NCC's co-operation in mitigating the ecological impact of the barrage. An expert on waders, John Goss-Custard, was commissioned to investigate its probable ecological consequences, and to propose substitute feeding grounds for the displaced birds. He made several suggestions: the barrage gates could be adjusted to provide some intertidal habitat in winter. Alternatively new tidal flats could be recreated nearby, or mudflats recreated over a wider area by the eradication of cord-grass. The idea that found most favour was the substitution of equivalent wildlife habitats nearby. A suitable site was found on part of the Gwent Levels at Wentloog, near Newport (ironically, as this was itself a proposed SSSI, as coastal marshland, the creation of lagoons and mudflats would involve 'damaging' it!). Conservationists complained that the this new site could accommodate only a fraction of the birds that used the Taff estuary, even if they were adaptable enough to exploit the opportunity.

By being allowed to present a private Bill to Parliament the developers found a way of circumventing the normal planning process. The decision would be made not by local representatives or after a public inquiry, but by a Parliamentary committee, to which petition could be made. Some 69 bodies, including the NCC and the RSPB, presented objections, centred on the inadequacy of the proposed substitute habitat and casting doubt on whether the lagoon behind the barrage would be as pure and shimmering as the developers projected it. On the contrary, without expensive artificial oxygenation, it could easily become a stagnant, midge-infested pond (they claimed). There was also the small matter of its international status as part of the Severn estuary, a potential SAC, which could involve (and the voluntary bodies would do their best to ensure it *did* involve) the European Commission.

The Barrage Bill was eventually passed in 1991. Full details of the habitat compensation scheme were not announced until 1996, when the Government unveiled its plans for a £5 million, 375-hectare nature reserve on the grounds of a former power station between Uskmouth and Goldcliff on the Gwent Levels. Saline lagoons and reed beds would be created by the dextrous use of pumps and artificial embankments. However, welcome as these are, it can hardly be regarded as habitat replacement since it is food-rich intertidal mud, not reeds and lakes, that has been lost. The Gwent Levels is in any case too far from Cardiff for site-faithful birds such as redshank to take advantage of it. In the meantime, conservation bodies led by the RSPB, WWF-UK and Friends of the Earth brought their case to the European Commission on the grounds that Cardiff Bay qualified as both an SPA and an SAC. They claimed that the developers' formal Environment Statement was flawed in that it failed to evaluate the ecological impact in terms of the actual numbers of birds that would have to find new feeding grounds. However, the Commissioner found 'no lawful impediment to the scheme', insisting that 'substitute habitat' was being created side by side with the construction of the barrage. That view seemed dubious after the Lappel Bank case in 1996, in which the EC ruled that it was unlawful to exclude an area from SPA designation on economic grounds.

Work on the barrage across the mouth of the bay began in 1993 and was completed on 4 November 1999. In the interest of the birds, it had been hoped that the Development Corporation would leave its gates open, and allow the tide to take its course for one last winter season. The Corporation thought otherwise, and the gates promptly slammed shut. The event was televised, amid scenes of confusion suggesting that the operators were still a little uncertain about how it all worked. Ultimately the enclosed bay will fill up with fresh water.

Cardiff Bay is the most prominent example of habitat substitution, a popular tactic used by developers in the 1990s, and one which the conservation agencies often went along with. It enables developers to honour their environmental obligations to the letter, but involves considerable risk to wildlife. It is too soon to form any definite conclusion about the fate of the birds of Cardiff Bay. Certainly the bay is now useless for wading birds except as a limited high-tide roost. Studies by the BTO using colour-ringed birds indicate that the redshanks have moved to the nearest estuary, that of the small River Rumney a few kilometres east of Cardiff (Burton 2001). Curlew and oystercatchers have also dispersed along the coast. Competition for limited food and space is likely to increase, but it is too soon to tell whether this will affect the survival rate of the displaced birds – though it is probable. In the meantime, the Severn estuary has been declared a Special Protection Area (SPA) by the European Union. Any future development will have to be judged by its impact on the estuary as a whole.

Against the odds: the case of Rainham Marshes

Wildlife protection and urban regeneration make uneasy bed-mates. Perhaps every expanding town has at least one major clash, from the housing estate and link road built on Aberdeen's best wildlife site, Scotstown Muir, to the unfortunate newts of Peterborough and the shrinking intertidal mudflats of Teesside. The struggle to preserve wild open spaces in Greater London is a story in itself, but in the 1990s one of them dominated the headlines: the long-running saga of Rainham Marshes.

The 'grazing marshes' of the Thames estuary, immortalised in novels such as *Great Expectations*, once extended from the Isle of Sheppey to the Isle of Dogs and beyond into Essex. Over three-quarters of it was drained

'Is *this* SSSI?' Rainham Marshes: an object lesson in our treatment of wilderness. This part is reinstated marshland, following a court order. (English Nature/Peter Wakely)

or developed during the twentieth century, and the section on the north bank between Rainham and Purfleet in the London Borough of Havering now marks the last wild wetland within London. In 1986, the NCC scheduled 480 hectares of it as the Inner Thames Marshes SSSI, the largest in the London area. The main ditches form breeding grounds for water vole and a variety of insects, including the scarce emerald damselfly. Lagoons have formed where silt from the river has been dumped and here teal gather to feast on sea-aster seeds. Hen harriers and short-eared owls hawk overhead searching for voles and mice, and botanists can still find some of the special plants of brackish grazing marsh, such as divided sedge and brackish water-crowfoot. All the same, Rainham Marshes had a battered, unkempt look. Standing near the huge landfill site that marks its western boundary, breathing in the fumes from nearby industries, you gazed on an unloved landscape of fly tips and illicit camps, unknown fluids leaking from rusty drums, overhead wires and muddy scars left by motorbikes. Bad things happen there. An ecological surveyor found a man's body, hog-tied and shot through the head. Police are sometimes seen poking the undergrowth with sticks. Even the NCC's own director had his doubts: 'is *this* SSSI?' were his incredulous words. Wildlife often thrives in grotty areas, but Rainham Marshes had been allowed to become much too dry, and much of it has not been grazed for years, apart from the odd itinerant peddlar's horse. The far end, at Wennington and Aveley Marshes, is in better condition since it was looked after by the Ministry of Defence which had a rifle range there. But it, too, is drying out. Most local residents were sceptical about its conservation value. Conserving the oceans and rainforests was all very well, but, as Bill Adams (1996) noted, 'when the issues were really in their own back yard, people became critical of the ideological freight train of conservation rhetoric, as indeed they did of the glossy portfolio of the developers'.

The developers were numerous, but one stood out from the rest: MCA, the Music Corporation of America, were interested in the site for a huge development involving a theme park, film studios, shops, offices and 2,000 homes. At £2.3 billion, it would have represented the second largest ever foreign investment in Britain, and provide a claimed 30,000 jobs. MCA were able to exploit a bizarre loophole in the planning system following the abolition of the GLC and the consequent suspension of the development plan for Greater London (which had protected the site). Havering Borough Council gave the MCA an unusually open-ended planning permission with no time limit. The Secretary of State decided not to call it in for a public inquiry even though the development would have led to the greatest loss of an SSSI to development since the Wildlife and Countryside Act was passed in 1982.

MCA offered conservationists a mitigation package of unprecedented size. In Britain such compensation is normally grudging and minimal, offered on an acre for acre, tree for tree basis. In this case, £16 million was eventually on the table, representing half of the NCC's entire annual budget, more than enough, theoretically, to buy and manage every grazing marsh in the Thames estuary. To its credit, the NCC did not grab the 'blood money' and run but maintained its objection. As it turned out, the theme park and the film studios were never built, and so the cash offer was illusory. At the end of 1990, MCA was bought by a Japanese company and sold on to a Canadian one. Five years on, it was revealed that the theme park, complete with dinosaurs, was to be built not in London after all, but in Japan.

After MCA disappeared from the scene, the site continued to degrade. The A13 road was re-routed across the edge of the marshes, and, contrary to assurances given at the public inquiry, spoil was dumped within the SSSI. Parliament allowed the construction of the Channel Tunnel rail link, complete with marshalling yards, across the site. English Partnerships, the urban regeneration quango, won itself outline permission for an industrial complex on the site, despite its stated belief that open spaces are 'important contributors to the process of regeneration'. The case looked hopeless.

Then something happened. In 1990, local opinion had been in favour of developing the marshes, then regarded as a lawless wasteland. It suited developers to keep it that way. All the same, a few realised that the place was not without its redeeming merits, which could, with suitable investment and care, be turned into an asset. Its growing band of supporters formed the 'Friends of Rainham Marshes'. Public opinion was changing. When the MoD decided to withdraw from its part of the marsh, some local residents, without waiting for permission, blocked some of the main drains and flooded the site. Very quickly it was thronged with birds, demonstrating that, given the right management, there was still life in the marshes. A few years on, the RSPB acquired the eastern end of the SSSI with the help of a half-million pound private donation via the landfill tax credit scheme, and entered into an agreement with the Port of London Authority for the sustainable management of the silt lagoons. It is now controlling water levels to create shallow pools, ditches and wet grassland, and starting on the huge task of clearing unexploded ammunition. Soon it hopes to open the marshes to the public and build an education centre there. The western end of the SSSI, at Havering Marsh, remains derelict and could still be developed. But, given the sea change in public opinion since 1990, and improved protection for SSSIs as a result of the CROW (Countryside and Rights of Way) Act, not to mention Mayor Ken Livingstone's personal commitment to protecting the site, the odds are now against the developer.

What conclusions can be drawn about this astonishing turnaround in the fortunes of Rainham Marshes? That it isn't over until the bulldozers move in? That beneath the glitter of the multimillion pound corporations lies the common clay of broken promises and unfulfilled dreams? The NCC could try to defend the SSSI on scientific grounds, and Friends of the Earth make waves in environmental waters, but in the end what mattered was the involvement of the local community and a greater sympathy for conservation objectives, which swept on as the case for development crumbled.

Skiing in the Cairngorms

One of the advantages of living in northern Scotland is that so long as there is snow you can go skiing whenever you like. Most places north of Perth now lie within an hour or two of a resort, except in the far north and west where there is seldom enough snow or no suitable slopes. Scottish skiing does have its limitations. There is often a vicious wind, and if you leave the pistes you can soon find yourself marooned in tangled heather or teetering on icy rocks. In some seasons downhill skiers find themselves jampacked into the limited runs of snow, and you pay as much for the privilege as on alpine pistes of far superior extent and quality. Scottish skiing is mainly a weekend activity. There is only one developed resort with nearby hotels and après-ski facilities and that is Aviemore in Speyside. The ski slopes of Coire Cas on Cairngorm lie just a short car or bus journey away.

It was inevitable that downhill skiing and nature conservation interests would find themselves opposed. Areas where snow lies late enough to attract skiers at the peak season in spring, and are also accessible to roads, are limited. Because of their altitude and snow-trapping qualities, such sites often have special vegetation and wildlife. Practically all of Scotland's downhill skiing areas lie adjacent to, if not within, Sites of Special Scientific Interest. The development at the Lecht and the threatened one at Ben Wyvis lie among some of the best lichen-rich, high level blanket bog in Britain. Glenshee and Aonach Beag lie on mica-schist and limestone rocks noted for their rare flora. Skiing is the one development that penetrates the nesting grounds of mountain birds such as ptarmigan and dotterel, and the inevitable litter attracts predators such as crows (as well as innocent snow buntings, feeding on the crumbs). Constructing runs and facilities on these thin-soiled mountain slopes creates scars that take a long time to heal. The harsh climate and slow growth rate can cause gullying and flash floods, and attempts to revegetate the ground with pasture grasses create anomalous green fairways instead of the natural colours of heather, crowberry and lichen-covered rock. But although Scottish resorts look tacky compared with their sleek alpine counterparts, the direct environmental impact is confined to the skiing area along with their car parks and cafés. The greater problem at Cairngorm is that the chair lifts, which continue to run during the summer, bring visitors within a short distance of the summit, from where you can roam at will over the plateau. In consequence, tracks now radiate from the eroded summit of Cairngorm to nearby corries, the steep defile of Loch Avon and beyond to the distant summit cairn of Ben Macdui, Scotland's second highest hill. On summer weekends 1,000 or more people walk to the top of Cairngorm and often beyond. The issue goes beyond its obvious effect on wildlife to one of how we treat our finest wilderness – exploitatively, or with reverence and respect.

Skiing and conservation interests managed to coexist without serious conflict until 1981, when the Cairngorm Chairlift Company applied for planning permission to extend its activities westwards into the three northern corries, especially Lurcher's Gully. It was claimed this would double the size of the resort and greatly increase its profitability as well as easing overcrowding. The application was supported by the Highland Regional Council and the Highlands and Islands Development Board; ranged against it were the NCC and a number of prominent naturalists and hill walkers. The Secretary of State called in the application, and the result was a major public inquiry. For six weeks the eyes of the environmental world were on a village hall in Kingussie. I sat in on a few of its sessions. In the chilly but crowded hall developers and conservationists faced each other across the hall as in Parliament, with the inspector, the press and the public seats completing the square. The unfortunate witnesses were made to sit alone in the middle, where they had to answer as best they could the practised questioning of a QC from across the hall. 'It comes to this, does it not?' ... 'Are you seriously suggesting that ...' 'Come, come, Mr Morris, two plus two equals four, does it not?' I remember a lengthy cross-examination over whether the wear and tear around Cairngorm was 'tolerable' or 'not tolerable'. The witness wanted to say it was not, but unfortunately the silk was holding up a letter in which he had said the opposite. 'It's just about tolerable,' he conceded, *'but fast becoming intolerable'*. Back in the hotel, our

Skiing in Scottish condi-
tions creates eyesores: not
only pylons but eroded
channels, tatty fences and
kiosks, access roads and
litter. Would better planning and more investment
solve the environmental
problems? (English
Nature/Peter Wakely)

evening briefings held the excitement of a council of war. A professional
conservationist spends much of his life at dreary meetings or behind a desk.
Moments of drama are rare, and are cherished accordingly.

It helps, of course, to be on the winning side. The NCC did win that
inquiry, but it turned out to be only the first round of a protracted battle.
In the entrepreneurial atmosphere of the 1980s, no fewer than five separate ski-related developments on SSSIs were approved over the objections
of the NCC. One of them broke Scotland's National Planning Guidelines
by extending downhill skiing into a National Nature Reserve, at
Caenlochan. Pylons and humming wires now cross that botanically famous
'amphitheatre with its ring of green alpine flushes' and rare alpine grasses
and sedges, described in the New Naturalist volume, *Mountain Flowers*.
Meanwhile, the chairlift company still had its eyes on Lurcher's Gully. In
1989, it slapped in another planning application, once again supported by
the local council. However, the times were changing, and growing concern
about the future of the Cairngorms had found a focus in the Save the
Cairngorms campaign, an alliance of 14 voluntary outdoor recreation and

conservation bodies. With most of the Scottish media now on their side, and a record 7,000 letters of objection in the minister's post bag, the application was refused. The Secretary of State went further, announcing in 1990 that the National Planning Guidelines would be revised to rule out any further expansion to the westward of the existing runs on Cairngorm. This was a turning point. No longer could the protagonists of skiing development credibly demonise conservationists as a boffinish élite living outside the area, as they had in 1981. The Scottish Office had affirmed the international importance of the Cairngorms as a wildlife and heritage asset, and had it included in a tentative list of World Heritage Sites. Shortly afterwards the Cairngorms Partnership was established to produce a strategy for the area. In 2001, the prospect of a Cairngorms National Park was real – and near.

Even so, the Highland Regional Council was determined to build up Aviemore into a year-round tourist attraction one way or another. The answer was a funicular railway that could convey a much greater number of people to the upper slopes of Cairngorm than the existing chair lifts. It was the central attraction in a £17 million package fronted by the chairlift company, three-quarters of which would be met by the public purse, including a major contribution from EU funds. Unfortunately Scottish Natural Heritage chose to base its objection on the narrow grounds of consequential pressure on the summit plateau. The developer was able to meet this by amending its plans to exclude access to the summit from the terminus building. SNH felt it had therefore no alternative but to withdraw its objection, thus letting the minister off the hook. This effectively pulled

Out of sight of the pylons and runs, the heart of the Cairngorms in late May – Garbh Coire with Angel's Peak and Ben Macdui (left) in the distance.

the rug from under the feet of the voluntary bodies. The RSPB had commissioned an alternative, environmentally acceptable scheme based on the more resilient lower slopes in which a gondola would provide a 'mountain ride experience' through the pine woods and over the moors, with grand views of the Cairngorm corries beyond. On scenic grounds it would probably have been superior to the funicular railway. Unfortunately, like all carnivores, humans prefer to look down from the heights than to raise aspirant eyes to the summit. The RSPB and WWF asked the courts for a judicial review on the grounds that the proposed development area had been excluded from the proposed SAC at the Cairngorms. They lost, and decided against an appeal because it was felt that this was not the way to woo the new Scottish Parliament. EU funding duly came through in 1999, and construction work on the railway began the following year (it opened just before Christmas 2001). Whether dark warnings about ecological damage and public safety are justified remains to be seen, but those who seek wilderness and solitude in the Cairngorms may have to turn their backs on this place. Perhaps the most important conclusion to be drawn from the long battle over Cairngorm is that people do not necessarily value 'facilities' and easy access above wildness. The *idea* of the Cairngorms as a wilderness, as a place one visits to recharge the batteries and obtain a sense of a world apart from our own, is potent. The railway may give families a fun ride on clear, windless days, but as an idea it is an embarrassment.

10

Animals That Get In Our Way

I wonder what kind of nature conservation we would have had if bears, wolves and other large animals still roamed wild in Britain. Perhaps it would have developed more along the American pattern, where National Parks are about wildlife as well as access, or even along the lines of African countries, where big, fierce animals are protected as a major source of foreign income. If so, our nature reserves would have had to be a great deal bigger, and policy would have been centred on vertebrate mammals, not birds or plants. As it is, the few big animals are on the margins and Britain's conservation policies were based on the management of natural vegetation in relatively small sites. Birds became important because birdwatching is so popular, but in general animals are expected to fend for themselves. We were the last country in Europe to consider reintroducing beavers.

To be a big animal in Britain is to be a nuisance. Most of them have their own Act of Parliament. Seals eat fish, badgers spread disease (it is claimed), deer browse trees and raid crops, and foxes cause class warfare. In short, big animals get in our way. So do some small ones. As guests in this chapter on our uneasy relations with our most attractive mammals and birds, I include the bats, which sometimes take up residence in our homes, and so also get in the way. On the whole, we are kinder to these animals today than we were a generation ago. Perhaps we are even heading towards the other extreme. The public will soon need some persuading that we need to massacre deer to preserve woods, and might even need to be horrid to foxes to preserve peace in the land.

Not a problem after all?

In 1969, Dudley Stamp dedicated several pages of his book to what was then regarded as a problem animal, the grey seal. A large proportion of the world's population of grey seals breeds around the British Isles. In 1914, when a law established a close season for the first time, there were only a few hundred grey seals in Britain, and not many more in the world. The animal was permanently protected under the Grey Seals Protection Act of 1932, which extended the close season to include the entire breeding season. Since then, their numbers have grown exponentially from hundreds to thousands to tens of thousands, as have complaints that the seals were eating too many fish, especially salmon, as well as damaging fishing nets. In some places, such as the Farne Islands, it seemed that the grey seals might even be harming the environment when their dragging bodies stripped the natural vegetation and caused soil erosion. The 1960s answer was to shoot them, in licensed culls, to keep the population at an agreed level. Since it is difficult to shoot seals in the water, the answer was to kill the seal pups, born in autumn and confined to land during their first few months. Where the animals were numerous, as in Orkney, it was considered prudent to shoot about half the pups born each year.

The calls for grey seal culling came mainly from the fishing industry. The then Nature Conservancy became involved when two of the largest seal rookeries, at the Scottish island of North Rona and the National Trust's Farne Islands, became National Nature Reserves. As Stamp pointed out, many found it 'hard to reconcile conservation with the annual carnage of 360 seal pups' recommended for the Farne Islands by a commission chaired by the Conservancy's own deputy director. That particular conundrum was solved by the less science-led National Trust, which owned the islands and decided that such scenes of blood and gunsmoke were inappropriate on a nature reserve. However, culls of grey seals went on elsewhere. In 1977 a projected five-year culling programme began in the Outer Hebrides that was intended to reduce the population from about 50,000 to 34,000, the level of a decade earlier. Heavy seas often prevented the Norwegian sealers, commissioned by the Department of Agriculture and Fisheries for Scotland (DAFS), from landing, but they managed to shoot 47 adults and 145 pups at the Monach Isles, another National Nature Reserve. They also managed to dispatch 276 adults and 352 pups at Gasker, an island off Harris. In its annual report, the NCC reported this without comment. These shootings took place out of the public eye, but a similar cull planned for Orkney the following year caused a public outcry. Seals had been traditionally hunted around Orkney until 1962, when stricter controls were introduced. With that partial protection, numbers increased from around 700 to over 1,000 by 1978. During that time the Conservation of Seals Act was passed, which provided a close season for common seals for the first time and gave ministers reserve powers to protect an area throughout the year. Seals that threatened to damage fishing gear – or fish farms – could still be shot, but otherwise killing seals now required a ministry licence. The attempted cull of 800 grey seals at Orkney in 1978 'for the prevention of damage to fisheries' had no real scientific justification. Pictures of the sealers going about their gruesome business were printed on the front page of every family newspaper, and DAFS was forced by the public reaction to abandon the cull. Since then, the grey seal population has grown and grown at roughly 6 per cent per year. Whether they pose much of an economic threat depends on who you talk to, but on fish farms anti-predator nets seem to work so long as they are maintained properly (Berry 2000). Some farms also use ultrasonic scarers.

Since receiving protection in 1970, the common or harbour seal has also increased slowly, especially in The Wash, its main breeding ground (there are smaller numbers at Orkney, Shetland and the Scottish west coast). In 1988, however, disaster struck them in the form of an epidemic of the phocine distemper virus, which killed half the common seals in The Wash, and 10 to 20 per cent of their compatriots elsewhere on the east coast – around 12,000 seals in total. Dogs, failing fish stocks and other unlikely causes were brought in as suspects. At a North Sea conference held in April

Common seals hauled out on sandbanks at Blakeney Point, Norfolk. (English
Nature/Peter Wakely)

1994, Dutch scientists presented convincing evidence that the outbreak
was linked to chemical pollutants that had disrupted the seals' immune systems. Fortunately the epidemic blew over, and grey seals seem to have
escaped its effects altogether. Perhaps the common seal undergoes greater
risks because of its more frequent use of polluted east coast waters.

Since 1978 there has been no real long-term policy on seals. Local fishermen in Orkney and the Western Isles are licensed to kill seal pups where
economic damage can be demonstrated, but large-scale culling has been
abandoned. With an environment-friendly image and a healthy tourist
business to maintain, local authorities are not calling for its resumption.
Indeed, seal-watching trips around the Scottish coast are said to generate
£36 million a year in income. It was public opinion that made the difference, not the law, nor any new scientific evidence. The issue came to be
seen as a matter of conscience, not merely one of scientific calculation. A
six-year study by Aberdeen University ending in 1995 concluded that seals
are not responsible for the decline of cod or salmon in Scottish waters.
Common seals eat mainly herring and sprat in winter, sand-eels and
cephalopods in summer. It was not the seals that have ruined Britain's fishing industry but the fishermen themselves, and far more effectively and
ruthlessly than the poor seals ever could have done.

Persecuting a protected species

The legislation surrounding the badger is a good example of how even a
retiring, nocturnal animal that rarely gets in anyone's way can still attract
a lot of attention in a crowded country. Quite a number of different par-

ties are interested in badgers. Conservationists want to conserve them. They do not need to do very much in that direction since badgers are common. They have survived changes in the countryside remarkably well. Very few parts of mainland Britain are badgerless, and the animals even live in suburban gardens. Some desperadoes enjoy cornering and killing badgers with the help of dogs. Although this unpleasant activity has been illegal for decades, an estimated 9,000 animals are killed in this way every year (Morris 1993). Dairy farmers regard badgers with dread, as potential carriers of bovine tuberculosis. The Ministry of Agriculture (MAFF) granted itself the licence to kill a lot of badgers on these grounds. Over 30,000 badgers have been gassed or cage-trapped and shot by MAFF since 1971, with another 12,000 to 20,000 dead badgers targeted for the year 2001 before the foot-and-mouth crisis put operations on hold. Motorists have no particular interest in badgers, but nonetheless manage to run over another 50,000 of them every year. Occasionally a badger makes a nuisance of itself and so has to be caught, tranquillised and moved somewhere else ('translocated'). For example, in March 2001, a local authority in Scotland spent £30,000 rehousing a family of badgers whose sett had undermined a main road.

Many believe that the Ministry's campaign against badgers on behalf of cattle farmers is cold-hearted and unnecessary. It started in 1971 when a dead badger was found to be infected with bovine tuberculosis on a farm where some cattle had the same disease. Did the badger catch it from the cattle or vice versa? A quarter of a century and 30,000 dead badgers later, it was still uncertain whether the policy of slaughtering badgers was helping to control the disease. In 1996, Professor John Krebs, then head of NERC, was asked to look into the matter, and advise on the resolution of the problem. The result was the largest-scale biological experiment ever tried out in the field, where all badgers would be trapped and killed in ten different areas measuring 10 square kilometres each. The incidence of TB would then be compared with similar-sized areas where badgers were left unharmed. What was left unclear is what would happen if it *was* proven that badgers spread disease. Would it then be open season on badgers in cattle-rearing areas? Not according to the Agriculture Minister, who explicitly ruled out a 'national wholesale eradication of badgers'. In any case, the Krebs trial ran into difficulties. It was difficult, if not impossible, to trap every last badger over 100 square kilometres, and the more badgers that survive, the weaker the data will be. And the longer the trial goes on, the harder it becomes to isolate badgers from other variables. For example, the floods of autumn and winter 2000 confined many cattle to barns where they are less likely to come into contact with badgers. Conservationists such as Malcolm Smith of the Countryside Council for Wales argue that MAFF should be spending more on promoting better farm practices – such as not stuffing cattle so full of chemicals that they have little natural resistance to disease. The NFU indignantly denied that farmers are not taking enough care of their animals, turning a blind eye to the chronic lack of preventative measures against foot rot and diarrhoea in British herds (Lawson

2001). And so we are back to badgers. In 2001, the Commons Agriculture Committee continued to insist that the Krebs trial is 'the only feasible way of obtaining the information essential to establishing the relationship between bovine TB in cattle and badgers, and whether culling is a viable policy option'. Whether trapping and shooting badgers controls the spread of TB any more effectively than doing nothing will not be known until the full results are available, in 2004 – if then. As this book went to press, there were signs that Government was losing faith in the trial. Computer modelling has cast doubt on the efficacy of culling. The foot-and-mouth crisis put paid to the original timetable. In April 2001, MAFF stated that it was 'minded to develop a range of policy options' that could be tried out outside the area of the present trial. This could be the prelude to allowing farmers to cull badgers themselves (Harris 2001).

Despite all the dangers it faces from our own species, the badger is protected as fully as any wild animal can be. Past legislation protected the animal but not its home, thus allowing badger-hunters to escape prosecution by claiming they were digging for foxes, something that was hard to disprove unless they were caught red-handed. In 1992, the Protection of Badgers Act made it an offence not only to harm a badger but to 'damage, destroy or obstruct' a badger's sett. This means it is now illegal to put a dog into a sett. It is also illegal to keep an unlicensed badger as a pet, or even to have a dead one in your house unless you can show it was not killed illegally. In order to close the legal loopholes through which badger-baiters have often escaped, the law comes close to assuming guilt unless it can be proved otherwise. Someone found guilty of hurting a badger can now be heavily fined or thrown into prison for six months. The badger therefore enjoys a degree of protection equivalent to that of an endangered species. The reason is not because badgers are endangered, but because they are popular. As Pat Morris (1993) has remarked, 'the badger probably has a greater public following than any other British mammal, and its future is thereby as much a political issue as a biological one'. Whether all this protection makes much difference to badgers is doubtful. The threats they face from mankind are as great as ever. Most people today are more familiar with the pathetic, mat-like corpses on busy roads than with the striped face of the live animal as it emerges from its bank and sniffs the evening air.

The 'branchy beasts'

When Dudley Stamp was writing *Nature Conservation in Great Britain* there were perhaps 200,000 red deer in Britain, a situation he described as 'too many deer, chasing too little food on too little land'. He had in mind the emaciated carcasses littering the Highland glens after the severe winter of 1962–63, a sight far removed from the Arcadian beasts roaming Richmond Park. There are even more deer today: some quarter of a million red deer range the Scottish hills, plus an unknown but large number lurking in dense forestry plantations – perhaps 347,000 all told, plus about 12,500 more in England. Yet Fraser Darling reckoned that the most red deer that the land could support without suppressing regeneration and in other ways damaging the environment was about 60,000. In other words, in Scotland we have five or six times the ideal number of red deer. British deer no longer have natural predators. Long ago, the wolf used to do the job, but today deer have to be controlled with the rifle.

In Britain, advice on the conservation and control of red deer is the responsibility of a government commission (formerly the Red Deer Commission, now the Deer Commission for Scotland). There is also a British Deer Society, concerned mainly with welfare issues. The Commission's task is to promote good deer management by maintaining healthy herds and making sure they do not make an environmental nuisance of themselves. It can, at least in theory, cull deer on private land if their numbers are too high, and pass on the bill to the owner. In practice, however, the Commission's powers are very limited. Many Highland estates, eager to raise well-antlered trophy stags to attract wealthy stalkers, do not cull as many hinds as they should in the belief that more hinds will produce a greater proportion of trophy stags, which can then be selected by shooting the poorer, less well-endowed ones. More likely, if more hinds were culled, they would free up habitat for the stags, and both stags and their habitat would be the better for it. As it is, in the prevailing grossly overstocked hills poor, runty stags are the norm. Another reason is that hind shooting is often left to the stalkers, who have other claims on their time in winter, or maybe have just grown lazy. The situation has been made worse by artificial feeding in winter, to reduce natural mortality. Conservationists are, of course, less bothered about the size of a deer's antlers and more worried about overgrazing.

The traditional Scottish 'deer forest' was a good habitat for wildlife (or at least it would have been if keepers had been kinder to raptors and carnivores). In the nineteenth century, deer numbers were often low enough to allow some tree regeneration along the valleys and glens, along with thick heather and marshy hollows where the deer would wallow on hot days and find fresh herbage in the spring. With present-day numbers, added to the disastrously heavy stocking of sheep encouraged by CAP headage payments, regeneration has long since come to a virtual standstill except inside tall deer fences – which of course increase the pressure on the remaining unfenced land. Once lush stands of heather, blaeberry and moss are turning into wastelands of short, patchy heather and grass. A side issue is the hideous scars of vehicular tracks created to make stalking easier, while the public were denied access to the hill in case they spoiled the sport.

Red deer are free-ranging wild animals, which nonetheless become the property of an estate owner for as long as they remain alive on his land. However, overall numbers can only be reduced when estates agree and co-operate in a culling programme. If a single owner shot most of his deer, other deer would simply move into the vacuum created, much to the annoyance of his neighbours. This is more or less what happened when NCC staff shot out the deer on its newly acquired holding at Creag Meagaidh during the 1980s. Scientists have had little influence on deer management, despite internationally acclaimed studies of deer on Rum and elsewhere, since estates tend to think they know better. In the 1960s, Morton Boyd and Dick Balharry of the Nature Conservancy got representatives of five estates in Wester Ross to bury their differences and manage the local deer more sustainably as a voluntary co-operative – the Gairloch Conservation Unit. But that did not stop what Boyd described as a 'tidal wave' of new forestry plantations making a mockery of any hopes of permanently reducing stocking densities on the open hill. Nor did the Gairloch idea translate well to the Cairngorms, where neighbourhood relations were more frosty.

The situation went from bad to worse until the 1990s, when things suddenly started to improve. A series of mild winters forced some estates to shoot more hinds, since more animals were surviving the winter. The 1996 Deer Act, which created the Deer Commission for Scotland out of the clapped-out Red Deer Commission, also gave it more flexibility and powers to broker 'voluntary control agreements' over large areas. But the most significant event of all was the purchase in 1995 of the vast 30,000-hectare Mar Lodge estate in the Cairngorms by the National Trust for Scotland. Experience has shown that practically the only way a conservation body can deal with the red deer problem is to own the land. Hence the NCC succeeded in getting regeneration underway at Inshriach and Creag Meagaidh, which it owned and managed, but not in the rest of the Cairngorms where weak gentlemen's agreements held sway. The Mar Lodge purchase came with strings attached. The Trust was obliged to maintain sport-stalking – and grouse shooting – by its mysterious benefactor, the 'Easter Charitable Trust', thought to be a consortium of Highland landowners who stumped up most of the £5 million purchase price. Even so, Mar Lodge offered an opportunity to manage deer on a sustainable basis over much of the Cairngorms. Deer management is an unsentimental business and the guns have been busy at Mar Lodge, reducing the herd from an estimated 5,000 at the start of the Nineties to within a whisker of its target of 1,650 animals (950 hinds, 750 stags) by 2000. The Mar Lodge experience is being followed closely by deer forest managers all over Scotland. The Trust and its estate staff have entered a long-term management agreement with Scottish Natural Heritage based on the principle of minimum intervention (a little hard to swallow, this phrase, with gun-shots echoing around the glens). A decision was taken not to plant trees or use fertiliser for the time being. If it works, Mar Lodge may influence deer management over much of the Highlands, and bring fresh life to its much-diminished natural woods.

While red deer have been monarchs of many a Highland glen since the mid-nineteenth century, the increase of roe and other deer in lowland woods has been largely a postwar phenomenon. Between the later Middle Ages and the 1950s, roe deer were comparatively scarce animals, confined mainly to Scotland and northern England. Since then they have spread naturally or been released over most of Britain, to be joined by muntjac, sika and fallow deer originating from escapes from deer parks. There are

Oak-hornbeam woodland in Hertfordshire. The dominance of pendulous sedge and brambles may be a response to overgrazing by deer. (Derek Ratcliffe)

now about half a million roe deer, 100,000 fallow deer, and upwards of 40,000 muntjac and 12,000 sika roaming wild in Britain, and they are all increasing. The presence of much larger densities of deer has profound implications for woodland management, especially for coppicing, the traditional system of regenerating woods based on a cycle of cutting and regrowth. Famous woods such as Hayley Wood and Monks Wood in Cambridgeshire have been ravaged by deer, making a mockery of their management plans. Regeneration is possible only under 'dead hedges' of brushwood, or, more certainly, behind ugly, expensive fences. In farmed landscapes, deer tend to use the woods as a secure base, foraying out at dusk to raid the hedgerows and crop fields. They are slowly changing the character of the woods from dense, lush coppice rich in wild flowers to wood-pasture, a more open, less biodiverse landscape based on mature trees and grass.

What does the conservation-minded woodland manager do about it? He can try to shoot some of the deer, but without co-operative neighbours this is unlikely to be a long-term solution. And small, shy deer in dense woodland make difficult targets. Some members of conservation charities say deer should not be shot, at least not on nature reserves; and there shouldn't be any fences either. A widely adopted compromise is to fence parts of the wood and try to shoot or evict all the deer inside. But the deer often find a way back, especially when the lush regrowth tempts them in. In the meantime, the impact of deer on woods is an important phenomenon that is receiving enough attention. As the woodland authority George Peterken sardonically notes, 'If you can't beat them, learn from them' (Peterken

1996). By studying the phenomenon, at least you learn something about natural processes. The deer are forcing us to become cleverer woodland managers. The best solution, if there is one, may be an integrated plan that accepts the deer as a *fait accompli*, but attempts to control their numbers by a mixture of shooting and chemical repellents, combined with rotational fencing. In the meantime anyone contemplating the release of yet more deer from parks should be rather forcibly discouraged.

Strange passions

Anyone who knows the classic New Naturalist monograph, *The Herring Gull's World* by Niko Tinbergen, will remember the famous Ravenglass gullery. When Tinbergen was carrying out his studies of bird behaviour on this part of the Cumbrian coast, in the 1950s and 60s, at least 10,000 pairs of black-headed gull nested among the dunes, along with 800 pairs of sandwich terns and other species of gulls, terns and waders. Today, despite Ravenglass's status as a Local Nature Reserve, all the gulls have gone. The crash began in the 1970s, and after several successive poor seasons they deserted as a body in 1985. The site had not changed much physically, nor had the gull suffered a widespread decline. The reason was predation. And the main predator was the fox.

Foxes had raided the colony for as long as there are records, but the decline and eventual loss of the colony took place during a period between 1966 and 1983 when foxes were no longer controlled. Ironically, this was at the decision of the nature reserve managers, who, after the departure of Tinbergen and his team in 1968, adopted a 'hands off nature' approach (Simpson 2001). Nature, they thought, would find its own balance. Unfortunately events showed that without fox control, the ground-nesting birds at Ravenglass did not have a future. What the gulls needed was not laissez faire 'management' but a gamekeeper! By the time they started shooting foxes again, in the 1980s, it was too late.

One way or another we do kill a lot of foxes. About 100,000 a year out of an adult population of about a quarter of a million are killed by motorists – knocked down and squashed flat as we rush along crowded roads about our daily business. If the statistics are right, most of us will probably kill at least one fox in the course of our motoring career. Gamekeepers and sheep farmers kill many more, and about 15,000 foxes each year are successfully hunted with hounds. A lot of urban foxes probably die of mange, which wiped out most of Bristol's well-studied foxes in the mid-1990s. Yet, despite the likelihood that most foxes will meet a violent or unpleasant end before their natural span is out, they are in no need of special protection. Foxes are extraordinarily successful and adaptable animals, occurring in every habitat from remote fellsides to suburban parks and gardens (a favourite hide-out is under the garden shed). They can eat practically anything, from berries, insects and worms to any animal they are capable of catching and killing. Foxes also eat carrion, including leftover chicken dinners in restaurant bin bags. The fox is the commonest large carnivore, not only in Britain but in the world.

Most of the foxes I have seen were dead. I occasionally spot a live one snooping around the house near dusk, mostly in late spring when they raid the waterfowl nests, and sometimes later on when they stalk fledgling pheasants. Usually they see you first, but occasionally you surprise one, intent about its business seeking the daily half-kilogram of meat a fox needs to stay alive. Then it gives a cross, very dog-like bark, slips away through the shadows and halts a safe distance away, no longer a fur-and-tooth animal but a vague dark shape. Normally, however, all you see of the fox is what it leaves behind – droppings, often bejewelled with beetle's wing-cases, a musky smell around the dustbins, or bits of reddish hair caught in the barbed wire.

To the traditional countryman the fox is, or was, synonymous with the worst kind of vermin, next in popularity to a rat. They are accused of killing lambs (though they take mainly weak or already dead ones, and probably make little difference to the sheep-farm economy). They certainly take pheasants, duck and poultry – the one thing everyone knows about foxes is the orgiastic killing frenzy that often follows when a fox gets into a henhouse or pheasant-rearing pen. Close-up, foxes have unusual eyes with elliptical pupils, which give them a cruel appearance. Such things, and the stories we used to learn at school about wicked Reynard or Brer Fox, helped to create the legend of the cunning, irredeemably wicked fox. More recently, television has created a more benign image, a cuddly, laddish creature, like Basil Brush, a small, bushy-tailed doggy, inclined to be a bit naughty, but essentially loveable. Neither image has much to do with real foxes, but anyone who has spent a season trying to protect breeding terns experiences very similar feelings to a poultry farmer after seeing his efforts brought to nothing by a rogue fox.

Conservation bodies have been uncharacteristically quiet during the great debate about the rights and wrongs of fox hunting. While hunting has little effect on overall fox numbers, since more animals are killed by trapping and shooting than by dogs, the issue has become an emotional tripwire. The reason for the silence from the nature conservation end is that membership bodies dare not say anything for fear they will lose members. The country agencies say nothing because the matter is politically sensitive. In its comments to the Burns Inquiry, English Nature insisted that hunting was not a conservation issue since foxes were natural predators and nature has its own way of finding a balance. Essentially this is because foxes live in social groups that establish and defend a territory. Because of this behaviour, foxes have their own method of population control. In towns, however, the density of foxes is greater purely because of artificial feeding. In Bristol, about ten per cent of householders regularly left out food for 'their' foxes (Baker et al., 2001). What English Nature did not mention is that a ban on hunting, whatever its effect on foxes, whatever its possible ethical merits, would undoubtedly strain relations between conservation agencies and country landowners.

Before it was cancelled because of foot-and-mouth, an expected half-million people were expected to join a countryside march in London on 17

March 2001. A ban on hunting clearly concerns many who do not actually hunt themselves. Some believe it would be the thin end of the wedge, the start of a general assault on country sports, including angling, led by animal-rights groups claiming massive, if largely passive, public support. More broadly, the hunting issue symbolises a clash of sensibilities: of a culture based on modernisation and change pitted against one of custom and tradition. There is a great fear nowadays of being seen to be *incorrect*, a sense that old-fashioned liberality is being replaced in many quarters with an enforced conformity based on moral absolutes. It seems to make little difference that the Burns Inquiry was unable to make up its mind whether hunting was cruel (or, indeed, about anything else), when so many people are obviously convinced that it is. The two views are irreconcilable. A sensible government will try to calm things down.

The fox is potentially the joker in the nature conservation pack. He is only a common natural predator going about his business, but because he enters gardens, and eats the food put out for him, the fox has become for many a kind of outdoor pet. He has transcended wildness and become property. But if country, as well as town, foxes are to be given the status of a pet, then other forms of wildlife will suffer for our sentimental indulgence.

Wild geese

Wild geese are one of conservation's success stories. True, goose populations were relatively low in the postwar years when the first national counts were made, but the numbers of most species of geese overwintering in Britain multiplied by a factor of three or four over the next 30 years. The 30,000 pink-footed geese present in 1950–51 had increased to 101,000 by 1983. The bigger greylag goose underwent a similar increase from 26,500 in 1960 to just over 100,000 in 1984 (Lack 1986). On Islay alone, the rarer barnacle goose increased eightfold from around 3,000 in the 1950s to 24,000 in 1978, a quantity that local farmers, whose valuable grass they ate, considered to be much more than enough. With the exception of some of the greylags, and the introduced Canada goose, all these birds breed in the arctic tundra. Britain, and especially Scotland, becomes an important base for them in late autumn, a landing stage for geese from Greenland, Iceland and Spitzbergen. Geese that traditionally eat barley grain, such as the pinkfoot and the greylag, congregate in east and central Scotland. Grass-feeding geese, such as the barnacles and white-fronts, prefer the west coast, with its mild winters and early 'bite' of new shoots. For these months, Britain and Ireland become a place of world importance for wild geese: we are host to around three-quarters of the world's pink-footed geese, half the dark-bellied race of the brent goose, a third of the barnacle geese, and the whole of the distinctive Greenland race of the white-fronted goose. When I lived in Aberdeenshire, one of the great wildlife spectacles was the grey skeins of pink-footed geese coming in to roost on the loch as the autumn sky turned rosy-pink. The call of wild geese still ignites in me blood memories of cold, exhilarating mornings, waiting in the darkness amid the

Brent geese on the Blackwater estuary in Essex. (English Nature/Peter Wakely)

reeds for the moment when fresh, feathered food from Iceland passes over the gun barrels. Their great northern journeys excite wonder, as does their cleverness in capitalising on the way we are changing the land.

Why are wild geese doing so well? It seems to be through a combination of protection and food supply. The more geese that survive the winter in Britain, the more will return to breed in the limitless, mosquito-infested polar wilds. Wild geese in Britain had a hard war, followed by a period of fairly intensive wildfowling in the late 1940s. However, the conservation pioneers led by the Wildfowl Trust went to a great deal of trouble to establish sanctuaries where wild geese and other wildfowl could roost and feed undisturbed. The first formal Wildfowl Refuge was set up on the Humber in 1955, followed by others on the Ribble at Southport, Caerlaverock in Dumfries, Lindisfarne in Northumberland, Tentsmuir Point in Fife and elsewhere. Some of these places later became nature reserves, where the shooting was regulated by the Nature Conservancy. In the new climate of the 1950s, with the Protection of Birds Act, and the growing popularity of birdwatching, wildfowlers, give or take a few independent 'cowboys', broadly supported the scheme.

At that time, wild geese fed mainly on salt marshes on and around estuaries, or, in the case of the pink-footed and greylag geese, on waste grain in barley stubble, followed by raids into fields of turnips and potatoes. However, by the 1970s, their behaviour was changing. The sowing of cereals in autumn instead of spring might have robbed the geese of their stubble fields, but it substituted an almost limitless supply of fresh cereal shoots. Similarly, the fertilisation of pastures or their replacement with fresh-sown leys produced supplies of tender grass, especially along the landfalls of the geese on the Scottish west coast. The geese also approved of the big new fields created to accommodate combine harvesters, since it gave these nervous birds a much better view. They tend to stay in the same place longer than in the past. New reservoirs were requisitioned by greylag and pink-footed geese as roosting spaces, and gave them the opportunity to move further inland. Even the relatively conservative brent goose had discovered the potential of autumn-sown cereals and oilseed rape by the mid-1970s, and began to feed inland. Of course, these new opportunities brought wild geese into conflict with farmers. On the wealthy grain farms of eastern Britain, the loss of up to 5 per cent of the crop to geese might not have dented the farm income too seriously, but in marginal areas, such as the Western and Northern Isles, agricultural damage was more serious. In particular, the habit of the barnacle goose of feeding as a single vast flock, like a swarm of feathered locusts, made them very unpopular on Islay. In global terms, the barnacle goose is rare. It just happens that most of them visit Islay.

The problem came to a crux in 1981, when, in line with our international commitments, Britain extended full protection to the brent and barnacle goose – and, in Scotland, to the white-fronted goose – at a time when goose numbers were at an all-time high. The NCC wanted to designate certain key areas as SSSIs, and try to work out agreements with farmers by creating undisturbed refuges in the hope that the geese would congregate there. In the meantime farmers would be compensated for agricultural damage. But many farmers preferred the traditional method of inviting in the marksmen. This now required a licence, which the Scottish agriculture department seemed more than willing to issue 'to prevent serious agricultural damage'. It soon became clear that the system was being abused, and that some of the geese were being shot for sport, as in the good old days. On Islay, the flock of barnacle geese fell from a peak 24,000 in 1976 to 15,000 in 1983. 'Shooting was undoubtedly a contributory factor,' wrote the NCC in its annual report to the minister. The Secretary of State for Scotland took the line that until management agreements had been sorted out, he would continue to dispense licences. One of the sticking points, as far as some of the islanders were concerned, was that they didn't believe the conservationists' figures. There were *far* more geese than was claimed. There was also considerable doubt about the efficacy of alternative ways of scaring geese away, for example, by flying kites. The trouble is that wild geese are not stupid. They are capable of learning that kites are not hovering birds of prey, and even to judge when the close season arrives (with warmer weather), when the shotgun blasts go over their heads and are not aiming at them.

Since the 1980s, it has been government policy to compensate farmers on Islay whose land has been damaged by geese. By 1994, this worked out at £9.50 per goose per year, totalling around £300,000. Noticing this, farmers in other parts of Scotland visited by large numbers of wild geese demanded similar compensation. Scottish Natural Heritage, which had to foot the bill, questioned the legitimacy of such payments, pointing out that it was paying for damage to crops and pastures whose productivity was made possible only by subsidies for agricultural production. In other words, the taxpayer was in effect paying for the crops and paying for the geese. SNH favoured an alternative scheme, pioneered at the Loch of Strathbeg in Aberdeenshire, where farmers are paid to take land out of production altogether as 'sacrificial fields' for the geese. These fields are fertilised or reseeded to bring on sweet, early bites of grass for the geese, which has the added advantage of concentrating the flocks in a particular place and so improving the birdwatching. Stubborn birds can be persuaded to use them with the help of bird-scarers. This system is now in place in many parts of Scotland, with payments available from agri-environment schemes, such as Environmentally Sensitive Areas.

In 1997, the Government established a National Goose Forum, administered by SNH, to advise ministers on goose management (the new word for goose control) in Scotland. SNH continues to run compensation schemes on Islay, the Solway Firth and on South Walls in Orkney. In the latter, refuge areas have been created along the lines of the Loch of Strathbeg model. On Islay, on the other hand, SNH pays around £400,000 a year on 118 individual agreements to protect some 32,000 barnacle geese and 12,700 Greenland white-fronted geese 'from disturbance and shooting'. The lessons learned, it says, have 'contributed significantly' to the development of a National Policy Framework on how to live side by side with wild geese. The answer is that we do so by paying for them.

An endangered species in the loft: bats and their conservation

When I was in the NCC, in the early 1980s, I climbed into quite a number of loft spaces to look for bats, and well remember the atavistic thrill of entering their dark world, feeling the breath of their faintly flittering wings against my face. Since then I have shared a house with a large colony of pipistrelles, which tended to make their presence known in the heat of summer, when, despite claims to the contrary in the conservation literature, the smell of accumulated droppings and urine made parts of the house uninhabitable. (We told everybody that these droppings are just the thing for roses – rather rash advice, I fear.)

Unlike birds, bats, with the exception of the endangered greater horseshoe bat, received no legal protection until 1981. Then the Government made up for its previous neglect and protected the lot of them, common and rare alike. It was full protection, too, extending to their roosts and winter sites (hibernacula), and forbade people to so much as touch them without a licence (though you were allowed to evict them gently from any room you happened to be living in). The Wildlife and Countryside Act protected British bats more comprehensively than any other native mammal, and, at the time, than any other country (Morris 1993). What made it all the more remarkable was that bats frequently choose to live in our homes; nearly half of all bat colonies are in ordinary houses built since 1960. As far as a bat is concerned, a house roof is an attractive combination of cave and giant hollow tree. Thus protection brought every Englishman's castle within the law as bat habitats.

There was considerable justification for all the fuss. Probably the still common pipistrelle was included because it was regonised that most peo-

tell one kind of bat from another. But all British bats were
, in numbers. The greater horseshoe bat, which is relatively easy
ey because of its habit of congregating in a limited number of hiber-
ig places, had spiralled down from an estimated 300,000 individuals in
)0 to only 4,000 in 1993. The fundamental reason is that bats are finding it harder to find enough insect food. There are fewer hedges and
copses, and suitable stretches of water. Even cow dung, once an important
source of flies, is nowadays virtually sterile thanks to the drug Ivermectin,
used to get rid of internal bovine parasites. The two horseshoe bats experience a particularly hard time, since they feed on cockchafers and other
large insects dependent on permanent pasture. They are in danger of
using up more energy finding food than they gain from eating it. Forest
hygiene methods that insisted on removing all dead wood were also removing the homes of bats, such as the noctule, which use hollow spaces in
trees. Another danger to bats was the use by timber treatment firms of
highly toxic, lindane-based sprays to prevent dry rot or woodworm. These
are fatal to any bats unlucky enough to come into contact with them.

For all these reasons, bats have become one of the great conservation
crusades of the late twentieth century. Legal protection for bats does not
achieve a great deal by itself. To bring a case against a householder or timber treatment company is difficult, and the magistrates usually let them off
with a token fine. But the handful of cases that came to court helped to
publicise the problem, and persuade the companies to contact the NCC or
a 'bat worker' first. Soon the NCC was receiving about 2,000 inquiries a
year, most of which were passed on to a growing band of licensed bat
experts. The result would be a call, and an opportunity to put the case for
bats. A survey conducted during a 'National Bat Week' in 1990 suggested
that while three-quarters of home owners did not like bats very much, most
of them were won round after being 'given some facts about bats'. The
NCC gave away 60,000 copies of its leaflet, *Focus on Bats* (featuring a particularly cute long-eared bat). If a home owner insisted on eviction, a way
could normally be found without harming the creatures, for example by
identifying and blocking the exit holes once they had dispersed. Among
the more awkward cases were churches, where bats made themselves
unpopular by soiling the altar cloth with their droppings. At least one vicar
tried to scare them away by revving his motorcycle up and down the aisle.
But the majority of people agreed to share their homes with the bats,
which, I think, does them credit.

The image of bats has mellowed considerably since 1981, thanks partly to
sympathetic television programmes but, above all, to the dedication and
domiciliary visits of members of the local bat groups. The Bat Conservation
Trust was founded in 1991, and now has over 2,000 active members – more
than the Mammal Society – and 90 local groups, one for nearly every county in England and Wales. Our knowledge of where and how bats live has
vastly improved as a result, helped by the invention of hi-tech 'bat detec-
tors', which identify bats by their distinctive high-frequency calls. Bat detec-
tors helped to discover two new species of British bat in the 1990s:

Nathusius' pipistrelle, which is probably a migrant and recent colonist, and another, now called the soprano pipistrelle, distinguished by its high-pitched call. The latter species is now known to be widespread, turning up in many bat boxes, especially woodland. University research is also refining knowledge of bat behaviour, for example, by showing that the Daubenton's or water bat prefers placid, sheltered, open waters, and finds it difficult to detect food among dense waterweed or ripples.

Since bats are sociable animals, and highly faithful to a particular roost or hibernation site, it has been possible to protect some of their homes. Many cave entrances have been covered with grilles to protect them from disturbance, with the co-operation of caving societies. (On the other hand, safety legislation compelled some local authorities to seal the entrances of disused mineshafts, incarcerating several colonies of horseshoe bats in the process, until they were persuaded to substitute a grille.) Specially designed bat boxes have proved effective for some species, such as pipistrelles and long-eared bats. Important sites such as Greywell Tunnel on the Basingstoke Canal have been made Sites of Special Scientific Interest and the subject of management agreements. Perhaps the most interesting example is the derelict Gothic mansion in Woodchester Park, Gloucestershire, never lived in by humans, but now home to greater horse-shoe bats, which roost in the chimneys. Another horseshoe bat roost, in some crumbling farm buildings in Devon, has also been acquired and restored, with the help of the Vincent Wildlife Trust. One even hears talk of electrically heated roosts; perhaps the elusive Bechstein's bat would appreciate these, for only the second colony ever found in Britain was in an unexpected place – an airing cupboard!

British bats and their estimated numbers (from Harris et al. 1995)

Species	Adult population
Greater Horseshoe	4,000
Lesser Horseshoe	14,000
Whiskered/Brandt's	70,000
Daubenton's	150,000
Natterer's	100,000
Bechstein's	1,500
Barbastelle	5,000
Pipistrelle	2,000,000
Soprano Pipistrelle	Unknown, but lots
Serotine	15,000+
Noctule	50,000
Leisler's	10,000
Brown Long-Eared	200,000
Grey Long-Eared	1,000

Raptors: the problems of success

The recovery of Britain's birds of prey since the banning of organochlo-rine pesticides in the 1970s, as well as the reduction in persecution, has been another of the brighter corners of the nature conservation world.

Release of captive-bred red kites at a woodland site in England in 1997. (English Nature/Peter Wakely)

Our population of peregrines has more than recovered its pre-war level, and this regal falcon now nests on cliffs on the southern coast from Cornwall to Kent, and increasingly in quarries inland. Around 30 pairs nest on tall buildings or pylons, sometimes in towns; a pair was reported even using the Millennium Dome. The marsh harrier, once reduced by pesticides and persecution to just half a dozen pairs, now follows the ploughs in Essex fields and is a frequent sight over East Anglian marshes. The buzzard returned in triumph to much of England during the 1990s. (A noisy pair nests in a poplar tree just a few score yards from my Wiltshire house.) The sparrowhawk, which has fully recovered its former range and often hunts in suburban gardens, is now common enough to be controversial: it has been blamed, probably wrongly, for decimating garden birds (your cat will take more, but the real cause is the absence of seeds and grain in modern farm fields). Thanks to reintroductions and escapes we now have red kites and goshawks in numbers unknown for centuries.

For some species, recovery has been rapid, for others slow or non-existent. Since 1990, however, the overall recovery rate has slowed down, and peregrine, hen harrier and sparrowhawk numbers have fallen slightly. For the first two the cause must be persecution. In its 1991 report, *Death by Design*, the RSPB produced evidence of widespread use of poisoned bait and pole-traps in pheasant-rearing areas and on upland moors. The documented cases, which accounted for 40 golden eagles, 65 peregrines, 57 hen harriers, 39 goshawks, 24 red kites and 367 buzzards, probably represent just the tip of the iceberg. Between 1995 and 1999, a further 100 incidents were documented in Scotland alone, of which 39 involved poison (Scottish Executive 2001). On the moors of northern England the situa-

Ian Prestt, late director of the RSPB, by a crow trap on a Borders grouse moor.
(Derek Ratcliffe)

tion is said to be even worse; successfully breeding hen harriers in England
fell to just five pairs in 2000, yet on English Nature's estimate there is
enough habitat for 240 pairs. Egg theft is also rife, for this form of trophy
hunting still attracts its share of fanatics. Scottish Executive's report, *The
Nature of Scotland*, cites a particular eagle's nest in Perthshire that had been
guarded around the clock, only for the successful fledgling to succumb to
poisoned bait shortly afterwards. The incidents of poisoning show no sign
of declining, and, in the case of golden eagles and hen harriers, were higher in 2000 than in previous years.

The reason why raptors are unpopular in some quarters is, of course,
because they take birds that have a value and are regarded as property,
specifically grouse and racing pigeons. Whether raptors are to blame for
the falling numbers of red grouse has long been hotly debated. In 1992, a
grouse moor at Langholm in Dumfriesshire, once celebrated for its game
bags (2,523 grouse were shot there in a single day in 1911 – a Scottish
record), was chosen by the Joint Raptor Study Group, representing
ornithologists and moorland managers, for an experiment. Over five years,
the effect of raptors on grouse numbers was carefully monitored. The
resulting report seemed to confirm what the keepers had been saying all
along. The raptors increased, but the grouse did not. Hen harriers killed
30 per cent of breeding grouse every spring, 37 per cent of the chicks each
summer and 30 per cent of the surviving chicks the following autumn. The
seasonal bag of grouse declined from 4,000 to 100, at which level the management costs of the moor could apparently no longer be met. In 1998,
grouse shooting at Langholm was mothballed. However, it is more compli-

cated than that. The estate had reduced sheep numbers, allowing the grass to grow rank and harbour higher vole densities than usual, and more voles meant more harriers. At the same time, half of the heather present in the 1940s had disappeared, producing a long-term decline in grouse. On other, more heather-dominated moors the predation of grouse was much less. Langholm, in other words, seems to have been an exceptional case. As for what should be done about it, some parties to the report advocated the licensed removal of harrier chicks from grouse moors, releasing them on other, grassier moors where they would feed harmlessly on pipits and voles. Another possibility discussed was an agreed 'quota' of harriers, enforced by pricking surplus eggs. The RSPB opposed such schemes, preferring an alternative in which 'diversionary food' would be offered to the harriers, just as dovecotes have been set up on some grouse moors to divert the peregrines. The issue is a sensitive one. The Scottish Gamekeepers Association is adamant that in some areas hen harriers must be 'culled' to allow 'a sustainable harvest of grouse'. But of course the hen harrier is a protected species, and although the Scottish Executive could license such a cull, it would be sure to be strenuously opposed. Some moorland managers nonetheless destroy harrier nests outside the breeding season, which is arguably legal, and burn out the tall, leggy heather that harriers prefer. A few, as we know, resort to illegal poisons or shooting, presumably with the tacit approval, if not the connivance, of the laird.

A second cause of controversy is racing pigeons. In some districts peregrines are blamed for taking an unacceptable number of these birds, especially inexperienced youngsters; sparrowhawks are also accused of killing

A peregrine falcon celebrates its recovery from pesticides and persecution.
(HarperCollins Publishers)

homing pigeons. Mysteriously, peregrines are declining in parts of the country. A recent report to the DETR suggests that the peregrines are mainly picking off stray birds. The UK Raptor Working Group has suggested re-routing and delaying the races to reduce predation, which reportedly works quite well.

Why make such a fuss about raptors when, compared with many species, they are not faring too badly? The Scottish Executive's forthright answer is that persecution is illegal, and that all wildlife deserves to be treated with respect 'as part of what makes our country special'. Just as racing pigeons are the property of their owner, so, at least in a legal sense, wild birds are the property of the nation. However, while willing to enforce the law, and, if necessary, increase the penalties in the few cases of persecution that stand up to legal proof, Scotland's institutions prefer to work out a solution by consensus. At the time of writing, the Scottish Executive is considering the recommendations of a report produced by the UK Raptor Working Group in 2000, which calls for stronger powers to convict offenders combined with greater support for moorland managers – and (of course) more monitoring and research. Although this fragile alliance of managers, bird experts and administrators is beset by tensions, it is hoped that the kernel of common ground – the desire for large areas of healthy, regenerating heather – may provide a way forward.

11

Biodiversity

Biodiversity became one of the buzzwords of the 1990s, just as Ecology was in the 1970s. The word ecology was coined by the German biologist, Ernest Haeckel, from the Greek *oikos* or home – for ecology was about the study of an animal or plant's 'home' or habitat. 'Biodiversity' was invented, or at least popularised, by the American ecologist, E.O. Wilson. It is a characteristic modern compound-noun, combining 'biological' and 'diversity', and means what it says: living things in all their infinite variety, including genetic variation within species. When traditional English was still spoken we would have referred to it as 'the variety of life'. Biodiversity is not just about species, but about the communities and habitats they form together. In his well-known book, *The Diversity of Life*, Wilson demonstrated that the variety of life on earth is greater than anyone imagined, but it is also threatened as never before by the rate of consumption of the world's natural resources. According to him, the probable extinction rate is a hundred, perhaps even a thousand times as high today as it would have been had people not existed (one has to say 'probable' because we have still discovered and described only a fraction of life on earth). Not only will this make the world a duller place, but it risks damaging our own interests. Some forms of life, perhaps unknown to us, will be 'keystone species', says Wilson, whose loss would drag down an entire ecosystem. For example, a large number of species depend on the nests of the wood ant. If something happened to the ant, we would also say goodbye to a whole swarm of moths, beetles and hoverflies. Equally, a single ill-judged action can quickly kill off a whole tribe of wildlife. On the Hawaiian island of Oahu hundreds of endemic snails were driven beyond the brink into extinction by an imported mollusc: introduced to control an African pest, it ate the native snails instead. Yet the place still *looks* like a tropical island paradise. It has lost its special species invisibly. Most extinct species are very small because nearly all life is very small. But being small does not make you insignificant. The most significant predator in the British countryside is not the fox or the eagle, but the ant. The environmental health of soil or fresh water might depend on some protozoan or micro-fungus that only a handful of experts could even recognise.

As the godfather of 'state of the planet' studies, Wilson's books command a respectful international readership. Hence it was with fresh memories of *The Diversity of Life* that participants at the 'Earth Summit' conference in 1992 started to use the term 'biodiversity' instead of 'species diversity'. The UN conference on Environment and Development was held at Rio de Janeiro in Brazil, in a specially designed conference centre, casually – and ironically – built on an erstwhile wetland, rich in departed species.

A keystone species? Hills of the yellow ant dotting ancient downland turf at Wylye Downs NNR are themselves a mini-habitat for many species of plants and insects.

The leaders of 178 nations came together with varying degrees of reluctance to talk environment. It was apparently the largest-ever summit gathering, and seemed to signal that the whole world had suddenly cottoned on to the fact that damaging the environment harms our own future. What made this conference different from less well-attended predecessors was that it bound signatory nations to the principles of sustainable development and biodiversity. Each country agreed to draft its own 'Agenda 21' – a blueprint for sustainable growth intended to try and make the twenty-first century more environment friendly than the twentieth. Among the documents signed was the Convention for Biological Diversity.

To everyone's surprise, the UK Government seemed to take all this seriously. Civil servants in the then Department of Environment took up their pens and, by the start of 1994 (which by Whitehall standards is almost reckless haste), had produced a quartet of White Papers. The first three, on sustainable development and climate change, need not detain us. The fourth was the beginnings of the UK Biodiversity Action Plan. At this stage it was more of a review than a plan, concerned mainly with itemising current policies and procedures in education, science and nature conservation that related to biodiversity. But it did set Government a goal: 'to conserve and enhance biological diversity within the UK and to contribute to the conservation of global biodiversity through all appropriate mechanisms'. Specifically, it would produce action plans for a wide range of species and habitats that had either declined or in some way incarnated environmental quality. In the usual way, Government set up a steering group, composed of officials from government departments and agencies, and also representatives from scientific institutions and voluntary bodies. Meanwhile a consortium of the voluntary bodies (RSPB, the Wildlife Trusts, WWF-UK, Friends of the Earth, Plantlife, Butterfly Conservation), gingered up at Rio by a mini-conference of their own, threw down the gauntlet by publishing their own ideas just ahead of the Government's. The document was called *Biodiversity Challenge*.

Biodiversity Challenge is one of the milestones on the journey of the voluntary bodies from amateur natural history societies to partners in environmental policy-making. The challenge was to halt the decline of biodiversity in Britain by rescuing threatened species and habitats, and place Britain at the forefront of world conservation policy. The report drew up a formidable list of species and habitats, and set out in some detail what needed to be done, who should do it, and how we should know when we had done it. In the first place, argued Sir William Wilkinson in his foreword, 'we must have clear objectives'. The goals should be ambitious but attainable. Each mini-plan on a species or a habitat was to be confined to a ten-year period, and nailed down with costings and a 'realistic' target. Wilkinson and his team were keen to demonstrate that the voluntary bodies could be as hard-headed as any businessman when it came to investment analysis and the delivery of defined goals. They were also anxious to force the pace before the self-congratulatory glow of Rio had dimmed and people had found something else to talk about. In Britain, 'lack of information is not the major problem', claimed the voluntary bodies. They could tell us which species are threatened and in many cases state in fairly exact terms what action is needed to save them. But conservation bodies could not do this work alone: putting biodiversity planning into practice had to be a partnership, using the far greater resources of national, regional and local government, as well as private business. *Biodiversity Challenge* reminded Government that species are a good way of measuring the sustainability of your policies. If they make species die out, your policies are not sustainable. Since they are interdependent, to sustain the environment you have to sustain species. Without a Biodiversity Action Plan up and running, they concluded, the other Rio documents on sustainable development would be worthless. Cheekily, the voluntary bodies added that if Government considered their 'rather modest targets' overambitious, it should do the sums itself, and then submit them to 'public debate'.

The constant refrain of *Biodiversity Challenge* was 'ambitious but realistic'. It proposed a vast programme of work that would not only require research but oblige bodies to work together to a common programme,

We are still discovering new species. This is an overlooked native tree, service or whitty pear, *Sorbus domestica*, with its proud finders, Mark and Clare Kitchen.

instead of in the traditional watertight compartments and client groups. Behind the swarms of rare species and archipelagos of beleaguered habitats, it set out a series of broad principles, a kind of environmental Ten Commandments. All uses of 'biological resources' should henceforth be sustainable. Non-renewable resources must be shepherded and used wisely and cautiously. Damage to the environment was only justifiable where 'the benefits outweigh the environmental costs'. Important projects should be appraised environmentally as well as financially. Subsidies leading to the loss of biodiversity (for example, in much of current forestry and agriculture practice) should be withdrawn. 'Critical natural capital' must be conserved. *Biodiversity Challenge*'s optimistic supposition was that a crowded island of 58 million human inhabitants – 3.6 persons per hectare in England – can sustain most of its material aspirations without damaging the environment. A challenge indeed. But as the very first paragraph points out, at least two million out of those 58 million were members of one or more of the bodies that produced *Biodiversity Challenge*.

The Government's response, the Biodiversity Action Plan (BAP), is Britain's blueprint for conserving the full variety of natural communities and habitats together with all their constituent species. It identifies a series of objectives for conserving biodiversity and commits the Government to take '59 steps' in that direction. The Plan is essentially a digest of consensual conservation philosophy as it stood in 1994. Its underlying principles include ideas of sustainable use, of due precaution when making decisions, and the need for sound knowledge to guide policy and action. It implicitly demands policy changes in every activity dependent on natural resources – farming, fishing, energy, forestry. The Plan also aims to involve more people in conservation activities, and to contribute British know-how to Europe and elsewhere in the world. Above all it recognises that conserving biodiversity should be 'an integral part of Government programmes, policy and action' (this aspiration received its 'statutory underpinning' in law six years later, in 2000).

The BAP is stuffed with fine, hopeful words such as 'healthy', 'enhance' and 'encourage', but nevertheless is rather short on specific ideas. Many of the '59 steps' take the form of general exhortations ('encourage the regeneration of woodland') or restatements of present activity ('continue English Nature's species recovery programme'). The official BAP Steering Group leaned heavily on *Biodiversity Challenge* for information, and the bulk of its recommendations are concerned with just one of the 59 steps, though arguably the biggest one: prepare plans for threatened species and their habitats. After all, if we continued to lose wildlife despite all the fine words, then the rest of the plan would be worthless. This is what is truly radical about the BAP: it has fastened a grand edifice of sustainable landuse policy to the backs of a collection of wild plants and animals that we are hardly even aware of, as though our future depends on our ability to sustain and enhance flamingo moss, narrow-headed ant and thistle broomrape. Needless to say, those of us that know and cherish these obscure things were overjoyed.

The Steering Group, charged with taking the plan forward, published its first 'tranche' of action plans commendably quickly, in 1995. Most of its 116 Species Action Plans (SAPs) and 14 Habitat Action Plans (HAPs) were cherry-picked from *Biodiversity Challenge* and then worked up by the JNCC. To qualify, a species had to be rare ('found in fewer than 15 ten-kilometre grid squares'), or fast declining ('by more than 25 per cent in the past 25 years') or for Britain to contain a quarter or more of the world population. Legally protected species would also qualify. The action plans are written to a formula. They are brief – a couple of pages and a map – and so necessarily generalised. They outline the 'current status' of a species, suggest why it is rare or declining, what has been done to conserve it and what needs doing. The rest of the plan farms out the proposed work to different 'lead agencies' and assigns a target for the first ten years. For example, the plan for the nightjar is to increase the number of 'churring males' from about 3,400 to 4,000, and begin to re-establish the bird in parts of its former range. To achieve even this limited objective would require the restoration of heathland, enough clearings in large forests and, more vaguely, 'promoting extensive agriculture systems in the wider countryside'. Everybody is given a job: Forest Enterprise must bear in mind the nightjar 'when considering felling and restocking proposals', the agriculture departments must 'support extensive low-intensity agricultural systems', English Nature must broaden its wildlife grants to cover 'key areas of lowland heathland', and the JNCC is made responsible for monitoring nightjar numbers. While much of the plan is kept nice and vague, a zealous interpretation would imply quite far-reaching changes. And all this just for the nightjar. Fortunately, efforts made on behalf of the nightjar should also benefit the woodlark and other BAP species.

Species bias: the proportion of species with action plans
(Source – Biodiversity: The UK Action Plan/JNCC Annual Report '96/'97). We have not yet made any plans for conserving species of protozoa, bacteria or viruses.

Taxonomic group	Approx no. of native species in UK	No. of SAPs published 1995–2000	Percentage of species with SAPs
Flowering plants and stoneworts	1,500	67	4.5
Lichens	1,500	32	2.1
Bryophytes	1,000	44	4.4
Ferns and fern-allies	80	4	5.0
Fungi	15,000	13	0.1
Insects	22,500	163	0.7
Arthropods other than insects	3,500	5	0.14
Non-arthropod invertebrates	4,000	15	0.4
Freshwater fish	38	5	13.2
Amphibians	6	3	50.0
Reptiles	6	1	16.7
Birds	390	26	6.7
Mammals	48	11	22.9
Total	**49,568**	**389**	**0.78**

UK Biodiversity Action Plan: terrestrial Habitat Action Plans

Ancient and/or species-rich hedgerows	Lowland heathland
	Lowland meadows
Aquifer-fed, naturally fluctuating water bodies	Lowland raised bog
	Lowland wood pasture and parkland
Blanket bog	Machair
Cereal field margins	Maritime cliff and slopes
Chalk rivers	Mesotrophic standing water
Coastal and flood-plain grazing marsh	Mudflats
Coastal salt marsh	Native pine woodlands
Coastal sand dunes	Purple moor-grass and rush pastures
Coastal, vegetated shingle	Reed beds
Eutrophic standing waters	Upland calcareous grassland
Fens	Upland hay meadows
Limestone pavements	Upland heathland
Lowland beech and yew woodland	Upland mixed ash woods
Lowland calcareous grassland	Upland oak wood
Lowland dry acid grassland	Wet woodland

Of course, the Biodiversity Action Plan was not the first time rare species had attracted attention. The NCC had funded research on endangered species from otters and red squirrels to snow gentians and lady's slipper orchids. The gauntlet was taken up by its successor bodies, who, to provide some extra originality, lumped the projects together in special programmes. By the 1990s the voluntary bodies had started joining in as 'lead partners'. Bats, butterflies and wild flowers now had parent conservation societies in the form of the Bat Conservation Trust, Butterfly Conservation and Plantlife, who were eager to take part, not least because Biodiversity projects gave them an expanded *raison d'être*. Similarly mammals, birds and 'herptiles' had respective active societies in the Mammal Society, RSPB and British Herpetological Society. For the less well-known orders, such as bees, lichens and snails, lead partners were harder to find. Fortunately the study of rare species in the field holds enormous appeal for many naturalists, both amateur and professional. For example, the ecologist Terry Wells took on the saving of the ribbon-leaved water-plantain, and Fred Rumsey and Mary Gibby of the Natural History Museum the Killarney fern. Laboratory work such as DNA analysis and micropropagation was farmed out to leading museums, botanic gardens and universities. Captive rearing

Attempts to conserve the marsh fritillary butterfly have been largely unsuccessful. Decline has continued through poor management and habitat fragmentation. Releases of captive stock usually fail. Nature reserves are too few and too small. (Natural Image/ Bob Gibbons)

of wartbiters and field crickets became the responsibility of London Zoo. Biodiversity spawned a series of PhD theses on rare butterflies and moths, part-funded by the conservation agencies. Experts were fairly numerous. A species like the oblong woodsia fern had a whole group of enthusiasts hatching plans to strengthen existing sites and bring it back to old ones. Cash is harder to find. The then Environment Minister John Gummer's hope was that private businesses would rush to champion endangered species in order to advertise their environmental credentials. But relatively few did. At the launch of the first tranche of plans, Gummer could only chalk up a handful of private sponsors. It is noticeable that, while private funds could sometimes be found for skylarks (Tesco) or butterflies (ICI, Wessex Water), it was hard to attract much commercial interest in tiny lichens or weevils. Center Parcs and the Sainsbury Trust have supported some of the plant projects. Perhaps the sponsorship of two water companies and the Environment Agency will cheer up the 'depressed river mussel'. But most of the money still comes from the conservation world, from the country agencies or the wealthier charities such as RSPB, and even so tends to be channelled towards the more glamorous species.

BAP 'Species Champions'

Anglian Water	Pool frog
Anglian Water/Thames Water/	Depressed river mussel
Environment Agency	
Center Parcs	Bullfinch
	Deptford pink
	Shore dock
	Convergent stonewort
	Lesser-bearded stonewort
	Slender stonewort
	Starry stonewort
	Tassel stonewort
	Great tassel stonewort
	Churchyard lichen
Co-operative Bank	Bittern
Glaxo Wellcome	Medicinal leech
Frizzell Financial Services	Song thrush
ICI	Large blue butterfly
	Pearl-bordered fritillary butterfly
Mileta Tog 24	Stag beetle
Northumberland Water	Roseate tern
Shanks McEwen	Corncrake
Tesco	Skylark
Water UK/Regional Water	Otter
Companies/Biffaward	
Fund/Fina	
Wessex Water	
	Early gentian
	Speckled footman moth
	Heath fritillary butterfly

The most important Biodiversity-related project was the Millennium Seed Bank, built at Wakehurst Place in Sussex by the Royal Botanic Gardens, Kew, with the help of the Heritage Lottery Fund and Wellcome Trust. At its heart is a bomb-proof bunker where seeds of the world's rarest plants are stored in temperature and humidity-controlled conditions. Among them is seed from most of Britain's native flowering plants, gathered by naturalists during the 1990s, and now available for approved reintroductions and related conservation activities. In the process, the technology for preserving genetic stock of wild species has improved, although a few species with big fleshy seeds are still almost impossible to store (interestingly, the field work showed that certain plants hardly ever produce fertile seed in Britain). The seed bank includes exhibitions devoted to 'the magic of seeds' as well as some beautiful 'wild gardens' and grounds, and is open to the public.

A second, larger 'tranche' of Species Action Plans was published, in four fat volumes, in 1998, 'the culmination' (reads the blurb) 'of many months of work involving Government departments and their scientists, agencies, voluntary conservation groups, owners or managers of land and academic bodies, to set more challenging-but-achievable targets for hundreds of species, large, small and minute'. This time the majority were indeed small

The Wellcome Trust Millennium Building at Wakehurst Place, home to a British and world seed bank of wild plants, opened its doors in August 2000. (Royal Botanic Gardens)

and minute. For example, there is *Cliorismia rustica*, 'a stiletto fly', which makes its home on sandy river banks 'especially where sand shoals have built up at flood level'. Or 'icy rock-moss', *Andreaea frigida*, which lives on boulders irrigated by melting snow, or 'knothole moss', *Zygodon forsteri*, which is confined to rain-tracks on large beech trees furnished from 'a reservoir held in the trunk cavity'. Even these sound quite catholic in their requirements compared with a 'pin-head lichen' called *Calicium corynellum*, which grows only on 'the damp, north-facing wall of a church tower in Northumberland': most of the British population was lost when one of the stones was replaced. For species such as these, 'action' consists mostly of finding out more about their distribution and weird and wonderful ways of life – a paid hobby for some solo lichenologist or dipterist. Probably there will never be enough funds to put all these plans into practice. As Nick Hodgetts of the JNCC points out, the plans are really 'a wish list'. For the more obscure species, survey forms the essential first step, and whether one goes beyond that step depends on whether the beast or plant really is as rare and threatened as was thought.

The annual meetings hosted by the country agencies to review progress on biodiversity were highlights of the conservation calendar in the 1990s. We learned, for example, that helping one species to survive often benefits others. Digging shallow pools for the natterjack toad at Woolmer Forest in Hampshire also helped the woodlark, several rare spiders and an endangered water beetle. Managing downland to create the 'small-scale mosaics' required by the wartbiter grasshopper contained useful tips for anyone looking after a grassland nature reserve. Slides of biodiversity in action suggested that conservationists are willing to contemplate fairly drastic action to save a rare species: mechanical diggers, toppling trees, burning reed stubble and bare ground raked by tyres and treads. Much of it tends towards the same end: since many rare species are confined to early seral stages, and become threatened by scrub invasion or dense swamp, saving them is a constant battle against natural succession. Biodiversity action encourages nature reserve managers to do what perhaps they should have done in the first place, to create and maintain ponds, sandy banks, short grass and flower glades. Behind all the details, these occasions created a buzz of excitement quite different from the formulaic aridity of the action plans themselves. With its focus on species, the biodiversity business has helped to revitalise natural history.

Let's take three examples of work in progress.

Three species

I have chosen three species (or sets of species) with their own action plans to illustrate how these work out in practice. The bittern has one of the biggest budgets of any species, and has become something of a masthead for the whole BAP. The pearl mussel has little hope of attracting the millions spent on bitterns, but in Wilsonian terms it may be a more significant species. Not only does Scotland now hold a sizeable proportion of its collapsed world range, but the mussel may be a 'keystone' species without which a whole clear-water ecosystem may find itself in trouble. Finally there are the stipitate hydnoids or tooth-fungi, a group of little-known species that, among other things, illustrate the problems of planning on less than adequate information. Between them they pull into sharper focus some of the strengths and limitations of Species Action Plans in practice.

A boom in the reeds

The bittern is a stocky, brown-streaked bird about the size of a small goose. It lives almost entirely on fish, caught in pools and ditches within dense reed beds. It is a shy bird. At the approach of a human being, a bittern will go rigid, swaying as the wind catches the reeds, with its long neck extended and glaring eyes swivelled forwards. Since they are the same pale brownish colour as the reed stems and fly only with reluctance, bitterns are rarely seen. For that reason, and also to avoid trampling the reeds, bittern surveys are based on calling male birds (although this has resulted in overoptimistic counts, since bitterns move around and different calling locations may belong to the same bird!). Experienced researchers can recognise

Skulking in the reeds: the bittern's hunched, seemingly dejected posture is justified by its recent history in Britain. (HarperCollins Publishers)

individual birds by subtle variations in their eerie, booming call, which fen men called a 'bottle-bump'.

The bittern has had a troubled history in Britain. In times gone by they were common enough to appear on the menu at lordly feasts. But their reed-bed habitat was much reduced when the fens were drained for agriculture. Persecution from shooting, egg collecting and the plumage trade even led to the temporary loss of the bittern at the end of the nineteenth century. By the 1950s we had about 80 calling birds, most of them in East Anglia. But then numbers began to fall again: fewer than 50 in 1976, around 25 by 1987, no more than a dozen or so by the early 1990s. What was going wrong for the bittern? Pesticides probably contributed to the initial decline, but the birds went on declining even after the worst pesticides were banned or restricted. The probable reason was staring us in the face in nearly every nature reserve: reed beds were drying out and becoming scrub-invaded. The open reed beds with their fish-filled ditches and pools had been maintained by reed-cutters. Wet conditions are needed to grow good quality reed for thatching, and the cutter also needs a network of broad ditches to navigate his punt. Every time a reed-cutter went out of business another set of reed beds fell into disrepair. Conservationists did their best to step into their shoes, but they seldom did enough. Essentially the poor bittern was running out of places to find fish.

As one of the most endangered British birds, the bittern was an obvious candidate for the first tranche of Species Action Plans in 1995. The circumstances were exceptionally auspicious. Some of its strongest sites were nature reserves. The 'lead partner' – naturally enough the RSPB – could bring enviable resources and experience to bear on the project, and workers could also draw on the considerable Dutch expertise in creating wetlands. It also helped that the bittern's habitat was itself a conservation priority. A further significant factor was prestige. While Britain adds virtually nothing to world bittern numbers – we hold only 0.2 per cent of Europe's bitterns – British bitterns are nonetheless the best studied in the world, and so we can at least offer expertise. British bittern experts have advised wildlife managers from more bittern-rich EC countries, including Spain, Germany and France. RSPB could not afford to let such a famous bird die out (lest a million angry bird-lovers cancel their memberships). For these reasons, the RSPB went all out to save the bittern.

The battle was fought on two fronts: improving habitat conditions on nature reserves, and creating new wetlands that might attract a passing bittern or two. Creating new and better reed beds is expensive, often requiring mechanical diggers, planting and elaborate ways of managing water, not to mention land-purchase and catchment safeguard. On the other hand, once the ground is wet enough, reeds spread quickly, springing up like rice paddies, often from rhizomes buried in the peat. Part of the RSPB's own effort went into restoring the reed beds at its famous reserve at Minsmere on the Suffolk coast, the main stronghold of British bitterns, by excavating a series of shallow, bunded lagoons. The bittern responded reasonably well. In 2000, at least seven hen-birds were present in the breeding season, compared with only one in 1990, and Minsmere is once again somewhere where you have a reasonable chance of spotting a bittern.

The plan is also to increase the natural range of bitterns by creating more reed beds. Habitat creation, especially of wetlands, is popular with potential sponsors, for creation and 'enhancement' are more exciting words than mere maintenance. English Nature part-funded much of the work on nature reserves, but substantial grants, amounting to millions of pounds, have also come from the Heritage Lottery Fund, the European Union's Life project and from private businesses. Old reed beds within the bittern's former range in North Wales are being restored. More reed beds have been created on derelict land, on former opencast coalmines in Montgomery and Northumberland, and in gravel pits at Attenborough, Nottinghamshire. At Wicken Fen, some 70 hectares of reed bed and water are being created in the hope of attracting just a single pair. At Needingworth near St Ives on the edge of the Fens, Hanson Aggregates are to help the RSPB transform a further 1,000 hectares of gravel pits into wetlands over the next 30 years.

This action plan seems to be working. The overall number of booming males showed only a modest increase from 18 to 24 birds between 1990 and 2000 (see below), after a setback in winter 1996–97 when roughly half of our bitterns died of cold or starvation, or emigrated. The BAP target is 50 birds by 2010. This seems fairly realistic. In *The State of the Nation's Birds* (2000), Chris Mead rates the bird's prospects as good, so long as key reed beds are kept in good repair and there are not too many hard winters. The bittern has also been a useful figurehead for wild wetlands, its well-known stooped, beaky profile attracting funds less likely to be forthcoming for copepods or stoneworts. It also provided a stimulus for conservationists to abandon in-built caution and adopt drastic measures – and they were drastic. It probably needs good nerves – and good public relations – to temporarily convert parts of a famous nature reserve to the appearance of smoking blackboards, or to watch 20-tonne diggers play mud pies with prime habitat. Drastic measures may be needed to undo decades of neglect, but, given the completeness of our reliance on machines, they may also be the only ones available.

'Booming Bitterns' in the UK 1990–2000 (RSPB figures)

	1990	1991	1992	1993	1994	1995	1996	1997	1998	1999	2000
No. of booming males	18(20)	16(17)	18(19)	15(17)	15(16)	19(2)	22	11(12)	13(18)	19(22)	24(30)
No. of sites	11	12	10(11)	9(11)	10	10(11)	10	7 (8)	9(12)	11(14)	14(16)

NB: the figures are conservative; those in brackets are maximum estimates.

Pearls in the river

The freshwater pearl mussel, *Margaritifera margaritifera*, is a quiet, unobtrusive animal that spends its adult life half-buried in gravel in shallow-water riffles in the beds of fast, clean rivers. It looks a little like a dark Roman

Left: Living pearl
mussels in the bed of a
Welsh river. Below left:
A large mussel, 15
centimetres long and
several decades old.
(Anna Holmes/
National Museums &
Galleries of Wales)

amphora sticking out of a submerged wreck, the brim and handle formed
by the extruding siphons by which the mussel breathes and filters suspended matter from the current, extracting anything edible. The pearl
mussel is one of the few species of British animal that may be older than
you. They can live for over half a century, and centenarian mussels are not
unknown. A single mussel can filter about 50 litres of river water every day.
A recent study of a related species in Belgium estimated that the colony fil-
tered water equal to that of a purification plant for a city of 150,000 people (Oliver 2001). What interested our ancestors, however, was not their
ecology, but the fat pearls that mussels occasionally produce when
annoyed by a parasite or a bit of grit. Unfortunately, fishing for pearls
means killing a lot of mussels: you need to open dozens, maybe hundreds,
to find a pearl. Even so, they were plentiful enough to support a sizeable,
and presumably sustainable, trade (the pearly costumes of Tudor kings
and queens owed a lot to *Margaritifera margaritifera*). In 1861, a German
dealer offered to buy all the British freshwater pearls that could be found.
Fifteen years later the British pearl fishery had collapsed. A few hardy souls
continued looking for pearls in the diminishing number of suitable rivers,

The River Dee near Braemar: one of the last strongholds of the freshwater pearl mussel.

such as the Tay or the Spey, but in 1998 the freshwater pearl mussel was legally protected, necessitating a regrettable end to an ancient trade (Cosgrove et al. 2000).

The freshwater pearl mussel is considered to be endangered worldwide. In England and Wales, it is almost extinct – but nonetheless still poached by would-be pearl hunters: a pile of shells photographed recently 'represented 64,000 years worth of growth'. The last healthy population in Wales, in the Afon Ddu in Snowdonia, was casually extirpated by a land drainage operation in 1997. The clear, soft-water rivers of Scotland are now a world stronghold for the species, but even here serious decline has set in. Out of 155 rivers known to produce pearl mussels in 1990, they survive in only 52, and are still common in only ten. What is killing off the pearl mussel? Pearl fishing is only partly to blame. Much more difficult to tackle are fundamental changes to its environment. Mussels need clear, well-oxygenated water. They are vulnerable to nutrient enrichment, which reduces oxygen and creates silt, which clogs up their siphons and buries them under a suffocating blanket of mud. River engineering also takes its toll by destroying mussel-rich shoals of gravel whenever the river is straightened or deepened. Pearl mussels do not like floods either, which is unfortunate because since the late 1980s Scottish rivers are flooding more often, and depositing yet more silt on the river bed. Mussels also depend on healthy stocks of salmon and trout. Tiny baby mussels drifting in the plankton survive by clamping onto the gills of these fish, where they feed on blood and mucus for a few months, before letting go and starting a new life on the river bed.

This behaviour helps, of course, to disperse the mussels. It does not seem to harm the fish, but the falling numbers of salmon and brown trout in our rivers is certainly harming the mussel. Investigators have found ominously few juvenile mussels; superficially healthy looking colonies turn out on closer inspection to consist mainly of geriatric individuals. Without recruitment they are doomed. And without thousands of pearl mussels in their gravel beds, each cleaning and filtering their 50 litres of river water every day, what happens to the water quality?

So what is to be done about it? The freshwater pearl mussel is listed on annexes and appendices of various international conventions and is now protected in British law. Of course, protecting a species without protecting its habitat does not help it much. It does, however, ensure the mussel a place in the BAP. In 1995, a Species Action Plan was prepared for *Margaritifera margaritifera*, which, as usual, itemised a programme intended to lead to recovery. This is what it amounts to:

• Identify the mussel's water quality requirements, and 'seek to ensure that these form the basis for setting Statutory Water Quality objectives' for mussel sites. (*In practice, this means monitoring water quality, which is a process, not preventative action.*)
• 'Seek to ensure that catchment management plans' and other relevant plans 'take account of the mussel'. It is hoped that local biodiversity plans will take this forward by 'encouraging favourable catchment management'. (*A bit vague?*)
• Designate the best sites as SSSIs and/or SACs. (*This doesn't necessarily help the mussel. It is dying out on SSSIs too.*)
• Consider reintroducing them to former sites. (*But we do not know how to do this. Every reintroduction attempt so far has been a complete failure.*)
• Provide advice to water bailiffs and other water folk and encourage the police to target mussel poachers. (*But mussel poachers are not the main problem.*)

This rather hopeless list of prescriptions illustrates both the strength and weaknesses of Species Action Plans. On the positive side, they make resources available – in this case, money for survey and research, by which means a lot was learned about the status of pearl mussel and its life-threatening problems through the efforts of Dr Mark Young and his team at Aberdeen University, and Graham Oliver at the National Museum of Wales. Without grants from conservation agencies there would be no research on threatened species in Britain, and without such research universities will lose touch with natural history. On the other hand, it is one thing to elucidate the problems, but quite another to solve them. The SAP for the pearl mussel seeks recovery in bureaucratic terms – more plans, policies and designations – and better policing. But these are all means to a solution without actually being one. The catch-22 is that no plan will work unless the fundamental causes – in this case low fish stocks and dirty water – are addressed, and if they were we would probably not need a plan at all; the mussel would go on looking after itself, as it has managed to do unaided since long before *Homo sapiens* invaded its watery territory.

Fungi with teeth

Despite their huge diversity and fundamental importance, fungi receive little attention from British conservationists. This may be because experienced mycologists are not, in general, much interested in conservation,
while most conservationists are distinctly unenthused by mycology. They
are rarely mentioned in formal descriptions of SSSIs except in the most
general terms. As yet there is no Atlas of British Fungi, nor an accepted
Red Data list, nor is the conservation of fungi high on the priorities of any
research institution. There is not even a comprehensive check list of
species: the last one was published in 1923 (another is currently in preparation). Yet fungi are incredibly numerous – as the New Naturalist volume
Plant Diseases has shown, there are all kinds of pathogenic and microscopic fungi as well as mushrooms and toadstools – more species in fact than
all the wild animals, birds, flowers and mosses put together (an average
wood may have three times as many *large* fungi as wild flowers; over 3,000
species of fungi have been found in less than a square mile of Esher
Common in Surrey). Without fungi every land ecosystem would collapse;
trees would die, soil would choke on its own rubbish and the land would
pile up with debris and dead animals. We would probably get by for a while
through our reliance on chemicals, but everything else would die out.

In many other European countries, by contrast, fungi are seen as quite
as important as flowers or ferns. Conservation in Poland, Denmark,
Germany, Holland and the Scandinavian countries is years ahead of
Britain, with fungal mapping and monitoring schemes in place, as well as
databases, red lists and university research. Probably the reason is that
fungi have more commercial importance in these countries than here,
and so there are regulations to maintain wild stocks. In Italy, for example,
elaborate local laws deter 'poachers' from stealing the villagers' truffles
and *porcini*. In Holland the decline of the *girolle* or chanterelle has been
monitored and mapped, and the reasons adduced from scientific study.
But in Britain we do not have the slightest idea whether the numbers of
truffles, chanterelles or any other fungi are going up, down or staying the
same.

The Biodiversity Action Plan put a few rare fungi in the spotlight for the
first time in Britain. After a good deal of list-shifting based on very inadequate data, some 40 species were singled out for conservation action.
There was a strong element of tokenism in their choice. The chosen few
are mostly large and attractive, relatively easy to recognise and, in some
cases, are rare throughout northern Europe. Indeed some were selected
because they are rare in *other countries*. The largest group of BAP fungi are
the tooth-fungi or stipitate hydnoids (= stalked hedgehog fungi). As the
name implies, they have teeth or spines where other mushroom-shaped
fungi have gills or pores. Most of the 15 species covered by the BAP do not
have accepted English names, although a recent handbook tried to offer
us 'Concrescent Corky Spine Fungus' and 'Green-footed Spiny Cap-
Fungus'. Even so, they are quite attractive, with their range of funny
shapes, contrasting colours, and, for some species, blood-like blobs of red

A 'stipitate hydnoid', *Phellodon melaleucus* or 'the black and white scented spine-fungus'. It cannot compete successfully with higher plants and requires soil of low natural fertility.

juice. Some also have the agreeable habit of growing like melting wax, engulfing twigs and bits of grass as they expand.

The Species Action Plan thinks it will cost about £55,000 to find out something worthwhile about the stipitate hydnoids. Most of the money has been stumped up by English Nature and Scottish Natural Heritage. In Scotland, a small team of enthusiasts led by Adrian Newton at the University of Edinburgh has systematically mapped their distribution and found out more about their habitat preferences. In England, surveys have been carried out in the New Forest and the Windsor area, the main strongholds of stipitate hydnoids in the south, augmented by a desk survey by myself, pulling together what is known about them, and what needs to be known before we can conserve them properly (Mistakes will happen. The best site in the New Forest was damaged by felling in 2001). The Plan also speaks of *ex situ* cultivation of one stalactite-like species, *Hericium erinaceum*, to produce material for translocation attempts (the BAP shows a faith in translocations and reintroductions that is not often justified by results).

Where has all this survey taken us so far? First, we discovered that none of the 15 are as rare as they were thought to be. This is interesting: their red-listing was based on their status in the Netherlands and other European countries, not in Britain. We also have a much better understanding of the kinds of places they like: wood banks, the sides of tracks, stream banks, even old sandpits, usually in long-established woodland of oak, sweet chestnut and Scots pine, often, but not always, on free-draining, acidic soils. It seems they like rather arid soil, where there is less competition from grasses and other tall vegetation. They always occur near trees,

but not necessarily mature trees. It may be that they survive because areas
on light infertile soils were often preserved as woods, parks or commons.
Their relatively healthy state in the Scottish Highlands mirrors that of
Norway and Sweden, where they are also still frequent.

Low-power, low-budget survey work like this often produces incidental,
unplanned benefits that may outweigh the original aims of the project. In
this case, it gives Britain's growing band of competent field mycologists a
project to get their teeth into (not meant literally, though many of us do
indeed enjoy the comestible side of our hobby). It also flags up a group of
fungi that tell us things about fungi in general, for example, that mycor-
rhizal species (those that grow on the roots of plants, especially trees) are
sensitive to chemical pollution, especially active nitrogen. They also form
another good reason for preserving wood banks and natural woodland
soils, and restoring woods infested with rhododendron. A further benefit
is that it helps to build bridges between the hitherto separate worlds of
mycology and conservation, and may even help to secure a higher priority
for mycology in scientific institutions and conservation work. Perhaps the
'concrescent corky spine fungus' may one day be as familiar as a prickly
pear or a hedgehog.

Biodiversity replicated

The 1990s saw a considerable outbreak of biodiversity planning, particularly among local partnerships, where local knowledge and skills help 'to
deliver national biodiversity targets'. These are known as Local Biodiversity
Action Plans or LBAPs. The lead has been taken by planning departments,
with the help of the conservation agencies, and often involving others in a
formal partnership. For example, the LBAP for North-east Scotland enjoys
the support of a government department, three local councils, the
National Farmers' Union of Scotland, Aberdeen University and 'an education consultant', among others. It is 'a locally-driven process' with a slogan:
'*people achieving local action for their local wildlife*'. What it boils down to is a
commitment by communities to look after their wildlife, with special reference to species and habitats that contribute to the local character. The
plans highlight an area's special wildlife and landscape qualities, and the
action needed to conserve them. For example, the Leeds LBAP recognises
the value of its remaining hedgerows and field margins ('a critically important habitat type for many once-common birds'), reed beds and magnesian
limestone grassland ('only a handful of small, isolated fragments left'),
together with their threatened bats, crayfish, pasqueflowers and harvest
mice. The Edinburgh LBAP proposes special action for swifts ('prepare
development-control guidance on provision of nest spaces in new development') and cornflowers ('include cornflower in specifications of seed-
mixes for annual meadows created as a condition of new developments').
Such action is largely voluntary, although it could be, and in some cases has
been, made a planning condition for new developments. In the view of the
Commons Committee reviewing Biodiversity progress in 1999, local
wildlife sites should be registered, and planning guidelines revised to

ensure that 'there is a general presumption against development on these sites unless no suitable alternative can be found'. Many LBAPs are accompanied by glossy booklets with enticing pictures of wild flowers and healthy habitats – lessons in local biogeography, with a display of 'biological correctness' thrown in.

Are Biodiversity Action Plans working?

By 2000, 436 species action plans had been prepared. However, only a minority had been put into practice, and it is too early to judge their overall effectiveness. For the record, however, the Biodiversity Challenge Group assessed the interim results as follows:

The status of BAP habitats and species

	Habitat plans	Species Plans
Insufficient information	8	17
Declining	1	11
No change	0	13
Signs of recovery	5	17
Recovered (i.e. targets met)	0	0
'Lost'	0	1

The lost species is Ivell's sea anemone, which is presumed extinct in Britain, a species that just missed the Biodiversity bus.

The fact that not a single target has yet been reached (2001) suggests that, even with the right kind of action, species and habitats take time to recover. Most 'signs of recovery' so far were due not to a real increase, but to better survey effort. For example, the early gentian, *Gentianella anglica*, is deemed on the way to recovery not because of what anybody did, but because several large new colonies turned up in the course of investigation. Interestingly, recent molecular analysis at Cambridge suggested that the early gentian is not a species at all, but an endemic early flowering form of the common autumn gentian (though whether that means it is not worth bothering about any more, I am not sure). It has proved easier to help species restricted to a particular habitat or geographical area, such as stone curlew or cirl bunting, than widespread ones such as skylark or song thrush. You can target cirl buntings with a special agri-environment scheme (see Chapter 6), but to stop skylarks declining you need to address fundamental agricultural policy.

Conservationists have tended to talk up the BAP in non-scientific terms. It has kept conservation groups focused on goals. It has helped to promote the health of the natural environment as a national goal, comparable with health and education. It involves parts of the community hitherto on the fringe or looking the other way, such as public utilities and businesses. It has provided a new focus to the way the planning system addresses wildlife and nature conservation issues. In their review, *Biodiversity Counts*, the voluntary bodies claim that Biodiversity delivers a better 'quality of life' by promoting environment-friendly policies: 'The greatest challenge is [in]

The relatively few species confined to Britain have evolved in isolation, probably
during the past 10,000 years. This means they are only slightly different to their closest
relatives. This is one of them: Marshal's eyebright, *Euphrasia marshalii,* confined to the
coast of northern Scotland and so vulnerable to climate warming. Should the
preservation of endemics be a conservation priority? (Natural Image/Bob Gibbons)

greening government policy: for access, the economy of rural areas and
people's health and quality of life.' By means of a hive of talk-shops and a
mountain of plans, the Biodiversity industry has capitalised on public opinion, brought nature conservation more firmly into the national agenda,
and produced more co-operative ways of working.

Is there a downside? According to *Biodiversity Counts,* some departments
and agencies have engaged with the BAP more than others. High marks
are awarded to the Environment Agency, English Nature, the Forestry
Commission and environment departments in Scotland and England. On
the other hand, cabbages are offered to agriculture departments, the water
regulator Ofwat, NERC and the Home Office for their lack of zeal. More
importantly, perhaps, the BAP has so far shown only limited benefits for
wildlife. It has not yet rescued a single species, nor has it stopped further
declines from happening, even among much-loved animals such as the skylark or the water vole. Another, rather obvious concern is that Biodiversity
may soon sink under its own bureaucratic weight. Some fear the process is
promoting a species-centred approach to nature conservation which is
wasteful and logistically daunting. Taken to its logical extreme, the BAP

Threats to listed Biodiversity habitats (from *Biodiversity Counts* (2001))

Type of threat	Number of habitats threatened
Land use change	
1. Habitat destruction	31
2. Agricultural intensification	27
3. Lack of appropriate management	21
4. Habitat fragmentation	17
5. Water abstraction, drainage or inappropriate river management	16
6. Coastal development and management	16
7. Changes in agricultural management	13
8. Afforestation	13
9. Changes to woodland management	5
Environmental pollution	
10. Climate change/sea level rise	19
11. Water pollution	18
12. Air pollution	13
Other	
13. Recreational pressure	13
14. Fisheries management	13

would require hundreds of intensively managed nature reserves, and intrusive means such as translocations which, if widely adopted, would blur the boundaries between conservation and gardening. It is possible to argue that conservation BAP-style rather misses the point. In a long-farmed environment like Britain, most species have adapted to traditional forms of human husbandry and harvest. There would be no need to address the

The skylark: not only a biodiversity target species but a government indicator of the quality of *our* lives. (Nature Photographers Ltd)

plight of individual beetles or mosses if we managed the land sustainably and did not pollute the environment. By trying to help some declining beetle or bug, we are therefore addressing a symptom rather than the cause. Biodiversity planning appeals to those, perhaps the majority of conservationists, who are wrapped up in the *means* to an end and feel at home in the planner's world of plans, targets and zones. What one misses in all the literature devoted to it is any readiness to stand back from the details and look at where this self-replicating mountain of plans may be taking us.

12

Sea Eagles and Parrot's Feathers:
Invading and Settling

This chapter is about newcomer species in Britain. As an island, Britain's terrestrial wildlife is isolated, although it is visited regularly by long-distance migrants, mainly birds, but also some insects with a powerful flight. Wild plants seem to find it impossible to bridge the channel. Most new species are therefore introduced by humans, usually by accident. Urban and artificial habitats, such as quarries or railway lines, offer open, sheltered conditions not easily found in nature, and often acquire a cosmopolitan flora of wild plants. New species are interesting, and create a dynamic in British wildlife, epitomised by the spread of Oxford ragwort, which followed the railways and thereby colonised the towns. Problems arise when species invade the countryside and start taking over, reducing natural diversity by outcompeting other plants. Later in this chapter I describe some of the plants and invertebrates that create conservation problems, as well as the ongoing activities of introduced animals such as the American mink. First, however, let us deal with an entirely different kind of introduction, by invitation as it were, of animals and birds that were once widespread in Britain, but which, for one reason or another, died out or were reduced to a tiny, endangered population. The stories of the carefully planned reintroduction of sea eagle and red kite, the accidental reintroduction of wild boar and the ongoing, slow-moving tale of the beaver, are interesting in themselves, but they also, I think, cast light on our relationship with wild animals. We simultaneously seem to want them to be wild and free, and yet also under our control. In the case of the sea eagle, our care at times became almost parental. Conservationists have been less celebratory about the wild boar, perhaps because it has escaped their control. Reintroductions are popular, and we are likely to see more of them, maybe including long-lost beasts such as moose and lynx, or even reconstituted wild oxen (perhaps with modified genes for increased placidity?).

Successful reintroductions: sea eagles and red kites

The naturalist Robert Gray, writing in *The Birds of the West of Scotland* in 1871, considered the white-tailed or sea eagle to be 'a much commoner bird than the golden eagle', and one that had 'never been at any time in the same danger of extinction'. On the second point he was, unfortunately, about to be proved wrong. Within little more than a generation, the sea eagle no longer bred in Britain. Magnificent though it was, the sea eagle was, like all predators, vermin. It was less shy and so easier to shoot than the golden eagle. One Skye laird hired a man to kill as many eagles as he could by luring them to the guns using sheep carcasses. Others sought their nests on beetling cliffs and stole their eggs. The last known breeding attempt by native sea eagles was on Skye in 1916. The last bird, an ageing, pale-feathered female, was shot on Shetland two years later.

The sea eagle is a tremendous bird, the largest raptor in northern Europe, with broad, vulture-like wings and a long, powerful bill. It is noisier than the generally silent golden eagle and noted for its raucous yapping in the breeding season. Its ornithologically-correct name, the white-tailed eagle, commemorates the distinctive white tail feathers of older birds. Sea eagles are now rare worldwide; the only European country to have them in healthy numbers is Norway, where the great birds build their vast nests on the cliffs and trees lining the fjords. There they can be still seen as one nineteenth-century ornithologist remembered them, 'watching the water, sometimes sunning [their] outstretched wings like a cormorant' (Coward 1920).

In 1959, and again in 1968, ornithologists released young Norwegian sea eagles at Glen Etive in Argyll and Fair Isle. These attempts at reintroduction were unsuccessful, probably because they involved too few birds. In 1975, the NCC decided to use their island nature reserve, Rum, as the base for a planned reintroduction programme, involving many more young birds from Norway's apparently inexhaustible supply. This was justified on the grounds that the west coast was still suitable for breeding sea eagles, and that it was human persecution, not natural change, that had killed them off. It was thought unlikely that the bird would ever recolonise naturally, for since 1918 the bird has been a very infrequent visitor to Britain. Besides, a successful 'Operation Sea-eagle' (as it was named) would be a great coup for those involved. Although he was initially opposed to the idea, Morton Boyd came to see it as 'a much-needed banner for nature conservation – an example of hands-on work on an epic scale' (Boyd 1999). Against internal resistance – some members of the NCC's Scotland committee were worried that it might annoy sheep farmers, while others

John Love releasing
a young sea eagle
on the island of
Rum in the late
1970s. (NCC)

were against reintroduction 'stunts' on principle – the NCC decided to go ahead. In June 1978 Boyd was at Festvag on the Mistfjord near the Arctic Circle, with Harald Misund, the Norwegian ornithologist who collected most of the chicks: 'Harald knew this eyrie and while he climbed to it using my shoulders and secured an eaglet into a canvas bag, I had time to watch the distraught parents plaintively calling with a high pitch '*kak-kak-kak* ... *kak-kak-kak*', circling against the snow-capped mountains of Nordland ... The young bird secured, the bag was handed down to me, and I descended with the precious load to the waiting boat'. In all, eight young eaglets were brought back to RAF Kinloss that year, crated up inside a Nimrod, before being transferred to Rum by Land Rover and then ferry. John Love, whose job it was to rear the eaglets until they were old enough for release, could not resist giving them names: Keiran, Fingal and Conon were the males; Risga, Danna, Ulva, Aida and Shona the females.

Between 1975 and 1985, a total of 82 sea eagle chicks were exported from Norway, to be reared, tagged and released. Food – mainly locally caught fish – was left out for them until they became fully independent and began to disperse. Several birds from the earlier batches perished prematurely, although only one died in captivity, a tribute to John Love's skill and care. At least one was poisoned with phosdrin, its body found close to the baited carcass of a hare; another collided with an overhead cable. Operation Sea-eagle was presumably a fairly stressful experience for the birds themselves: the deaths suggested that the birds were struggling to survive. It is hard to know what goes on in the consciousness of an eagle, but a large degree of bewilderment would, in the circumstances, be understandable. But the welfare aspects of the project were played down: in conservation it is numbers that count; one mourned a setback rather than a dead bird.

When the BBC was making its film *Return of the Sea Eagle* in 1981, the older birds 'were beginning to adopt territories over a wide area of the Inner Hebrides' from Mull to Skye (oddly enough, they all deserted Rum in the end). Two years later, three pairs had built nests. Interestingly, they were not very good at it. The first pair to nest on Skye made a particularly feeble effort from clumps of grass and wool, and others chose precarious places where there was no hope of the nest lasting more than a season. Once they started to lay, eggs were crushed as the birds clumsily changed places on the nest, and where eaglets did emerge they sometimes starved. Well-meaning ornithologists, watching this domestic disarray, tried to help by rebuilding the nests, supplying boiled duck eggs to give the learner eagles some incubating experience, and, in at least one case, clambering up the tree to stuff food down the throat of the starving chick. Evidently sea eagles need time to get things right! Not a single baby sea eagle was reared successfully until ten years after the start of the project. It is tempting to suppose that captive-reared sea eagles lack the natural savvy of their wild Norwegian parents (though carelessness with nest-building seems to be common with sea eagles). Some of the birds were also confused about their sex lives: two females were spotted vying for the attention of a single male, while another male tried cohabiting with a golden eagle. Experts feared that the original release programme might not be sufficient to maintain a viable sea eagle population in Scotland, and so a further 58 Norwegian eaglets were released on the mainland between 1993 and 1998. As a result, by 2000 there were 23 breeding pairs of sea eagles along the west coast, mainly in Argyll and West Inverness, which have reared 100 chicks between them. One pair on Skye had alone reared 21 chicks over the past ten or so years. Immature birds range more widely, and have been seen in places as far apart as Shetland and Northern Ireland. Most have chosen to nest on cliff ledges, generally low down, but a few have opted for trees. Fortunately the sea eagles have shown themselves capable of coexistence with golden eagles, and even of taking over their ranges (Nellist & Crane 2001).

Was it worth all the effort? Boyd certainly thought so: to him, it represented 'a tangible piece of positive practical conservation in a life burdened with negative bureaucracy and political manoeuvrings'. Probably most birdwatchers would agree with him, if not for quite the same reasons: a sea eagle soaring over the lochs and isles is a sight to gladden the heart. Operation Sea-eagle made a small contribution to saving the species by extending its range, and its success has encouraged similar reintroduction experiments both at home and in Europe, most notably the spectacularly successful Red Kite Project (see below). Perhaps even more importantly, breeding sea eagles have the makings of a tourist attraction. In 2000 a viewing area overlooking a sea eagle's nest on Mull was opened, and live close-circuit television pictures of another nest on Skye are beamed straight to the heritage centre in Portree. Even closer views are possible when sea eagles follow fishing boats, tempted by scraps thrown to them. The possibility of eyeballing Europe's largest eagle, with barn-door wings exceeding two metres in span, is sure to set more wildlife seekers on the road to the Isles.

Returning the red kite to England and Scotland was, by comparison with Operation Sea-eagle, speedy, efficient and successful. Everyone involved was amazed at how easy it was. The kite was 'one of the very few bird species which fulfil all the criteria for reintroduction', declared the NCC and RSPB. The bird has a relatively restricted world range, mainly in Spain, Germany, France and Sweden, and so a recovery in Britain would contribute to its conservation as a species. Once widespread – it was a familiar town bird in Shakespeare's England – persecution had since driven the red kite to the upland fastnesses of Wales. It is a predator of young domestic fowls and game birds, which made the kite unpopular on country estates. They were wiped out, it seems, with ruthless efficiency. By the 1830s, the kite 'was gone', wrote Morrey Salmon, 'before anyone thought to record its disappearance'. It survived, just, in southern and central Wales largely because there was little interest in game, and therefore no keepers. Although the bird was targeted by egg collectors, it was one of their number, Professor J.H. Salter, who was the leading figure in the Kite Protection Committee, who probably saved it from extinction. DNA testing has shown that until recently all Welsh birds were descended from a single female. Inbreeding may be one reason why the kite failed to recolonise England

Red kite, once on the verge of extinction in Britain, now well on the way to becoming one of our less rare birds of prey. (HarperCollins Publishers)

naturally, though it has increased its range in Wales in the last 20 years and might eventually have done so.

In 1989, when the red kite reintroduction project began, there were only about 80 breeding pairs, all in Wales. That year ten young birds, taken from their nests in Sweden, were transferred to Britain by courtesy of the RAF. This time there was no nonsense about giving them names. Contact with humans was kept to a minimum, and the young kites were left the kind of food they would find in the wild ('locally available species', including rabbit, squirrel, fox, rook, magpie and salmon) via a flap in their wooden aviary. The birds were released in batches of three to ten when about 12 weeks old, after being fitted with coloured wing-tags and a miniature radio transmitter fastened to the tail feathers. Subsequent releases consisted of birds from Spain, which were released mostly in England, and from Sweden, which were released in Scotland. By 1994, when some of the earlier releases had already begun to nest, some 186 kites had been released, half in England, half in Scotland. From 1995 yet more young birds were released in two further areas, in the English Midlands and central Scotland, with the hope of establishing 'at least five expanding population centres' of red kites in Britain.

It was only to be expected that some of the releases would end their days on a poisoned rabbit carcass, as our indigenous Welsh kites regularly do. And, indeed, two birds out of the very first release were quickly and efficiently poisoned, as were many more later on. Particularly toxic rat poisons were in use in the 1990s because rats had developed a resistance to warfarin, the traditional rodenticide. Even so, the surviving kites began to pair up. Five bred successfully in 1992, four in England, one in Scotland, producing ten young between them. From then on, their increase was almost exponential. By 1999, there were no fewer than 429 pairs of breeding kites in Britain. Wales is still the heartland, with 259, but in England and Scotland, where no red kites had bred for 150 years, there were now 131 and 39 pairs respectively. In his prospectus on British birds, Chris Mead rates their prospects as 'brilliant'. If the present 30 per cent annual increase continues, there could be 2,500 pairs breeding in Britain by the year 2010. Already kites are a familiar sight in many parts. It has been dubbed the bird success of the century.

There is no reason why red kites should not become one of the more familiar birds of prey in twenty-first century Britain. They eat almost anything meaty from rabbits, mice and birds to earthworms and beetles, including carrion. In winter they often pick over rubbish dumps, and in Wales large numbers of kites have been lured to a viewing platform by butchers' offcuts. The last time I visited Aston Rowant, one of the original release sites in the Chilterns, I saw more than a dozen red kites within about two hours, some of them at close range. There is a (surprisingly little known) viewing place close to a car park where you are practically guaranteed a kite. It seems amazing now that for many years the red kite was seen as the bird most likely to become nationally extinct. The story of its successful reintroduction is related in a colour leaflet, issued free by English Nature.

Red Kite numbers in Britain 1997–99 (figures are breeding pairs).

	1997	1998	1999
Wales	151	163	180
Chilterns	52	71	75
West Midlands	4	4	7
Northern Scotland	23	24	30
Central Scotland	–	2	4

Alternative ways of establishment: plans and escapes

Most of Britain's land mammals are smallish and good at keeping themselves hidden. In the distant past, of course, there were more large beasts. Stone Age dwellers would have known wild horses and cattle, and trapped brown bear and lynx for their fur. In the Middle Ages huntsmen pursued wild boar and wolves, and perhaps caught the last few beavers. The value of flesh and fur made life dangerous for a big mammal. Once Forest Law fell into abeyance, we even managed to thin out deer to well below present-day numbers. Until recently, there was no concerted rush to bring any of the extinct animals back. A century ago two Scottish lairds did try to introduce Canadian beavers to western Scotland, but without success. Such new animals that have become established, such as edible dormouse or muntjac, have done so through their own efforts, by escaping from their park. Some of us dream of reintroduced wolf packs roaming the Highlands, but, not surprisingly, sheep farmers are much less keen. It was only in the 1990s that serious consideration was made about bringing back a long-vanished wild mammal.

There are two kinds of beaver. Thanks to Walt Disney, the better known one is the American beaver, which builds dams, and led trappers, in pursuit of its fur, to explore the interior of North America. The European beaver is of similar size, but has paler fur and is a shy, rare beast that often lives in a burrow rather than a dam. Until recent reintroductions got underway, it was confined mainly to the former USSR, Scandinavia and parts of the valleys of the Rhone and the Elbe. Elsewhere, as in Britain, it was hunted to extinction, both for its fur (which made excellent hats) and because the Pope allowed the faithful to eat beaver tails on Friday (being scaly, they counted as fish). Apothecaries were interested in its musky secretions, on which modern aspirin is based. Beavers have an obscure history in Britain. That they did occur over here once is proven by their excavated bones, but those that have been carbon dated all belong to a relatively distant period between 7,700 and 1,160 years ago. The historical sources are equivocal – more so than reintroduction enthusiasts would have us believe. The inclusion of beavers in medieval lists of fur-bearing animals does not prove that they still occurred here, since the lists include definite non-natives, such as the sable. The chroniclers Gerald of Wales and Hector Boece, who mentioned beavers at different times, were prone to mythologise, and both admitted they had never seen one. Beavers were confused with otters; indeed the Gaelic name for beaver, *dobhran losleathan*, translates as otter with a broad tail. The most we can say is that a few

A European beaver takes a snack of young birch sprigs. We may be the first generation since the Middle Ages to see wild beavers in Britain. (Natural Image/Bob Gibbons)

beavers survived into historic times, certainly into the Saxon period and possibly through the Middle Ages to as late as the sixteenth century. But by the time people started studying wildlife there were none left. Since apparently suitable habitat for beavers still occurs, in parts of Scotland at least, it is probable that the animal was pursued to extinction by trappers.

The project to reintroduce the beaver to Britain came about through the recent vogue, encouraged by the European Union, for reintroducing lost species. The EU Habitats Directive encourages member states 'to investigate the possibility' of reintroducing extinct animals. The conditions are that suitable habitat must remain, and the introduced stock must be genetically as close as possible to the original strain. On the European mainland attempts to bolster endangered mammals such as the European bison and the Spanish lynx by captive breeding and introductions have long been on the conservation agenda. Both the European and the Canadian beaver have been naturalised in various places (sometimes the same place). Given the right environment and full protection, the European beaver can still thrive – in Europe it is now officially 'out of danger', compared with its status as 'vulnerable' only ten years ago. For example, it has been successfully reintroduced to Holland, which had been beaverless for 150 years.

Reintroducing the beaver to Scotland is being done 'by the book', that is, slowly and methodically. While someone like Roy Dennis could start reintroducing sea eagles almost at a whim, Scottish Natural Heritage has been anxious to ensure the widest possible public consultation, and to take no step without the support of the Scottish Executive. The project began in 1995 with a feasibility study, which found no reason why the beaver should not become naturalised in Scotland, as it has in several other European countries. Indeed, the study suggested that Scotland has room for an estimated thousand beavers, more than enough for a self-sustaining wild population. On ecological grounds, the proposal was sound. The next stage, completed by 1998, was to ask people likely to be affected by it what they thought of the idea. Some did not like it at all, on grounds of potential harm to salmon rivers and spawning grounds, of the risk to growing timber and of the possibility that beavers would erode river banks with their burrows or cause flooding by building dams. There was also an angry response from a prominent entomologist, fearful that the beavers would cut down too many aspen trees, noted for rare flies. WWF, on the other hand, thought that Scottish beavers would delight the tourists and help ecology by creating more habitat diversity. The wider (Scottish) public seemed to welcome the scheme.

The next step, following the feasibility study and the public consultation, was for a pilot study. The Forestry Commission volunteered a suitable site in Knapdale Forest, Argyll, where SNH proposes to release about 12 Scandinavian beavers in a large compound and see how they get on. 'The beavers will be studied to within an inch of their life,' said SNH's George Anderson. 'We're fitting them with radio-tags so we'll know what they're up to. It'll be like *The Truman Show*' (Beardsall 2001). If the necessary political support is forthcoming, the first beavers could be out of quarantine and into their Knapdale enclosure by late 2002, followed, if all goes well, by a full-scale reintroduction attempt starting in 2007. But 'if it is decided that the species should not be reintroduced to Scotland, the project would be dismantled and all the animals removed', says Scottish Natural Heritage. It all seems to be taking a long time, but they are slowly getting there.

Proposals to release beavers in England were still tentative in 2001. Five breeding pairs of Norwegian beavers kept at a wildlife centre are ear-marked for release in a nature reserve where they will live under semi-wild conditions within an electric fence and, it is hoped, contribute to its management by cutting down invasive bushes. If successful, beavers may be released on other wetlands in Kent and elsewhere. The ultimate goal would be to remove the fences. In the meantime, 'beaver-bulletins' on their progress are being issued at the Wildwood centre at Herne Bay.

The beaver project is justified by its promoters in terms of its strong likelihood of success and its evident popularity. It also helps that the beaver can be projected as 'a natural waterway engineer' whose activities help to slow river flows, control floods and preserve wetlands. As a sales pitch this all sounds very attractive, but is it nature conservation? Bill Adams (1996) has warned about the seductive allure of popularism, of doing what focus groups say the public wants, which will always be more attractive, cuddly animals rather than rare flies that live in aspen trees. Reintroducing the beaver will capture media attention, impress or amuse politicians, and, if successful, will be talked up as a major achievement. Perhaps it will also bring in more tourists to remote places that need all the help they can get. On the other hand, a project that takes at least a dozen years to get going must be taking up a lot of time and resources from conservation bodies who are constantly complaining they are overstretched. Objections could perhaps be made on welfare grounds (after six months in quarantine the animals will have become used to being fed), and that it blurs the distinction between nature conservation and zoo-keeping. Perhaps the future for large animals in the countryside should lie not so much in naturalising wild beasts but in the opposite approach, the ranching of primitive breeds, such as long-horned cattle, Przewalski's horses and Exmoor ponies, as 'management tools' on nature reserves. Judging from the eco-propaganda made on its behalf, the European beaver would slip into this role rather well.

Meanwhile, the wild boar is naturalising itself without any help from conservationists or anyone else. The popularity of wild boar farming – some 40 such farms were registered with the British Wild Boar Association in 1998 – made it almost inevitable that animals would escape. A study in 1998 estimated that at least 100 adult boars were on the loose by then in the High Weald on the borders of Sussex and Kent, with another 12 to 20 animals lurking in Dorset. A mass escape followed the great gale of 1987, which knocked down the fences, rather like the storm in *Jurassic Park*. The escaping porkers found perfect piggy habitat close by, with dense woods to lay up in by day and delicious crops to forage among later on. Most of these boars are probably hybrids with domestic pigs rather than pure-bred wild boars, but to anyone except a zoologist they look pretty convincing all the same: big, bristly and long-nosed with razor-sharp tusks. Releasing wild boars or allowing them to escape is an offence under the Dangerous Animals Act, though no one has yet been prosecuted for it. A MAFF report in 1998 recommended a review of the legislation covering boar farming, but, as with the similar cases of signal crayfish and mink, it is far too late to redesign the stable now: the horses left some time ago. At the average rate of population growth, there could be 500 or more adult boars living wild and free by 2010, even if no more manage to escape.

Should we rejoice at the re-establishment of another large extinct animal or tear out our hair at another potential pest on the run? The chances of anyone encountering a wild boar in an English wood are not great. Despite their ferocious appearance they are shy and wary, and like a quiet life. They will, however, have an ecological impact. There will be soil disturbance where boars wallow or dig for roots, and that is no bad thing. They also enjoy eating the bulbs of bluebells, wild daffodils and orchids, which may be more problematical. Boars rarely attack humans, but dogs and newborn lambs are fair game. They were also on the MAFF hit list during the 2001 foot-and-mouth epidemic, as potential carriers of the disease. As a former native animal, the wild boar will be more in tune with its environment than mink, coypu or grey squirrel. And their story suggests that the quickest way to restore lost animals to the countryside is not to write deliberative action plans but to farm them, and then wait for a high wind.

Plant and animal invasions

One of the consequences of farming shellfish or fur-bearing animals is that some will always escape, and may find that they like it here. The same is true of temperate plants, especially aquatic ones. From early times, foreign plants and animals have become naturalised in Britain. The corn cockle, for example, probably came in with imported seed grain and made itself at home in Iron Age cereal crops. Sycamore was widely planted from the seventeenth century onwards. For most of this time it seemed to behave itself, but in the twentieth century it became a serious invader of woods and ravines, outcompeting native oak with an alarming rain of shade-casting saplings. Another species to have changed its habits is the rosebay willowherb, formerly a rather delicate rock plant, latterly a mass invader of urban waste spaces and burnt ground. This is a relatively benign invader, which forms a kind of emergency bandage on fire-damaged heaths and commons while the natural vegetation recovers. One could not say the same about sterile brome, which has changed in my own lifetime from a harmless weed of open ground to a persistent invader of cultivated fields that even herbicides cannot fully control. Presumably it has mutated into a more aggressive form, or maybe we have killed off its herbivores. Another famous transformation from a formerly stable, placid species was the little pine beauty moth, following the planting of peat flows in northern Scotland with lodgepole pine in the 1970s. The moth underwent an unprecedented population explosion, probably because it had found what every herbivore must dream of, unlimited food, and not a predator or parasite in sight.

Foreign invaders are becoming a serious problem for native wildlife. We are fortunate in Britain that our wild plants and animals are true Europeans that reached us before Britain became an island, a scant 10,000 years ago. Floras and faunas that evolve in isolation, such as on Madagascar or New Zealand, are more vulnerable to takeover from aggressive colonisers. If Britain was like New Zealand, perhaps half of our animals and birds would be extinct by now, and isolating the rest on off-shore islands, such as Scilly or the Hebrides, would be our number one conservation priority. As it is, apart from a few well-known examples, such as grey squirrel versus red squirrel, or, more recently, mink versus water vole, foreign invaders still tend to be regarded more as a novelty than as a serious threat. There are even political connotations. The word 'alien', long used by botanists to mean newly-arrived non-native plants is now offi-cially frowned upon. The preferred word is 'neophyte'. But whether aliens or neophytes, the impact of these colonisers is not always benign. On the warm heaths of the New Forest it is hard to find a pond that has not turned into a monoculture of swamp stonecrop. A recent survey of Poole Harbour found that over 60 per cent of the biomass (weight) of marine species consisted of invaders, mainly slipper limpet, cord grass and an oriental seaweed, curiously known as Japweed.

Invaders seem to occur among some groups more than others. We have seen some spectacularly invasive mammals and freshwater fish, but rela-tively few birds; quite a few molluscs, beetles and flatworms, but not many moths or spiders; lots of flowering plants, but few mosses and liverworts. Most of them are harmless, perhaps even to be welcomed, such as the tree frogs of south-east London, or the powder-blue slender speedwell (though not all lawn owners would agree). A few cause a scare and make the head-lines, such as the escaped flatworm that is said to be stealthily eating its way through our earthworms. In most cases they are probably here to stay. The coypu was eradicated, but only after a very determined campaign by MAFF that might not be possible today (the traps would be sabotaged by coypu liberationists). What follows tries to sum up their impact on native wildlife at the start of the twenty-first century.

Invasive plants

For every native wild flower in Britain, there is a foreign one that has escaped into the wild. A recent report (SNH 2001) lists 830 introduced flowering plants that occur wild in Scotland. They reached us in a variety of ways, for example, as seeds or propagules in the ballast of ships or stuck to imported fleeces, but most of the more recent ones probably originate from gardens or plant nurseries. Most of them are rare, harmless, and even a delight to field botanists. Of the 830 species in Scotland, only 73 are wide-spread, and of those fewer than 20 cause problems. Unfortunately those problems can be serious, and are becoming more so, perhaps in response to a slowly changing climate. Invasive plants have a bad habit of moving into places with a high natural biodiversity and reducing them to a virtual monoculture. So we have river banks overrun with giant hogweed from the

Australian swamp stonecrop, also known as New Zealand pygmy-weed, a threat to the natural flora of shallow ponds. It was sold as an oxygenating aquatic plant by a single UK supplier until the 1970s. First recorded in the wild in the 1950s, it had become widespread by the 1980s, perhaps spread by birds. (Natural Image/Bob Gibbons)

Caucasus or Indian balsam from Central Asia, natural woods stuffed with sycamore or rhododendron, limestone headlands smothered in *Cotoneaster* or *Carpobrotus*, lowland heaths dotted with spiny, aromatic shrubs of the genus *Gaultheria*, and sand dunes with pirri-pirri bur from New Zealand. Most pernicious, perhaps, are the invaders of fresh water. Their prototype was Canadian pondweed, which spread like wildfire along Britain's canal

Giant hogweed, an awesome weed, once sold to gardeners as 'one of the most magnificent Plants in the World' (and it is), now feared as a 'Triffid' that burns you when you touch it. A quiet garden plant until about 1970, it suddenly broke cover and appeared on river banks and roadsides all over Britain. Its towering stature gives the giant hogweed an oddly proprietorial air. (Natural Image/Bob Gibbons)

network in the mid-nineteenth century and found its way into natural ponds and lakes. Scottish Natural Heritage recently spent £7,500 trying to get rid of it in just one loch in the Outer Hebrides. Loch Davan, in Aberdeenshire, once noted for its natural waterweeds, is absolutely choked with it. A strong current contender for the title of the worst weed of all is swamp stonecrop, *Crassula helmsii*, from Australia, which is sold by the aquaculture trade as an ideal oxygenator for garden ponds. It was new to me when I found it by a New Forest pond in the early 1980s. I should have risked offending Forest bylaws and dug it up at once. Today it has taken over that pond and many others like it, smothering the rich former vegetation with a mattress of stems and tiny, prickly leaves. So far, swamp stonecrop has been considerably more successful at invading warm, shallow heathland ponds than Britain and the European Union have been at preserving them, despite their prominence as SSSIs and SACs.

With diminishing amounts of good wildlife habitat, invasive plants are a problem we could well do without. Once established they can be extremely difficult and expensive to eradicate. The problem is not peculiar to Britain. The United States, for example, is said to spend £100 million every year fighting them, and on some Pacific islands they have all but taken over. The traditional 'big three' invasives in Britain are giant hogweed, Japanese knotweed (including the closely related giant knotweed) and rhododendron. The latter is a well-known invader of woods, but is an even greater nuisance on west coast islands with a mild climate. On Colonsay in the Inner Hebrides, where rhododendron had been introduced to provide cover for game and some colour in the spring, there is now a 20-year,

It likes islands. Rhododendron has become a major nuisance on Lundy, Colonsay and other places with a mild climate and moist, peaty soil.

£160,000 project to get rid of it, with the full support of the islanders. On Lundy, where rhododendron seems to have spread from the manor garden after a fire in the 1920s, the policy is containment, but even that is hard work. The frequency of young saplings in the island's peaty, granitic soil suggest it could spread much further unless controlled.

Meanwhile, the watery world of swamp stonecrop is attracting more exotic invaders. Parrot's feather, *Myriophyllum aquaticum*, looks disarmingly attractive in an aquarium, with its whorls of stiff, capilliary leaves. Floating pennywort, *Hydrocotyle ranunculoides*, forms masses of thick, round leaves that can form a green carpet on the surface of field drains and ponds. Both probably originate from garden throw-outs. Both seem to like nature reserves and other protected sites, where they block watercourses and crowd out the native frogbits and water crowfoots. According to a report by Plantlife, their sins include deoxygenating the water, killing fish and invertebrates, causing flooding and even drowning cattle! The cost of controlling this pair is £300,000 a year and rising. That of the more firmly established escapee, swamp stonecrop, is estimated at £3 million – assuming it can still be controlled. Methods using herbicides such as diquat or glyphosate, or even liquid nitrogen, tend to kill every plant in sight, and swamp stonecrop will recolonise the ground faster than most natives. To my knowledge, the only successful eradication to date is in certain shallow ponds in sand dunes, where every last scrap of the stonecrop has been laboriously raked up and removed by enthusiasts bent on saving the breeding pools of the natterjack toad. The only sure way to stop the rot is at source, by preventing the sale and distribution of invasive plants. Here, unfortunately, we run into an absurd legal catch-22 situation. At present the law requires that any plant must be *shown* to be invasive before it is proscribed – yet, by the time that has been proved, it may be too late for effective control measures. The only two invaders currently banned from sale

A ditch blocked by floating pennywort, first recorded in 1990, but already a serious pest of lowland waterways. (Natural Image/Bob Gibbons)

A less aggressive, perhaps even welcome, invader: Nootka lupins from Canada rubbing along with native flowers, including spignel, on the banks of the Royal Dee.

are Japanese knotweed and giant hogweed. As this book went to press, a DEFRA 'working group' was considering what to do next.

Sowing 'wild flowers'

A growing problem in the British countryside is the sowing of seeds of what are purported to be wild flowers, but more often are fodder crop varieties from overseas, especially eastern Europe. Among the popular ones are non-native strains of bird's-foot trefoil and kidney vetch from southern Europe, and cultivated alsike clover. The drifts of white 'ox-eye daisy' flourishing on the ring roads of Oxford in June are a large tetraploid strain of our diploid native plant originating in the plant-breeder's laboratory. The varieties of sainfoin, salad burnet, yarrow, knapweed and even grasses used in commercial seed packets generally differ genetically from the wild British equivalent. One commonly used grass, Hungarian brome, *Bromus inermis*, does not even occur here (except as an undesirable escapee). A large market for wild seed has developed through agri-environment schemes such as stewardship and long-term set-aside, and they have tended to be sown in an agricultural way, as 'instant carpets' of wild flowers. They are often sown cheek by jowl with native vegetation and with scant

Wild flower seed-sowing aims to replicate the colour and variety of natural meadows like this. Unfortunately, competition generally ensures that a few species soon become dominant, and the result is disappointing. Moreover many of the seeds are of non-native plants or of cultivated origin. (Natural Image/Bob Gibbons)

regard for what is appropriate to the site. Similarly tree-planting schemes in wild places often use non-native trees such as Italian alder or horticultural strains of willow or hawthorn, or alternatively use native trees in places where they do not occur naturally, such as field maple in western Cornwall. There are even examples of authorities planting *rare* trees, such as small-leaved lime or wild service, thus diminishing their real value as trees with mysterious and interesting natural distributions. The authors of a recent county flora are concerned that the sheer scale of amenity sowings and plantings 'weakens the integrity of the flora and threatens the very fabric and character of the landscape' (French *et al.* 1999). The solution advocated by the Flora Locale project and by many local wildlife trusts is to use only locally derived material and to allow the plants to spread naturally. For example, the Wiltshire Wildlife Trust collects its own seed from nearby downs and meadows for grassland restoration work, and sows it in small patches from which flowers spread naturally, choosing where they grow.

Unwelcome guests

The East Anglian population of coypu, which reached an estimated 200,000 animals at its peak in the early 1960s, originated from fur farms. In the Norfolk Broads they found a reedy waterscape resembling their native South America, and if it was a lot colder – the long, freezing winter of 1962–63 killed a lot of coypus – at least there were no jaguars. They became an economic nuisance, raiding fields of root crops and under-mining farm tracks with their burrowings. Consequently the Ministry of Agriculture declared war on them. By 1965 coypu had been eliminated from most areas outside their heartland in the Broads, and after two more decades of intensive trapping the last coypu was caught and shot in 1987, four years ahead of schedule. The effort required the services of 24 full-time trappers and cost £2.5 million. The coypu is only the second invasive mammal to have been eradicated. The first, the muskrat, another escapee from fur farms, was hunted down in the 1930s, mainly by leg-hold traps (which also caught a lot of other things).

The American mink is a much more serious ecological problem than the coypu, and, being more widespread and harder to trap, it is probably beyond our power to get rid of except on isolated islands. Mink began to escape from fur farms between the World Wars, but it was not until 1956 that they were found breeding in the wild, in south Devon. By the 1970s, the mink population had exploded all over Britain, but with the highest densities in the west. Given the regularity with which mink escape, or are released from captivity by half-witted animal liberationists, mink farming should have been banned long ago. It certainly should never have been allowed on islands naturally free of ground predators, such as the Outer Hebrides and Shetland. Undoing others' mistakes is hard work. There are now an estimated 10–15,000 adult mink in the Western Isles. Scottish Natural Heritage has started an eradication programme in the Uists and Harris, expected to last 20 years and cost £1.65 million. Unlike the stoat or polecat, the mink hunts along watersides, and swims well. Its most promi-nent victim is the water vole, which seems defenceless against mink. Consequently the vole's population has plummeted disastrously since the 1970s. Conservationists monitored its fall in detail – the water vole's plight first became apparent during a survey – but could do little to stop it. The

Water vole – threat-ened by American mink. (Nature Photographers Ltd)

animal was already under pressure from insensitive drainage operations that destroyed river-bank vegetation, and from overstocked farm animals that trampled and grazed it flat. Whether the water vole can be saved by 'mitigation measures' and better habitat management remains to be seen, but as long as mink are around it may be destined to remain one of our rarer mammals. The one hope is that recovering otter and polecat numbers may eventually suppress mink numbers.

Two introduced deer, muntjac and Sika, have also had a significant impact on native wildlife. Muntjac, like the native red and roe deer, browses woodland foliage, and large numbers of them can put paid to plans for natural regeneration in woodlands. They also eat a lot of woodland flowers. The oxlip woods of the East Anglian boulder clay, once so abrim with the creamy-yellow flowers in April, are largely a thing of the past, thanks to the muntjac. Controlling the animals by shooting is difficult, if not impossible, since these russet, dog-sized deer are barely visible in long grass. They are also very good at penetrating supposedly deer-proof fences. Sika pose a different, but equally intractable problem: since they are genetically compatible with the larger native red deer, there has been a widespread and progressive genetic introgression that replaces pure-bred red deer with 'hybrids'. It is now hard to distinguish them, making nonsense of efforts to 'shoot out' the Sika deer. Possibly the only places where pure-bred red deer may survive are islands such as Rum, assuming the Sika does not find its way there too. Thus in this gradual, largely unsung way, we are losing our largest land mammal. A similar problem afflicts the native Scottish wildcat through

Will we ever see this again? The famous oxlip woods on the boulder clay of East Anglia have been devastated by a population explosion of deer, which have a liking for the delicately scented blossoms. You can still find the flowers, but not the spectacle. (Derek Ratcliffe)

interbreeding with feral domestic moggies. Interestingly, while the wildcat has been a protected species since 1988, the hybrid is officially a pest. If, as some experts believe, wild cats and domestic cats belong to the same species, the legal distinction breaks down altogether. No one knows how many pure-bred Scottish wildcats remain, nor whether some can be saved from the consequences of wholesale genetic introgression.

Another American mammal that is obviously here to stay is the grey squirrel. Its spread has slowed down since the 1950s, when it quickly displaced the native red squirrel over most of England, but it is now taking over the red squirrel's woods in the Lake District and southern Scotland, and looks set to do so in Ireland too. Barring serious epidemics, the red squirrel can probably hold its own in the mature pine forests of the Highlands, and on islands such as the Isle of Wight and the odd mainland refuge such as Formby Point, where the greys are rigorously controlled. But the grey squirrel has defied attempts by foresters to control it using baits laced with warfarin, and has long had a refuge in gardens and urban parks where it is usually welcomed. We do not have many cuddly, confiding mammals; the grey squirrel is not as attractive as the red, but it is better than nothing.

The ruddy duck is another newcomer from America, originating from the Wildfowl Centre at Slimbridge. The birds had proved tricky to rear artificially and so it was decided to let the parents bring up their own young. Unfortunately some of the young birds flew away before they could be caught and pinioned, and the result was a substantial feral population centred on large lakes and reservoirs in the West Midlands. Although the brood females are aggressive and may compete with native ducks, the ruddy duck seemed a harmless and even welcome addition to the list of British birds. The West Midlands Bird Club even adopted the duck as its flagship species. Unfortunately the ruddy duck went on spreading, and has now started to breed in Europe. There are ominous implications for the white-headed duck of Spain, with which this species will compete, mate and produce fertile hybrid offspring. The white-headed duck is regarded as a threatened species, and its survival has been made even more precarious by increasing sightings of ruddy duck in its breeding grounds in Andalucia. For that reason, MAFF, under international pressure, approved a cull of the ruddy duck in England. The measure is controversial, though it has the backing of English Nature and the RSPB, and those involved seem anxious to say as little about it as possible. In the first six months, 448 duck were shot at a cost of £223 each. The Welsh have so far refused to shoot any ruddy ducks, and no one seems prepared to say how large a cull would need to be to satisfy the Spanish, nor whether the authorities are prepared to countenance its complete eradication.

Another headache for wildlife managers is animals that become established on islands that were previously free of them. This is the problem at South Uist in the Outer Hebrides, where hedgehogs were introduced in the 1970s, apparently by a gardener keen to try out a 'natural' way of controlling his slugs. Today about 6,000 of their descendants live there in old rab-

A drake ruddy duck, currently on the wanted list for threatening the genetic integrity of the rare Spanish white-headed duck. (HarperCollins Publishers)

bit burrows. Birds' eggs, especially those of ground-nesting species such as dunlin, lapwing and ringed plover, form part of their diet. With some 2,500 pairs of both dunlin and ringed plover, South Uist supported one of the highest breeding wader densities in Europe. Since the hedgehog arrived, their numbers have halved, and may fall further unless the hedgehogs are all rounded up and removed. On North Ronaldsay, where hedgehogs were also ill-advisedly released around 1970, some 10,000 of their offspring were trapped and deported to the Scottish mainland in 1986. Other places where some bright spark decided to cheer things up by introducing hedgehogs include Sark in the Channel Islands and St Mary's in the Isles of Scilly.

Rats are another unwelcome guest on bird islands. Probably introduced accidentally with lighthouse supplies, rats can survive on comparatively barren rocks by raiding seabird colonies and scavenging along the tide line. A single pregnant rat can in theory produce a population of 15,000 rats in a single year. Fortunately they can be targeted by using warfarin, which has a low toxicity to birds. Recently the Scottish Wildlife Trust succeeded in eradicating brown rats from Handa Island, with immediate benefits to the seabirds. In Wales, Flatholm, Puffin Island and Ramsay Island have also been cleared of rats. Eradicating them does not guarantee increased numbers of birds, but where there are no other ground predators, it is generally regarded as a good idea (Ratcliffe & Sandison 2001). The much rarer black or ship rat poses more of a conundrum, since it is now more endangered than its prey, with only three known self-sustaining populations, all of them on islands (it may be our rarest long-established wild mammal). On Lundy, where, appropriately enough, some of them live on Rat Island, the policy is to preserve a sustainable number, but to control the surplus using live traps.

A new, spectacular but most unwelcome addition to the list of naturalised animals in Britain is the American bullfrog. This is an enormous frog, four or five times the size of our native one when full grown. It has a tadpole to match, that grows over several seasons to the size of a sardine. The bullfrog was widely sold in the pet trade, and occasionally (inevitably) turned up in ponds and marshes. In 1997, their sale was banned to counter the potential effect on native wildlife, but it may be too late. In 1999, tadpoles and froglets found in two Sussex ponds seemed to represent a naturalised population, and not merely a recent release. The ponds were drained and efforts made to eradicate every last bullfrog, froglet and tadpole, but the episode shows how easily this large amphibian could become established, and how difficult – turning rapidly into impossible – it would be to control. More than the European marsh frog, which has become firmly established in southern England since 1935, bullfrogs are aggressive predators potentially capable of suppressing native frogs and newts. They can also carry a frog-killing virus (Banks *et al.* 2000).

Anglers and other fish enthusiasts have long modified the natural distribution of freshwater fish by restockings and introductions, both of native fish and of imported species such as carp and rainbow trout. In at least two cases there have been unfortunate consequences. That small perch relative, the pope or ruffe was introduced to several upland lakes, probably as escapes from anglers' live-bait tins. Unfortunately they feed on the eggs of native whitefish (*Coregonus* species), such as the famous powan of Loch Lomond. Whitefish are not only rare and threatened in themselves, but are keystone species in the intricate food web of upland lakes. They may survive the ruffe, but with a reduced population, which will inevitably cause kickbacks in the lake ecosystem. Another growing problem is the zander or pike-perch, a predator from eastern Europe, introduced from Woburn stock into the Great Ouse in 1963 at the whim of the local river authority. By the 1980s, the zander had colonised extensive river catchments, including the Severn, with the help of the canal network and further illicit releases. The zander is a top predator, which, in the fisheries minister's recent words 'can have far reaching and undesirable ecological consequences' (Smith & Briggs 1999). The cost of controlling it is estimated at £30,000 to £40,000 per year, with no certainty of success. Left uncontrolled, the zander will probably spread throughout Britain.

The casual decimation of one of our largest native invertebrates, the freshwater crayfish, is another sad example of the vulnerability of aquatic animals to carelessly introduced foreign invaders. The many crayfish farms established in Britain from 1976 onwards were stocked mainly with the larger (meatier) American signal crayfish. Unfortunately – as was known at the time – this species carries 'crayfish plague', a fungal infection that is invariably fatal to our native crayfish, without itself suffering from the disease. Moreover, as the more powerful species, the American crayfish overcomes the native one in clawed contests, killing the males and mating with the females. In rivers such as the Hampshire Avon or the Kennet, where native crayfish were once common, only the invaders are left. Eradicating them is no longer practicable – and perhaps even undesirable, since the only certain way of getting rid of them is to poison the water. Containment, including a ban on further crayfish farms, may preserve the native crayfish in northern England, but, worryingly, there are reports that crayfish plague is spreading to areas not stocked with signals. In Ireland, which had the sense not to import foreign crayfish, the disease has arrived by some other means, perhaps on rubber waders and fishing equipment. The ecological consequences of this change-over of crayfish is still uncertain, but there will be some, and there is no doubt who is to blame. It was the Ministry of Agriculture that licensed crayfish farms in the 1970s and 1980s without a thought about what might happen when the animals escaped (and one look at some of these farms would have told them it was when, not if).

In general, most invasive invertebrates have become economic pests rather than threats to native wildlife, although any ecological effect may become more evident with time. For example, the honeybee brood mite has devastated the hives of the honeybee throughout the world, and since 1992 has become firmly established in Britain. Fortunately it does not seem to attack wild bees to the same extent, but since the honeybee is an important pollinator of wild flowers, there may be a knock-on effect on our flora. Some invasive invertebrates pose a more direct threat to our freshwater and marine fauna. An American flatworm, *Dugesia tigrina*, is spreading rapidly in rivers and canals, and, like the signal crayfish, it seems to compete with and eliminate its native rivals. Another New World flatworm, *Phagocata woodworthi*, has been found in Loch Ness where, wonderful to relate, it was probably introduced accidentally by American scientists searching for the Loch Ness monster! In the Norfolk Broads, an Asiatic clam, *Corbicula fluminea*, is reported to be spreading on the bottom of the shallow lakes. The problem here is that the clams live in dense mats of up to 2,000 per square metre, and so they physically crowd out native animals and plants living there. Again, experience in America, where this clam once closed down a nuclear power station, indicates there is practically no way of controlling them without resorting to chemicals that kill innocent animals too. Another potential nuisance is the Chinese mitten crab, which has colonised several broad river estuaries in south-east England, including the Thames. Apart from potential disruption to natural food chains, their burrowings threaten to undermine river banks causing subsidence and flooding. Without doubt, invaders will be a major wildlife issue in the twenty-first century, and to control them we will have to be wiser about releases and cleverer at containment than in the past.

13

Summing Up:
Whither Nature Conservation?

A changing environment?

St Boniface Down, on the southern tip of the Isle of Wight, was one of the more ecologically interesting corners of Britain at the turn of the Millennium. Formerly an open sweep of chalk downland overlooking the sea – clean bands of green, white and blue – it is changing into a landscape characteristic of the Mediterranean coast: dense scrubby woodland on hot, dry soil, dominated by holm or evergreen oak. The oaks have been spreading for several decades and now occupy some 16 hectares. The National Trust has installed some goats there to try to prevent the wood spreading further. This represents a new type of natural vegetation in Britain. Inside the wood, naturalists have found species all but unknown elsewhere, including a big white toadstool, *Amanita ovoidea*, and a striking purple cup-fungus, *Sarcosphaera coronaria*. It is also rich in saprophytic flowering plants that like deep shade: a strange, sparse flora of orchids and yellow bird's-nest. Recent records of a new British moth, the oak rustic, whose larval

Like giant clay ovens, the towers of Ferrybridge power station gently heat the atmosphere. However it works out, global warming will make life more difficult for many species of wildlife. (Derek Ratcliffe)

foodplant is holm oak, may originate from here. Holm oak is native just
across the channel in Brittany. Is the oak wood at St Boniface Down a har-
binger of global warming? And if so, which is most important, a new,
dynamic but rather species-poor habitat, or the much-diminished, species-
rich chalk grassland habitat it is displacing? Should we welcome it or worry
about it?

Climate change is one of the great environmental bogeys. It is probably
real, but hard to distinguish from normal fluctuations in the weather. In

Heralds of climate warming? Lesser red-eyed
damselfly, *Erythromma viridulum*, (right) and
lesser emperor dragonfly, *Anax parthenope*, (left)
began to breed in Britain in the late 1990s.
(Natural Image/Bob Gibbons)

Increasing: the elegant
little egret began to
colonise Britain in the
1990s. 'Hundreds in
dozens of colonies' are
predicted by 2010,
assuming our winters
remain mild. (Natural
Image/Bob Gibbons)

The long-winged conehead has greatly increased its range, probably in response to milder winters and taller grass. (Natural Image/Bob Gibbons)

Britain its recent symptoms have been warm autumns, mild winters, disappointing springs and wet summers. Many of our insects would prefer a cold winter and a sunny spring. Mild winters mean fewer frosts, and frosts are one of our best defences against the spread of invasive species (frost-free coastlines are often marked by masses of *Carpobrotus*). But so far, at least, the biological evidence for climate change in Britain is unspectacular. There has been a slight northward extension of the range of some well-recorded species, such as the comma butterfly or the long-winged conehead grasshopper or the Dartford warbler. Three species of tongue-orchid (*Serapias* species) native in southern Brittany, turned up in southern counties during the 1990s in circumstances that suggest natural colonisation. Orchids have dust-like seed that must drift for miles in air currents, but for most species, the English Channel forms an effective barrier. We have not yet lost any species through climate change. However, a recent report (UK Climate Impacts Programme 2001) predicts the extinction of some mountaintop species by the year 2050, and the decline of many northerners, such as globeflower and capercaillie. Habitats vulnerable to warming include mountain tops and raised bogs (which depend on cold, wet conditions), chalk streams (which are vulnerable to drought) and 'soft coastal habitats' (which shrink as the sea rises). Change is also in store for native pine woods, calcareous grassland and mesotrophic lakes.

Coping with faster rates of change will demand flexibility, and co-operation between users and preservers. The sea is rising. The RSPB predicts that some 100 square kilometres of tidal mudflats will be lost by 2013, and

The limits of uncertainty. Different climate
change scenarios predict a modest increase or
a severe southward decrease to the distribution
of the large heath butterfly (from English
Nature: *State of Nature* report, 2001).

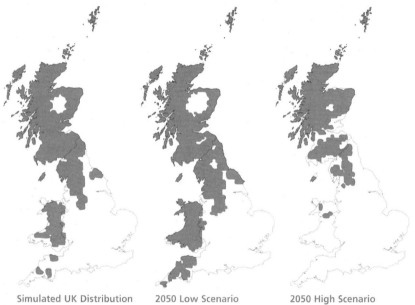

Simulated UK Distribution   2050 Low Scenario   2050 High Scenario

that by 2050 the English Channel will have risen by 65 centimetres. Other
world climate scenarios are more apocalyptic; it depends on how much of
the polar icecaps will melt. No one can prevent the sea from rising, but the
environmental consequences are predictable and can to some extent be
planned for. In the 1990s, English Nature campaigned for 'a living coast',
which advocated ecologically intelligent ways of defending the coast, that
involved rolling with the punch of the tide instead of hiding behind walls.
It also introduced the idea of 'set-back' or 'managed retreat', which entails
abandoning the old sea walls and allowing salt marshes and other tidal
habitats to develop inland. But there is an understandable reluctance on
the part of the authorities to surrender hard-won land. For that matter,
conservationists may not always be very keen either, since some of the vulnerable areas are nature reserves, assets often purchased at considerable
cost. Habitat destruction can be delayed by constructing protective barri-

ers, such as the fibre rolls anchored between wood piles at the mouth of Lymington River in Hampshire (Tubbs 1999), or by abandoning dredging near the coast, but in the longer term it is inevitable. Farmland has already been abandoned to the flood in Essex, on the Humber and elsewhere. Coping with the rising sea will be complicated. For example, no fewer than 60 organisations are represented on a forum to co-ordinate planning and management in the Solent. The one that matters is Associated British Ports (ABP), which has an expansionist agenda, including the development of Dibden Bay, one of the last natural corners of Southampton Water. The forum has made progress with a common strategy, but, not surprisingly perhaps, has not achieved any real consensus over the future of the Solent, and has no real power. This is in the hands of the shareholders of ABP.

Another change that affects the whole country is eutrophication (nutrient enrichment). Modern farming produces astronomical quantities of waste nitrate and phosphate, most of which enters the ecosystem via the soil or the drainage system. The majority of cattle and pigs are kept indoors for part of the year – in the case of pigs often all year round. A pig generates 11 times as much phosphorus a day as a human being, a cow 20 times as much. A dairy unit of 200 cows produces as much phosphorus as a town of 4,000 people (indeed, probably more, since human sewage is treated first before rejoining the environment). Slurry from livestock units also contains quantities of organic matter – 2,000 kilograms per year per cow (being biological carnivores, human beings average far less – about 43 kilograms a year) – and cattle slurry consumes far more oxygen than human excreta. This slurry, rich in toxins and active chemicals, often went straight into the nearest stream. In 1987 alone there were 3,870 reported water pollution incidents in England and Wales related to the discharge of farm waste, compared with 1,500 incidents in 1979 (George 1990). Only a small minority were prosecuted, and the fines doled out were scarcely a deterrent to wrongdoers, like the Somerset farmer who referred to his fine of £2,000 as 'cheap money' compared with the £200,000 it would cost him to fit new slurry tanks. On top of all this, tonnes and tonnes of chemically active, nitrogen-rich fertiliser used on all non-organic farms enter the watercourses every time it rains.

In the 1990s, the Environment Agency (and its Scottish equivalent) took a much tougher stance than had its predecessors on water pollution. 'Phosphate-stripping' plants have been built on sensitive waterways, such as the chalk rivers and the Broads. MAFF has tried to create 'nitrate free zones' in sensitive catchment areas, and offered farmers grants to clean up their farm waste. Although there has been some improvement in water quality, the scale of the problem is still wholly disproportionate to attempts to deal with it. The result of eutrophication on aquatic ecosystems is well documented (see Moss 2001 for a recent account), and devastating. Obvious signs are the decline of vegetation (pond plants need a firm bed and clear water), decreased oxygenation and increased turbidity. Animals that depend on clean gravel beds and well-oxygenated water, from pearl mussels to brown trout, struggle to survive. Less well known is the effect of

Nettle beds are spreading in river valleys, especially along watercourses where sediment rich in phosphate and nitrate from farm slurry, sewage and arable field run-off has accumulated.

soil enrichment on land. The Government's *Countryside Survey 2000* revealed the full extent of the problem – a rapid, nationwide fall of plant diversity, especially in vulnerable places such as riversides and road verges (DETR 2000). For example, roadside plant diversity fell by 9 per cent between 1990 and 1998, no doubt helped by active nitrogen from exhaust fumes. In the nutrient-sinks down in the valleys, where most of us live, colourful wild flowers have been replaced by a few aggressive plants of chemically enriched soil, such as cow parsley, hogweed and, above all, stinging nettle, perhaps the most successful wild plant in modern Britain. Eutrophication can only be reversed at source, by changes in agricultural support that require less nitrates and produce less phosphates, that is, by lower-input, organic methods of farming. Eutrophication is not a threat to the environment; it has gone beyond being a threat: it has happened.

GM (genetically modified) crops, on the other hand, are a threat. It could be bluff, and there may be few harmful side effects. It is a shame we have to risk it. On the basis of experience in the United States, where GM crops are grown routinely, contamination of non-GM crops by GM pollen will be inevitable. That should demand, at the very least, sensitive place-ment of GM crops, but one of the Government's farm-scale test trials is tak-ing place just two miles from the headquarters of the Doubleday Association, Britain's leading organic researcher. Contamination of organ-ic crops could mean that a farmer would lose his hard-won organic status. Thus, one government policy flies directly in the face of another. BTO research predicts that more effective weed control using GM sugar beet would eliminate fat-hen, an important food source for skylarks, and cause their populations to crash by 80 per cent in beet-growing areas; another government policy is to increase skylarks. English Nature was worried enough to stick its head above the parapet and call for more research before the commercial introduction of GM crops. Britain does not need GM crops. They are depressing confirmation that in agriculture, technology and Monsanto still rule supreme. Here, as elsewhere, unpopular policies breed direct action. Greenpeace's destruction of GM maize crops was non-violent, but there have been less gentle protests at Twyford Down and at Thorne Moor where £100,000 worth of peat-winning machinery was damaged by EarthFirst!, one of several New Age groups operating around the fringes of the law. It all seems a long way from the early days of nature conservation, revolving around field study and the appeal raffle.

The rise of conservationists and the slow death of natural history

The delight in attractive countryside expressed so vividly by past British artists is still something that seems imprinted on our bones, perhaps all the more strongly now that more of us have the means to reach it, but in the meantime have to live in surroundings that are far from delightful. Added to the desire for decent countryside there is perhaps also a sense of communal guilt or even anger that we have so mismanaged things. We are living at a time of declared rights, as in 1215 or 1789, and one of them is that land is communal as well as private, and that people should be allowed to roam over at least the open land. Some think wild animals, too, have rights. The opinion polls place countryside and environmental issues high among many people's concerns. People riot in the streets about unsustainable use of resources and the lack of respect for mother earth. None of this has helped to bring about a resurgence in natural history, however. From being the province of the amateur naturalist, field study is to a great extent now a paid activity or formalised as educational projects. If you see someone with a pond net, they are unlikely to be simply interested in what lives in the pond. Natural history as our forefathers knew it, incorporating elements of exploration, curiosity and museum making, seems to hold little appeal today. People have become content to be spoon-fed by the media. Nature, particularly animals and birds, translates well to television (though the words lag well behind the pictures). Wildlife tours are popular, but are essentially passive: you are taken places and shown things. Via the TV, wildlife has become popular entertainment. Britain's leading wildlife magazine is published by the BBC, but, with the usual exception of birds, no high street magazines cater for British natural history. Who, these days, owns a plant press or a microscope?

Why is this? We still like wildlife, but I think we have become less curious about it, less intellectually engaged. Nature conservation may be a crusade for many, but wildlife itself is less central to our lives. Perhaps people relate to it best in the privacy of their own gardens, just as the shepherds and ploughmen of yesteryear had their own pet names for the flowers they saw every day – shepherd's needle, pheasant's eye, devil's comb, gamber-griggles, slipper-sloppers. Wildlife gardening is popular, but a distance has opened up between people's media-fed perceptions of the countryside and the reality. You find crowded car parks, but not that many walkers. You can sit in a public place and see the most amazing things going on, and *no one is looking.* Visitor statistics to nature reserves are misleading. In may cases, people have come to picnic and sunbathe, admire the view or to walk the dog. One nature reserve received half a million visitors a year, claimed the NCC. Of course it did; it had a beach in front of it. Apart from the famous few, such as Minsmere or Wicken Fen, nature reserves receive no more visitors than any other patch of attractive countryside.

The growth of the nature conservation industry has been prodigious. When Dudley Stamp's book, *Nature Conservation in Britain*, was published in 1969, the annual budget of the Nature Conservancy was £1,320,000. That of the largest wildlife charity, the RSPB, was only £195,000. In 1999, the Nature Conservancy's successors, English Nature, SNH and CCW, received a total of £110 million in government grant. Hence the state contributed nearly 100 times as much to nature conservation in 2000 as it did in 1969. Allowing for inflation, it still amounts to a more than tenfold increase. The rise in income of the leading voluntary bodies has been even more spectacular, more indeed than the rise in membership, which is impressive enough. In 2000, the annual budget of the RSPB was £41 million, representing a 200-fold financial growth, while its membership is 20 times as large as it was in 1969. Other bodies, such as the county wildlife trusts, have seen comparable growth, if not quite such a meteoric rise in membership. I calculate that the combined annual income of the voluntary wildlife bodies in 2000 was well over £100 million. The Heritage Lottery Fund contributed £48 million to countryside and nature conservation projects in 2000 alone, and £321.8 million (for 968 projects) since it opened for business in 1995.

Mass memberships and multimillion pound budgets have given the official agencies considerable power of patronage through grants and other incentives, and the voluntary bodies have acquired considerable influence. As I hope this book has demonstrated, they have helped shape today's environment, and done their best for the natural world under very difficult circumstances. Even so, nature conservation remains a rather inclusive activity with an off-putting bureaucrat's language. Most of its many publications offer only a partial view, or are forbiddingly technical. After half a century the industry still lacks an informed, popular literature.

Bill Adams, one of the most perceptive and intelligent commentators on the conservation scene, accuses the conservation priesthood of treating their subject in a dreary, deracinated way. Look at what we have done with nature, he says: 'We classified it and located it, defined it, and tied it down as a set of objects, as species, sites or habitats. In order to protect nature from industrial rationality, we have increasingly used the logic and methods of industrial rationality itself ... We seek to use industry's weapons on nature's behalf, but in the process we substantially industrialise nature itself' (Adams 2001). In short, we are in danger of turning something wonderful into something boring. Children have a much better idea: they *enjoy* nature, climb trees, splash about in rock pools, play little games with daisies and plantains. It is wrong, thinks Adams, that conservation should be a sectarian activity. Rather, it should become a universal ethic, a set of rules by which we should try to live.

Maybe, but ethics are not what conservationists spend time on. I wonder how many readers will recognise Matt Ridley's hostile portrait of contemporary conservationists in action:

> 'The public – simpleton that it is – thinks that conservationists conserve nature. This is like saying footballers score goals: it is the aim, but it is a poor description of what most of them spend most of the match doing. Most conservationists can and do talk for hours to each other without mentioning an animal or plant. I have seen them do it many times. They talk about committees, guidelines, grant applications, advertising campaigns, legislation, conventions, protocols, conferences, secretariats, treaties, regulations, resources – just like the businessmen they affect to despise ... The rare few in conservation who remain naturalists and can tear themselves away from desks and photocopiers long enough to get their boots muddy and actually do something are soon doomed to impoverished obscurity, for little of the money reaches them. The suits and sandals [i.e. conservation bureaucrats] are far too good at intercepting it along the way. The route to fame and wealth is behind the desk.' (Ridley 1996).

Anyone who has wrestled with the annual reports, corporate strategies and other voluminous documents of the conservation industry will surely see some truth in this portrait. Of course, the embattled conservationist could claim that conservation is a business like any other, needing a strategy, targets, resources and advertising campaigns. And if this means spending most of your life behind a desk scavenging for money, so be it: 'Gardens are not made/By saying "Oh how beautiful"/And sitting in the shade'. But Ridley and Adams are right: the conservation industry *is* too inclusive – and far too inclined to place its faith in words rather than deeds. The UK Millennium Biodiversity report, published in 2001, found it necessary to remind us that 'action planning' is not enough: there must also 'be better delivery on the ground'. To read some conservation reports, you could be forgiven for mistaking species and habitats for a substance, like bubblegum, that will roll off the conveyor line in measured quantities once you set up the right inputs. This is what tends to happen when you spend too much time indoors talking to similarly-minded people. The corporate strategists, the hot-air merchants, rise to the top, and the real naturalists stay at the bottom. One wonders how many of today's conservationists are still moved by the beauty of nature, or have arrived at their calling from a sense of wonder, and the excitement of discovery.

Job opportunities in nature conservation

Does the future of the countryside lie in tourism? In the less intensively farmed parts of Britain, it already contributes more than farming, which survives only by public subsidy. According to Smout (2000), in Scotland the tourist industry contributes £2,500 million to the economy, with 70 million 'bed-nights' a year, and employs 180,000 people. The publication *Biodiversity Counts* estimates that 'environment-related activity' generates some £1,600 million and 100,000 jobs in south-west England – that is, some 5–10 per cent of its GDP. Hill walkers and mountaineers alone add £150 million to the economy of the Highlands every year. In the Hebrides, wildlife tourism is *the* growth industry. The whales, eagles and seabirds that helped contribute £57 million to the local economy in 1996 may already be worth more than sheep or spruce trees, at least when agricultural subsidies are deducted.

Of course, only a small part of the tourist industry is directly related to nature conservation. Back in 1985, an NCC-commissioned review of employment opportunities in nature conservation identified some 14,250 'jobs', including the following: public facilities, such as museums, zoos and botanic gardens (3,020); land-owning institutions, such as National Trust and local authorities (2,730), mainstream nature conservation bodies (1,600), media and publishing (1,400) and 'production and retail of associated goods' (for example, wild flower seeds, natural history equipment – 1,860). The reviewers, the Dartington Institute, concluded that an 'identifiable' conservation industry had developed that overlapped with established industries such as agriculture, forestry and tourism (NCC 1987). The industry has grown considerably since, but calculating actual jobs is problematical. On the basis of the expansion of key voluntary bodies since 1985, the number of jobs in conservation could be twice or three times as high, more still if part-time jobs are included, though what is meant by nature conservation is almost in the eye of the reviewer (for instance, I am not at all sure whether most of what I do counts as 'nature conservation', though some of it, like this book, presumably is).

The 'conservation industry' offers more varied opportunities than when I started out, in the mid-1970s. Then it seemed natural to apply for an advertised post in the official nature conservation agency, then the NCC. That is still an option. There are careers to be made in government and the prospect of a comfortable salary at the higher grades. You are close to important events and can influence decision-making. You are likely to be among like-minded colleagues and in a culture with a strong esprit de corps. On the other hand, working for a conservation agency can disappoint if your interest lies primarily in science and field-based natural history. You spend most of your working hours in front of a computer screen, or in meetings, and have to put up with mindless rules and hierarchies. Almost everywhere today, in conservation bodies, and even in research institutions, careers are based on managerial skills, not on the quality of your science.

For many, life in the voluntary bodies may be more attractive, especially now that the pay scales are closer to civil service levels and there are so

ιy more job opportunities. There is less of a career structure, but more
rsonal freedom and a less stifling atmosphere. And you will probably get
ιt more. For the more adventurous, there are career opportunities in the
media, especially as cameramen and wildlife film-makers, or organising
wildlife holidays in Britain and around the world. Another conservation
job is the independent consultant where success is rewarded by money,
although to be successful in this field you need to be competitive, with
good business sense and strong nerves – as well as a willingness to break
bread with the 'enemy'.

Most jobs in the industry require a measure of personal dedication and
some basic qualification, generally a degree in a relevant subject. Those
going into government research need a background of university postgraduate research, preferably leading to a doctorate. When I was starting
out, the main choices were the still running (and still good) postgraduate
conservation course at University College London, and two or three postgraduate ecology courses. Today there is a much wider choice; for example, Imperial College and Wye offer postgraduate courses in environmental management or sustainable agriculture and rural development, and
there are also full-time or part-time courses available in every imaginable
aspect of environmental studies, in many cases for a basic degree. But the
best training for conservation work is still to do some, as a volunteer (it also
looks good on your CV). Dudley Stamp thought that the most important
qualification of all is a love of the natural world: 'Nature conservation is a
vocation, not a job'. Many people today might want to substitute 'environment' for 'nature'. They amount to the same thing, although, again, one
cannot help noticing that a commitment to environmental values does not
always entail a serious interest in wildlife.

A vision for the future?

In the wake of the foot-and-mouth crisis of 2001, Britain's conservation
groups weighed in with an ambitious 'vision' of how harmony could be
restored between wildlife and countryside. The 'Rebuilding Biodiversity
Group', chaired by wildlife gardener Chris Baines, made a bid to place biodiversity at the forefront of rural development: wetlands act as flood
defences, organic farming works with nature to rebuild soil fertility, woodlands stabilise the land and wildlife makes us all healthier and happier.
Baines is convinced that the damage of the past half-century can be
reversed by a grand scheme of habitat restoration and creation. The public wants 'a new vision', the politicians are convinced that change must
come, and farmers do not want to go down the same road again. The bits
of the jigsaw are coming together. Britain wants a new start.

It sounds attractive, and keys into the aspirant language of modern politics. However, such hopes often rely on brushing previous experience
under the carpet: the wish becomes father to the thought. In fact, planning does not necessarily deliver a more diverse countryside, and has often
given us the opposite. Community forests and habitat creation schemes
provide only second-rate copies of lost originals, mass-produced paper-

backs in place of leather-bound books. It is the fallacy of planners to
assume that beautiful, life-enriching countryside can be created instantly
by design. Landscapes are generally created by individuals doing what
seems best to them. You can encourage beauty and biodiversity by the right
incentives, but all that drawing-board conservation will ever achieve is a
McDonald's version of Britain, much the same from top to bottom, without meaning or detail. Who needs it?

How will the near-future really play out? Is it really all good news? Most of
the doomsters of the recent past got it wrong. The oil did not run out,
species did not become extinct, oil spills and acid rain did not devastate
Britain's wildlife, intensive farming did not go on until the last hedge was
grubbed up (though air pollution really did give us global warming). On the
other hand, ecologists overstated the potential utility of wild animals and
plants. Wild British plants have not been very beneficial to agriculture.
Arable farming can get by perfectly well without hedges. Ecology was not the
big idea whose time had come. History, like evolution, is unpredictable. In
1945, Britain's greatest plant ecologist, Sir Arthur Tansley, considered it was
'scarcely probable that the extension of agriculture will go much further, for
the limits of profitable agricultural land must have been reached in most
places' (Tansley 1945). Unfortunately, as events were to show, you could
grow bananas on Ben Nevis with enough state support.

Integrated, 'holistic' rural policies are probably here to stay. The process
began with the 1990 White Paper, *This Common Inheritance*, and advanced a
step further in 1995 with its successor, *Rural England: A nation committed to
a living countryside*, and its nod towards Agenda 21 and the principle of sus-

Arable farming has reduced natural chalk grassland to isolated patches on steep
slopes, such as this one, the aptly named Handkerchief Piece in the Berkshire Downs.

Arguably, EU is finally moving towards a more intelligent land-use policy in which food production will be only one element, and mainly on the better soils. Subsidies seem here to stay; no one will take the political risk of ending them, and conservation will require them even more than agriculture. All the same, it worked in New Zealand, where agriculture is more important to the economy and social fabric than in Britain. In 2001, the central focus of EU agricultural policy was still to modernise the industry and make it more competitive. It might also want to be more environment-friendly, but it does not really know how to go about this, and certainly has not resolved the fundamental contradiction between efficient farming and conserving natural resources.

In 2001, an outbreak of a non-fatal disease, foot-and-mouth, resulted in the greatest slaughter of livestock ever seen in Britain. The farmers immediately received generous compensation for the loss of their herds, but the tourist industry received next to nothing. If any good is to come out of the disaster it must be to hasten the end of the heedless headage payments that result in far more animals than the land can bear – but no one is banking on it. Just before the outbreak, the Government had set up a 'task force' to identify ways in which hill farmers in England can diversify and change from open-air sheep factories to 'sustainable business enterprises that contribute to the upland economy, society and environment'. Significantly, the task force was not chaired by a MAFF official, but by the chief executive of English Nature, David Arnold-Foster. It found that many hill farmers were not benefiting much from existing agri-environment schemes, neither living in an ESA, nor being able to meet the demands of Countryside Stewardship. Also that agricultural fraud was widespread. We have reached the stage where subsidies on hill farms average £33,000 per year, but the farmer's net income is closer to £9,000. Not surprisingly, the farmer's sons and daughters are looking for other work. A quarter of hill farmers are now aged 60 and over. About a third left the business between 1994 and 2001, and, on a recent projection, perhaps another half – 100,000 farmers – will go between 2001 and 2005, mostly from the smaller farms. The family farm will be replaced by the business farm. On traditional farms, nature thrived incidentally. On business farms it will have to be planned for.

Can the catastrophe of modern farming be nature's opportunity? We are threatened by the NFU with the spectre of 'unsightly' scrub on abandoned hill land. We need lose no sleep over that. More low, scrubby woodland extending up the bare, denuded hillsides of northern England and Wales would be wonderful. We might even see the reappearance of an almost forgotten habitat, ungrazed subalpine scrub and that all-but-lost wonder, a natural tree line. Unfortunately it won't happen. Fewer farms does not mean fewer sheep (or deer). The quota levels remain unchanged. Pasture management may become even more intensive, involving yet more ploughing and reseeding as the more business-like farmers that remain use every available opportunity to increase their profitability. At no time in history have Britain's farmers and landowners simply abandoned the land.

However, some form of restructuring of hill farming does seem likely. There is at least a rare opportunity to get off the treadmill of cheap food production and redirect subsidies towards schemes that benefit the environment.

On the better soils, farming will be more intensive than ever. On the less good and the bad, there may be more room for wildlife. The greatest threats to wildlife will come not so much from habitat destruction as neglect. Methods of reducing eutrophication and coping with invasive species will be badly needed and are likely to be expensive (undoing often costs more than doing). With building programmes averaging two million

Projected changes to farmland and former farmland over the next half-century. (English Nature)

Potential farming and land use changes 1995 to 2045

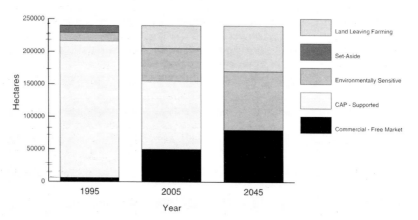

Potential uses of land leaving farming 1995 to 2045

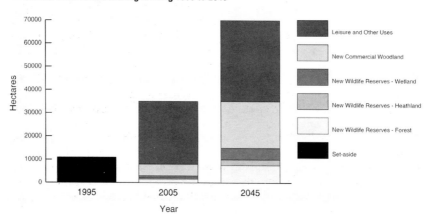

'homes' per decade – a city the size of Plymouth or Aberdeen every couple of years – the pressure on the countryside will be greater than ever. The countryside near the cities will become urban playgrounds. We must therefore take care that it remains countryside, and does not become suburbia. It will be hard for the authorities to resist the temptation to turn woods into 'safer' parks, or to plant red maples and daffodils on every roadside and picnic site, or shower us with streetlights and signs (including parking charges to pay for the daffodils and security cameras). Multimillion euro projects will come to the aid of some wildlife habitats, such as lowland heaths and acid bogs. The challenge will not be finding money so much as finding experienced stockmen and graziers. Created habitats, such as broad-leaf plantations and resown set-aside, will increase, which will be good news for the already common species capable of exploiting them.

Some traditional land uses that supported wildlife, such as grouse moors, will probably go out of business. If moor management continues, it will be to provide purple hills of blooming heather to attract tourists in early autumn rather than grouse for London restaurants. The rare breeds boom will continue, usefully grazing our wild places before being eaten by the better-off. I hope the conservation industry will go on encouraging traditional crafts – and new ones – that depend on the sustainable harvest of natural materials. In Britain the historical link between land use and nature is so strong that to break it would spell disaster. Sites managed sustainably by skilled reedsmen and charcoal-burners are, as a rule, better managed than by weekend volunteers, contractors or well-meaning idiots.

Protected sites will get bigger. European SACs are of a more sustainable, landscape-scale size than SSSIs (Portugal has only 50, but they cover 8.4 per cent of the entire country). The European Commission may possibly prove to be better at conservation than the UK government, and will almost certainly be more generous. As part of a much bigger pool, we will absorb European ideas and practices, and this is also to the good. There is, however, a danger that non-European SSSIs will lose out if conservationists jump wholeheartedly onto the European gravy train, and still more if they insist on going down the road of community planning and habitat creation. Other ideas floating around include English Nature's 'Lifescape' concept, which aims to 'join up the dots' of our special wildlife sites with corridors such as hedgerows and broad field margins 'through which species can move, making sites more ecologically viable' (English Nature 2000). It is hoped that local communities will contribute to the process, and that larger, less inclusive conservation areas will help build bridges between conservation and other land interests – and, most importantly perhaps, with educational institutions. An active generation of young field naturalists should be a number one priority; it does not yet exist.

The web and tangle of conservation institutions shows no signs of unravelling. Devolution triplicated it. The communications revolution makes better co-ordination possible, but spins its own webs. I suggest there is no chance at all of a single body emerging to administer nature conservation, access and rural planning in England. The investigators of a proposed merger between English Nature and the Countryside Commission in the mid-1990s concluded that it would cause more problems than solutions. The resulting body would collapse under its own bureaucratic weight. The splicing of the wildlife and countryside agencies in Scotland and Wales has not been a happy experiment. In both cases, wildlife has lost out, and by focusing on peripheral instead of core activities, the CCW and SNH have hazarded their own futures. In summer 2001, the Welsh Assembly was debating the question: Should CCW exist? The point is not the survival of that particular body but the probability that, if nature conservation becomes a dispersed activity, it will also become a low priority.

I doubt whether we shall see many fundamental changes among the voluntary bodies. As far as the county trusts are concerned, *vive la difference*. Local trusts are better at establishing relations with local authorities, farms and businesses than a remote headquarters would be. The county-based system works with the grain of human nature: the English, at least, are a nation of club members with a strong sense of local identity. There are, however, signs of division, between generations, between radical eco-activists and conservative naturalists. There is also a dangerous popular perception that wildlife means birds. On present trends, the RSPB may have five million members by 2020, but the other 99.9 per cent of wild animals will depend on crumbs falling from the bird table. As for wild plants, the BBC currently regards them as a minor component of its gardening

'Tomorrow's Heathland Heritage', funded by the EU, the Heritage Lottery Fund and landfill tax credits, was one of the most ambitious habitat restoration programmes of the 1990s. It aims to clear heathland of invasive scrub, and, where possible, introduce hardy cattle, such as this long-horn bull at Hartland Moor NNR in Dorset. (Natural Image/Bob Gibbons)

Environment-friendly river engineering on the Kennet at Ramsbury, funded by the Environment Agency. The idea is to create more river bed variety, with pools and riffles, along with more natural sloping banks.

programmes. The small charity Plantlife would have a glorious future ahead of it – if only we could manage to interest a tithe of the nation's millions of gardeners in wild plants.

Finally, what about species? Which animals and plants will fare best in the twenty-first century? 'Rebuilding biodiversity' is the goal. A few optimistic spirits have voiced the hope that the countryside of 2100 will be as rich as that of 1900. Their enthusiasm has overtaken their ecology. Under a planned programme of habitat restoration, the more opportunistic species among our wildlife, that is to say the common or invasive ones, should thrive, while those that require stable, semi-natural conditions will be confined to protected sites, namely the Natura 2000 sites and the larger SSSIs. Here, again, planning is not necessarily the solution. Global warming threatens biodiversity, come what may, especially the species of uplands and mountain tops. The current run of dull, wet summers may also harm the fortunes of many of our insects, irrespective of how well their habitats are cared for. The biodiversity process may ensure that extinctions remain few, at least among popular species. We will have to be careful to distinguish between glamour and substance, between quantity and quality. A disused opencast mine turned into a sort of wild garden with open access, laboratory-assisted wildlife and all the facilities may seem a more significant attraction than some midge-ridden bog miles away from anywhere, and of interest only to the specialist. In 1980, the great countryside debate was about the impact of intensive farming on wildlife. In 2001, it was about the future of farming with wildlife almost an afterthought. In books such as Graham Harvey's *The Killing of the Countryside* you get a sense that wildlife is brought in as supportive evidence – as prosecution witnesses so to speak. The debate is really about our own world, our own futures. Skylarks are one of the Government's 'indicators of environmental health', but it is our (spiritual) health that is meant, not that of the skylarks.

Perhaps the time has come to release field-based natural history from the belly of the conservation industry, where it has been confined these past 30 years. Is there not still something to be said for dumping the burden of care for a while and just enjoying nature for its own sake? A few years ago I was in the rain forest of Sabah, standing on a hillside overlooking a vista of glorious, virgin dipterocarp forest, with a sore on my hand, a tiger leech in my sock and strange, drunken sounds reverberating in the thick jungle air. It was a moment to savour, but my companion was unable to clear his mind of what was likely to happen to it: 'In a few years all this will be gone. It will be wall-to-wall oil-palm, like the rest.' Maybe, but concern about the future should not hold back delight in the present. In the 'wildscape' of the near-future, conservationists will have a considerable say. Many things, good, less good and indifferent, will be done in the name of nature conservation. But the objects of the great debate, the badgers and lizards and beetles, do not know that their names are on the annexes of a dozen conventions, nor that there is a five-figure Species Action Plan resting on their tiny heads. They just get on with their own lives, much of which are still mysterious. If we explore their world it should not only be because we need information for an action plan, but because we are intelligent beings with an innate curiosity about the natural world around us. To break free, naturalists will have to put the conservation industry behind them for a while, and rediscover that older quality embodied in the credo of the New Naturalist series, that 'inquiring spirit of the old naturalists' that James Fisher set out to reinvigorate, half a century ago. We should resist seeing wild animals as pets or 'targets' and respect their indifference to us, and the complete lack of personal contact every time a beast or bird looks us in the eye. We should affirm that there is more to nature than nature conservation. It is good for us to be reminded that nature is an infinitely more complex and tested scheme than anything we try to impose on it.

Appendices

APPENDIX 1: Main events affecting nature conservation in Great Britain, 1970–2000

1970 European Conservation Year. A Conservative Government was elected under Edward Heath. Third 'Countryside in 1970' conference was held. Conservation of Seals Act established a close season for common seals. Monsanto restricted use of PCBs after high residues were found in dead seabirds. The minister intervened to prevent fellings in the New Forest and preserve the unenclosed woods 'without regard to timber production objectives'. Nan Fairbrother's *New Lives, New Landscapes* was published, as was Max Nicholson's *The Environmental Revolution*.

1971 UNESCO 'Man and Biosphere' programme called for establishment of Biosphere Reserves with facilities for 'research, education and training'. A branch of Friends of the Earth was established in Britain. The Agricultural Development Advisory Service (ADAS) was set up. Dutch elm disease was detected in south-east England.

1972 *The framework of government research and development* (the Rothschild Report) established 'the customer/contractor principle' and gave notice of Government's intention to set up an independent statutory body, the Nature Conservancy Council. The Woodland Trust was founded by Kenneth Watkins. The Local Government Act set up two-tier district and county/regional councils. Government proposed to build a third London airport at Maplin Sands in Essex, noted for its wildfowl, then changed its mind. An endemic grass called interrupted brome died out in its last known site.

1973 UK entered the European Common Market and so joined the Common Agricultural Policy. The Nature Conservancy Council was established under Sir David Serpell (chairman) and R E Boote (director). The former Nature Conservancy's research arm became the Institute of Terrestrial Ecology. The badger (but not its home) was protected. The Water Act replaced local river boards with regional water authorities. The use of persistent aldrin and dieldrin pesticides for cereal seed dressing was banned voluntarily. Shell launched another 'Better Britain' competition.

1974 The county system was reorganised, introducing strange new places like Clwyd, Humberside and Cleveland. A Labour Government was elected, under Harold Wilson. Countryside Commission's *New Agricultural Landscapes* revealed dramatic changes to the landscape of England and Wales, resulting from the removal of hedgerows. The Control of Pollution Act made regional water authorities responsible for tackling pollution in rivers. The Sandford Report revealed weaknesses in our National Park system. National Nature Reserves were 'declared' at Swanton Novers Woods in Norfolk, Lathkill Dale in Derbyshire and Whitlaw Mosses in the Borders.

1975   The White Paper *Food from our own resources* advocated agricultural expansion. The Conservation of Wild Creatures and Wild Plants Act protected an assortment of rare animals, plants and insects. A Finance Act provided conditional exemption from Capital Transfer Tax for 'land of outstanding scientific interest' (LOSI). The Washington Convention on International Trade in Endangered Species of Wild Fauna and Flora (CITES) came into force. Membership of county wildlife trusts reached 100,000. RSNC launched the Watch group for young naturalists. Landlife, the urban conservation body, was founded in Liverpool. Local authorities cut back on road verge maintenance. MAFF started to gas badger setts hoping to control the spread of bovine TB. Dutch elm disease spread across central England. By 1977 most tall hedgerow elms were dead.

1976   North Sea oil began to flow. European Wetlands Campaign. Britain 'ratified its signature' on the Ramsar Convention on wetlands and designated the first batch of 'Ramsar' sites. During a long summer drought many heaths caught fire; two NNRs, Hartland Moor and Glasson Moss were badly burned. The National Trust acquired Wimpole Estate, Cambridgeshire. NCC acquired Gait Barrows in Lancashire 'after a decade of negotiation' during which it was badly damaged by quarrying. A Crofting Act gave crofting tenants the right to buy their land. The Government's Job Creation Programme provided a cheap source of employment on nature reserves. A Joint Otter Group was set up to investigate the decline of the otter. The BTO published the first *Atlas of Breeding Birds*.

1977   *A Nature Conservation Review*, an analysis and catalogue of biological 'key sites', was published. With that out of the way, the NCC started a *Geological Conservation Review*, 'a systematic basis for geological SSSIs' (the first of a projected 51-volume series of reports appeared 12 years later). NCC also published its first 'policy paper', *Nature Conservation and Agriculture*, proposing a strategy for farmland. The NFU/CLA countered with one of their own, called *Caring for the Countryside*, basically saying the countryside was safe with them. Planning guidelines on nature conservation were drawn up for local authorities. The first Red Data Book, on vascular plants, was published. In America, Greenpeace was founded.

1978   The Ribble Estuary was saved from reclamation at great expense and declared a nature reserve. In Exmoor National Park farmers turned moorland into fields. An application by Southern Water authority to drain Amberley Wild Brooks was turned down after a public inquiry. National Scenic Areas were designated in Scotland. The otter was legally protected in England and Wales (Scotland followed in 1982). Oil from the Amoco Cadiz tanker, grounded on the Brittany coast, killed a lot of seabirds. A cull of grey seals in Orkney was halted after a public outcry. W G Teagle's *The Endless Village* was published.

1979   The Conservatives under Margaret Thatcher were elected. The White Paper *Farming and the nation* offered a few sops to nature conservation. NCC and RSPB publicly disagreed over designating the Berwyn mountains in Wales; meanwhile the Economic Forestry Group afforested part of it. The EC Birds Directive required the setting up of Special Protection Areas (SPA) (the first batch were designated in 1982). The NCC published its report on *Nature Conservation in the Marine Environment.* Britain signed the Bonn Convention on the Conservation of Migratory Species. The large blue butterfly was declared extinct. BANC, the British Association of Nature Conservationists was founded. Wildlife Link was established, replacing the defunct Council of Nature. Attenborough's *Life on Earth* series was televised.

1980   The World Conservation Strategy, 'an exercise in confidence building', was launched. The NCC purchased Parsonage Down in Wiltshire. Administrative changes to Farm Capital Grants required SSSI owners to consult NCC. Horton Common in Dorset was mostly rotovated away. Permission to reclaim salt marsh at Gedney Drove End on The Wash was refused. Marion Shoard's *The Theft of the Countryside* was published, as was Richard Mabey's *The Common Ground.* RSPB acquired its half-millionth member. Sir Ralph Verney became NCC's third chairman just as the Wildlife and Countryside Bill began its tempestuous passage through Parliament. Richard Steele became its Director, replacing R.E. Boote.

1981   'The year when wildlife came to Westminster': the Wildlife and Countryside Bill was read, debated, amended and re-amended (June–July) before being assented to (October). Wildlife Link learned how to lobby. The National Heritage Memorial Fund helped NCC save the blanket bog of Blar nam Faoileag in Caithness. The National Trust for Scotland acquired the isle of Canna. The first Groundwork Trust was established at St Helens and Knowsley, Merseyside.

1982   After deciding it could not afford to save Romney Marsh, the NCC
notified West Sedgemoor in the Somerset Levels and was burned
in effigy for its pains. The first 'Section 29 order' was served – on
Baddesley Common SSSI in Hampshire. After a public inquiry, the
minister ruled against extending Cairngorm skiing into Lurcher's
Gully. DDT was finally banned under EC regulations 'except for
emergency use against cutworms'. Richard Body's *Agriculture: the
triumph and the shame* was published. Britain ratified the Bern
Convention on the Conservation of European Wildlife and
Natural Habitats.

1983   The NCC's chairman was 'sacked for doing his job' and replaced
by William Wilkinson. The UK published its response to the World
Conservation Strategy: *Earth Survival: A Conservation and
Development Programme for the UK*. The Government belatedly published its Financial Guidelines on SSSIs, and NCC staff knuckled
down to renotifying some 4,000 SSSIs to 30,000 owners and occupiers. Revised guidelines were also produced on what qualified as
SSSI. 'EC-funded Integrated Development Programmes' for the
western and northern isles failed to integrate nature conservation.
The John Muir Trust was formed to protect wilderness, and
Common Ground to promote cultural values in landscape.

1984   The NCC published *Nature Conservation in Great Britain*, a review
and national strategy recognising 'cultural values' as the main purpose of nature conservation. Shortly afterwards its head office
moved to Peterborough. The Wildlife & Countryside Act was tested and found wanting when proposed SSSIs were destroyed during the 'three month loophole' allowed for 'representations', and
also when attempted prosecutions failed due to poor drafting. In
Scotland NCC was obliged to buy off Fountain Forestry who
wished to afforest Creag Meagaidh SSSI. At Halvergate Marshes
protesters halted drainage by sitting in the excavator's bucket.

1985   A year of turning points. The Government's answer to Halvergate
was a Broads Grazing Marsh scheme in which farmers were paid to
maintain scenic character. The Forestry Commission published
Broadleaves in Britain, a significant policy shift in favour of native
trees and ancient, semi-natural woods. An EC Directive obliged
developers to make environmental impact assessments. The
Agriculture Act introduced 'Environmentally Sensitive Areas'
(ESAs); the first six were announced in 1986. The North Sea
Forum, later renamed the Marine Forum, was set up. There was a
stormy meeting between conservationists and islanders over the
future of Duich Moss SSSI on Islay. The Wildlife Trusts launched
an appeal under the slogan 'tomorrow is too late'. RSPB purchased Old Hall Marshes in Essex for £780,000.

1986   Nicholas Ridley became Secretary of State for the Environment. NCC published *Nature Conservation and Afforestation*, a policy paper arguing for more constraints on upland afforestation. The first Marine Nature Reserve was made, at Lundy. The Channel Tunnel Bill began its passage through Parliament. St Kilda was made a World Heritage Site. The British Hedgehog Society logged 5,000 members. Oliver Rackham's *The History of the Countryside* was published.

1987   European Year of the Environment (EYE). An unpredicted 'hurricane' lashed southern England, toppling millions of trees. The Brundtland Report on sustainable development, *Our Common Future*, was published. Government's 'Farming and Rural Development' package proposed spending more on attractive countryside, including more ESAs. The NCC's *Birds, bogs and forestry*, launched in Peterborough instead of Edinburgh, opposed forestry in the Flow Country. An expensive agreement preserving Elmley Marshes in Kent was concluded. The Dorset Trust bought Kingcombe Estate, noted for its wild flowers. Fisons started milling peat at Thorne Moors. *The Scotsman* thought the 'conservation pendulum has swung too far'.

1988   Sensing the public mood, Mrs Thatcher made a pale green speech to The Royal Society. Tax avoidance schemes on forestry were abolished. Ridley withdrew compensation for 'profits foregone' on forestry on SSSIs in England. Details of a voluntary set-aside scheme to reduce agricultural surpluses were announced. The NCC published *The Flow Country*, its second report on the Caithness peatlands. A Farm Woodland Scheme was set up to encourage tree planting on surplus farm land. The Broads Act gave the Broads Authority a new set of teeth, and the Broads the effective status of a National Park. The World Wildlife Fund (WWF) was renamed The Worldwide Fund for Nature (WWF).

1989   The Green Party received 14 per cent of the vote in the European elections, whilst a Gallup poll revealed that 84 per cent of the public were dissatisfied with the Government's record on the environment. The NCC published comprehensive guidelines on the selection of SSSIs, shortly before Ridley's announcement of its imminent replacement by three country-based wildlife agencies. The Freshwater Biological Association was swallowed up by ITE. The Water Act privatised the water companies, but set up a Watchdog, the National Rivers Authority. An unusually mild winter revived talk of global warming. The Countryside Commission advocated 'community forests' around cities, and a National Forest in the Midlands. It also introduced a Countryside Premium Scheme to encourage traditional land use. The first red kites were released. A new charity, Plantlife, was launched; so was a magazine 'for the modern naturalist', *British Wildlife.* The RSNC published *Losing Ground,* a report about habitat destruction. The RSPB celebrated its centenary by breaking the half-million membership barrier. The last coypu was eradicated from East Anglia. Peter Scott died.

1990   An Environment Protection Bill ('the Green Bill') had a messy passage, amid much public apprehension about the future of nature conservation after the break-up of the NCC. Chris Patten's White Paper *This Common Inheritance* launched a series of annual reviews under this title, but said more about how things are done than about what needs doing. More gales struck southern England. The Countryside Commission for Scotland made another call for National Parks in Scotland. Environmental Impact Assessments became mandatory for industrial developments. The Government embarked on a countryside monitoring survey (the results were published in 2000). The NCC completed unfinished business: an 'inventory' of ancient woodlands, a nature reserve at Fenns and Whixall Moss in Shropshire, publications on natural vegetation and estuaries, and an earth science strategy. It also notified its 5,000th SSSI and 2,000th management agreement. Skomer became the second Marine Nature Reserve. The RSPB purchased Abernethy Forest. The Shetland sand-eel fishery, blamed for seabird breeding failure, was suspended. The Department of Transport's 'Roads to Prosperity' promised a bonanza of new roads and bypasses, such as the one about to cut through Twyford Down, near Winchester. Bill McKibben's *The End of Nature* was published.

1991   Two new 'country agencies' took over the functions of the NCC in England and Wales: English Nature and the Countryside Council for Wales (CCW), together with a co-ordinating body, the Joint Nature Conservation Committee (JNCC). In Scotland, an interim body, the Nature Conservancy Council for Scotland, awaited further legislation. The Forestry Commission was reorganised into a Forestry Authority (grants and licences) and Forest Enterprise (managing the estate). A Countryside Stewardship scheme for England and Wales was launched. An EC Directive on wildlife habitats introduced yet another designation: 'SAC' (Special Area for Conservation). The RSNC report on roadworks, 'Death at Pooh Corner', estimated that some 40,000 badgers are killed by road traffic each year.

1992   The Natural Heritage (Scotland) Act launched Scottish Natural Heritage (SNH), combining the functions of the former NCC and CCS with a more 'softly softly' remit and a formal appeal procedure on SSSI designations. One of its first tasks was to pay a landowner over a million pounds for not wrecking an SSSI at Glen Lochay. CAP reforms reduced support for cereals and introduced compulsory set-aside in exchange for area payments. At the 'Earth Summit' in Rio, the UK signed a Convention on Biodiversity, a term borrowed from E.O. Wilson's *The Diversity of Life*, published that year. A new Badgers Act protected that animal's home. Wildlife Link published a 'health check' on SSSIs. An ITE research station was burned to the ground by animal rights protesters. The lagoon sandworm was declared extinct. A survey showed people walk an average of 250 miles a year but drive 5,000 miles.

1993 English Nature announced a string of new programmes and strategies, on 'natural areas', on the coast, on heathland restoration, on species recovery. More controversially it concluded a deal with Fisons over the future of peat bogs. More than a million hectares were declared eligible for ESA grants. Privatisation of the FC estate was considered and then rejected. 'Nitrate sensitive areas' to protect water supplies were designated, and more encouragement given to organic farmers. Thousands of tonnes of light oil were spilt from the wreck of the Braer, off Shetland. Stubble burning became illegal. Oxleas Wood in London, threatened by roadworks, won a reprieve. The National Trust banned deer hunting on its property. The RSPB published a new atlas of breeding birds. A survey showed Britons made two billion 'day visits' over the year, spending some £15 billion. Former RSPB director Peter Conder died.

1994 *Biodiversity: the UK Action Plan* was launched. Government published further White Paper 'responses to Rio' on sustainable development, sustainable forestry and climate change. The successful Countryside Stewardship scheme was taken over by MAFF. The minister decided against merging English Nature with the Countryside Commission. The nation rushed to access the joys of the 'information super-highway' via the Internet. The spectre of BSE threatened the future of beef farming as herd slaughtering began. New restrictions were imposed on out-of-town superstores. The RSNC was renamed the Wildlife Trusts Partnership. RSPB's attempts to preserve wild geese and corncrakes in the Western Isles caused friction. Glen Feshie was purchased by an unknown body called Will Woodlands. Membership of the National Trust reached 2.2 million.

1995 European Nature Conservation Year (ENCY). An Environment Act replaced National Park boards and committees with freestanding National Park Authorities (NPAs), now the sole planning authorities. It also merged the National Rivers Authority and HM

Inspectorate of Pollution to form The Environment Agency. The National Forest Company was set up to run the National Forest in the Midlands. Thousands of great crested newts were 'translocated' from a housing estate in Peterborough. The National Trust for Scotland purchased Mar Lodge Estate with the help of a mysterious trust. RSPB purchased its Forsinard nature reserve, protecting 7,127 hectares of flow country. The Countryside Movement (later renamed the Countryside Alliance) was formed to uphold fox hunting and other country matters. The CCW came under attack from minister Redwood. The White Paper, *Rural England: A nation committed to a living countryside* reflected a growing 'holistic' approach to rural development. Tributyl-tin (TBT) anti-fouling paint was banned, but substitutes proved almost as toxic. The Avon Gorge was polluted by toxic slag. Former RSPB director Ian Prestt died.

1996   Britain produced its first set of 'sustainable development indicators'. Work started on the Newbury bypass amid nightly televised scenes of protest. The Wildlife Trusts claimed proposed roadworks would damage 718 important wildlife sites nationwide, including 76 SSSIs and 31 ancient woods. A Wild Mammals (Protection) Act protected wild animals from gratuitous cruelty, but its main quarry, fox hunting, escaped. The Sea Empress ran aground on the Pembrokeshire coast, contaminating 200 kilometres of scenic coastline. RSPB purchased 242 hectares of carrot fields at Lakenheath to turn into reed beds. The Heritage Lottery Fund helped wildlife charities to acquire important nature reserves and restore lost habitat. The Countryside Premium Scheme was extended to Scotland. SNH withdrew its objection to a funicular railway on Cairngorm. The DoE projected a need for 4.4 million new homes over the next 20 years and set housing targets for county councils.

1997   New Labour under Tony Blair won the general election by a landslide. DoE was forthwith subsumed into a super-ministry, DETR, combining 'environment, transport and the regions'. The contentious Salisbury bypass was shelved in a policy U-turn on roads. A new regulation empowered local authorities to protect hedgerows of conservation or historic interest. After dry weather, some 89 SSSIs were said to be at risk from water abstraction. At the Kyoto conference, Government committed itself to reduce greenhouse gas emissions to 20 per cent below their 1990 levels by 2010. The Royal Botanic Gardens received a lottery jackpot for a Millennium Seed Bank of British and world plants. English Nature attracted critical attention after its less than zealous performance at Offham Down and Thorne Moors. Residents bought the isle of Eigg in partnership with Scottish Wildlife Trust and Highland

Regional Council. Basking shark and water vole received legal protection, but the viper's bugloss moth was deemed extinct (despite protection). RSPB membership reached one million.

1998 Scottish Natural Heritage submitted proposals to Government for the designation of National Parks in Scotland. A scheme to slaughter badgers in selected areas to test bovine TB control policy began in western England. The Raptor Study Group published its report on the Langholm experiment, concluding that hen harriers take significant numbers of grouse. In the West Midlands, ruddy ducks were shot to prevent them from mating with Spanish white-headed ducks. RSPB launched the Land for Life campaign calling for better legal protection for wildlife. It found many farm birds had declined by 50 per cent or more since 1975, including lapwing, skylark, linnet, bullfinch, turtle dove and tree sparrow. The budget of the BTO, which supplied this information, was cut. The last British pool frog died in captivity. An informal body, 'Flora locale' was formed to keep an eye on imported trees and wild flower seeds used in conservation schemes. The EU ruled that Britain was wrong to fail to designate SACs on grounds of imminent development. Overgrazing was revealed to be the most widespread cause of 'loss and damage' on SSSIs. Morton Boyd died.

1999 The Countryside Commission and Rural Development Commission were merged to form the Countryside Agency. The Government's 'rural development package' proposed to siphon off £1.6 billion of EU agricultural subsidies for agri-environment schemes over the next five years. A new strategy, '*A Better Quality of Life*', set out revised indicators of sustainability. Housing targets were revised and directed from 'greenfield' to 'brownfield' sites i.e. urban wasteland. Britain submitted 340 candidate SAC sites to the European Commission, but was told it was not enough. Some 8,000 birds 'were made homeless' as the waters closed over the Cardiff Bay mudflats. About 100,000 seabirds perished in oil slicks from the tanker *Erika*, which broke up off the Brittany coast. In Wales, an agri-environment scheme called Tir Gofal replaced ESAs. The National Trust bought Snowdon with help from the actor Anthony Hopkins. Environmental protesters hijacked a meeting of the World Trade Organisation in Seattle. Greenpeace activists cut down a Genetically Modified crop. The beef crisis was followed by a lamb crisis and then a pig crisis. Legislation protecting important wildlife sites was included in the Queen's Speech.

2000 The Countryside and Rights of Way Bill ('CROW Bill') became law, strengthening SSSIs and giving the public the right to roam on open land – but only after the Countryside Agency has completed its maps, which may take years. Rebellious Labour backbenchers were promised a free vote on hunting with hounds. A government survey revealed huge losses in biodiversity from nitrogen 'fallout'. Plantlife reported on local losses of wild flowers and a vast increase in stinging nettles. Ben Nevis was bought by the John Muir Trust. A Bill to set up National Parks in Scotland was passed by the Scottish Parliament. Two more National Parks were proposed for England: the New Forest and South Downs. A National Wildflower Centre opened in Liverpool. A jury found Greenpeace protesters had a 'lawful excuse' for destroying a crop of GM maize, and acquitted them. The RSPB purchased Hope Farm in Cambridgeshire and 352 hectares of Rainham Marshes in London. Climate change was a hot topic. Monthly temperature averages over the past two years were consistently higher than normal. The floods of October–December 2000 were the most widespread in 50 years, prompting the minister to advise us to 'wake up' to global warming. The RSPB urged us to put out grain for starving farm birds.

2001 Over three million cattle, sheep and pigs were slaughtered during an outbreak of foot-and-mouth disease. By the end of the year it had cost about £2.7 billion in compensation payments. Public footpaths were shut for most of the spring. After the General Election in June, MAFF was abolished and a new Department of Environment, Food and Rural Affairs formed (DEFRA – pronounced Deff-Ra). Scottish Executive published a policy statement, *The Nature of Scotland*, promising reform of wildlife laws, but not necessarily in the direction of conservation. *Biodiversity Counts*, a review by the voluntary bodies, chalked up some successes, but criticised some government departments for failing 'to fully engage' with the BAP process. The Hills Task Force's report proposed payments for environment-friendly farming in all Less Favoured Areas. Attempts began to eradicate mink from the Outer Hebrides. Planning permission was refused for a controversial bypass at Hastings. An inventory of England's trees found there are 25 trees for every person, more than at any time since the Middle Ages. A report on climate change predicted extinction in the next 50 years for some mountain species. The fossil-rich 'Jurassic coast' of Devon and Dorset was dubbed a World Heritage site. A government review report recommended the removal of the JNCC from the grasp of the country conservation agencies, and renaming it the UK Nature Advisory Council.

APPENDIX 2: Glossary of conservation words and abbreviations

Areas of Special Scientific Interest (ASSIs) The equivalent of SSSIs within Northern Ireland.

Areas of Outstanding Natural Beauty (AONBs) Areas in England and Wales designated for their attractive scenery, such as the Cotswolds, Chilterns and Mendips. Nature conservation is not among their formal objectives.

Biodiversity The variety of living things, including the habitats that support them and genetic variation within species.

Biodiversity Action Plan (BAP) The UK's commitment to maintain and preferably increase the variety of native wildlife. It entails 'holding the line' on species and their habitats through conservaton planning.

CITES (Convention on International Trade in Endangered Species of Wild Fauna and Flora) Commonly referred to as CITES (sy-teez), set up in 1975 and implemented by a European Union regulation in 1983. The regulation of wildlife trade in Britain is the responsibility of the JNCC (animals) and the Royal Botanic Gardens, Kew (plants). Enforcement is the business of HM Customs and Excise and the police.

Countryside Agency Formed in 1999 by merging the Countryside Commission in England and the Rural Development Commission. It administers National Parks and ancient monuments and 'works to conserve the beauty of the English countryside and to help people enjoy it'.

Countryside Council for Wales (CCW) The government agency responsible for countryside and wildlife in Wales, formed in 1991 from elements of the NCC and the Countryside Commission. It now reports to the Welsh Assembly.

Countryside Stewardship A Countryside Agency-run scheme offering English farmers flat-rate payments for promoting conservation on a range of natural habitats. In Scotland the equivalent is the Rural Stewardship Scheme, in Wales, *Tir Gofal.*

EC LIFE – Nature European programme that offers funding for 'innovative projects' to manage and enhance SPAs or SACs (Natura 2000 sites). Budget for 2000–2005: 300 million euros. The 'LIFE' bit stands for Financial Instrument on the Environment.

English Nature (EN) The government agency responsible for nature conservation in England, established in 1991.

Environmentally Sensitive Areas (ESAs) Scheme administered by the agriculture departments to encourage less intensive, more environment-friendly farming within designated areas. It has been up and running since 1986.

Farming and Wildlife Advisory Groups (FWAGs) Local associations that advise farmers on wildlife conservation and build bridges between the two interests.

Geological Conservation Review (GCR) A major review carried out by conservation agencies and scientific institutions over the past quarter-century, identifying the most important sites for earth heritage conservation in Britain. It is currently being published in 42 expensive volumes.

Groundwork Groundwork Trusts promote environmental action mainly in towns for the benefit of local communities. They are co-ordinated by Groundwork Foundation, established in 1985.

Habitat Action Plans (HAPs) Conservation plans for habitats listed in the UK Biodiversity Action Plan, such as fens, chalk grassland and oak woodland, that either are declining or are important for rare species, or are important internationally.

Heritage Coasts Non-statutory label used by the Countryside Agency and CCW for coastlines of scenic beauty.

Industry and Nature Conservation Associations (INCAs) Usually referred to as INCAs, an informal partnership between the two interests, analogous with FWAGs, originally set up on Teesside in the late 1980s.

IUCN What is now called the World Conservation Union, the IUCN has retained its original initials, which stood for 'International Union for the Conservation of Nature and Natural Resources'. It is the world forum on conservation matters, especially threatened species, with 541 corporate members from 116 countries. It has a Conservation Monitoring Centre in Cambridge.

Joint Nature Conservation Committee (JNCC) The body set up to advise government on GB-wide nature conservation issues since the formation of the three country nature conservation agencies in 1991–92. Shortly to be renamed UK Nature Advisory Council.

Less Favoured Areas (LFAs) Places, mainly in the uplands and the isles, where agriculture is marginal and special subsidies are available to support farming.

Limestone Pavement Orders (LPOs) As a non-sustainable resource, limestone pavements are specially protected, initially under the Wildlife and Countryside Act and later under the EC Habitats Directive made in 1994.

Local Biodiversity Action Plans (LBAPs) Local BAPs with their own priorities and targets, intended to contribute to the national BAP and 'deliver targets on the ground'.

Local Nature Reserves (LNRs) Statutory nature reserves run by local authorities, often for amenity purposes. They are not all SSSIs, but must be of special value locally.

Marine Nature Reserves (MNRs) Statutory protected area below the waterline, set up by the Wildlife and Countryside Act 1981, but scuppered by vested interests. Effectively superseded by SACs.

National Biodiversity Network (NBN) The developing electronic network that stores biodiversity information and biological records. Run by a consortium of bodies, currently administered by the JNCC.

National Heritage Areas (NHAs) The Scottish equivalent of AONBs and weak substitutes for National Parks, effective since 1980. Formerly known as National Scenic Areas. Nature conservation is not a formal objective.

National Nature Reserves (NNRs) A nationwide network of nature reserves owned or managed by English Nature, CCW or SNH to conserve wildlife or geological interest. All are important examples of natural habitats or geological formations, and are also designated SSSIs.

National Parks Ten mainly upland areas in England and Wales, established under the National Parks and Access to the Countryside Act of 1949 for amenity and public access. They are administered by Park Authorities under the Countryside Agency (England) or CCW (Wales). The Broads is a quasi-National Park with its own legislation. So, more doubtfully, is the New Forest. There are current plans for a series of National Parks in Scotland, starting with Loch Lomond and The Trossachs. Surprisingly, nature conservation has never been a major objective, except in SSSIs.

Natura 2000 The projected network of protected wildlife sites (SPAs and SACs) across the European Union.

Natural Areas English Nature's system of dividing the country into 120 'biogeographic zones' to provide an ecological framework for conservation planning. SNH has a similar scheme called Natural Heritage Zones.

NCC The Nature Conservancy Council, the government's advisory body on wildlife and geology between 1973 and 1991.

Nature Conservation Orders Orders made under Section 29 of the Wildlife and Countryside Act by the Secretary of State responsible for environmental matters to conserve the nature conservation interest of a site. They were repealed as no longer necessary in 2000.

Non-Governmental Organisations (NGOs) Slightly disparaging label given to the voluntary bodies by government and their agencies. Perhaps we should retaliate by calling them the Non-Voluntary Organisations.

Potentially Damaging Operations (PDOs) The embarrassing phrase used for activities on SSSIs that require permission from the conservation agency. In Scotland they are looking for a better way.

Red Data Books Lists of threatened species along with statements of their status and ecology. British Red Data Books are traditionally confined to Great Britain and the Isle of Man.

Reserves Enhancement Scheme English Nature's scheme to improve the state of SSSIs managed, usually as nature reserves, by voluntary bodies, and to increase their appreciation by the public.

RSPB The Royal Society for the Protection of Birds, Britain's most successful wildlife charity.

Scottish Natural Heritage (SNH) The government agency responsible for countryside and wildlife in Scotland, formed in 1992 from the Scottish NCC and the Countryside Commission for Scotland. It now reports to the Scottish Executive.

Sites of Special Scientific Interest (SSSIs) The statutory instrument of nature conservation in Britain since 1981. They represent the best examples of wildlife habitats and geological features over the full range of natural variation.

Special Area for Conservation (SAC) Potentially powerful Euro-designation made under the EC Habitats Directive for habitats and species of special interest. Candidate SACs where due consultation has taken place are referred to as cSACs. Those at an earlier stage are 'proposed' or pSACs.

Special Protection Areas (SPAs) Euro-designation for areas of importance for rare or decreasing birds or for migratory species. All SPAs are also SSSIs.

Species Action Plans (SAPs) Conservation plans for species listed in the UK

Biodiversity Action Plan that are threatened globally or are declining in the UK.

Species Recovery Programme English Nature's scheme to increase the numbers of rare or threatened species, such as red kite, starfruit and ladybird spider.

Sustainability A sustainable development is one that allows social and economic needs to be met without damaging the environment. (If we really lived like that, we wouldn't need nature conservation.)

Tree Preservation Orders (TPOs) Orders made by local authorities to protect trees or groups of trees for public enjoyment. Unfortunately TPOs are made only for landscape reasons, not for interesting wildlife nor even for rare and special trees.

Wildlife corridors Strips of habitat, such as hedgerows or field headlands, that link wildlife sites, allowing animals, plants and insects to move from one place to the next, and preventing genetic isolation.

Wildlife Enhancement Scheme English Nature's scheme to strengthen partnerships with SSSI owners and managers for the benefit of wildlife. It offers annual payments for managing the land beneficially, and fixed-costs payments for capital work.

Woodland Grant Scheme Forestry Commission grants for managing and improving woods, and for excluding stock from woods in hill areas. It is available for woods over 0.25 hectares with over 20 per cent canopy cover.

World Conservation Strategy (WCS) A call for the world to live within its resources. Subtitled 'Living Resource Conservation for Sustainable Development', the strategy was instigated by the IUCN and the United Nations Environment Programme (UNEP) and launched in 1980. The UK's response, three years later, was essentially to claim we had done all that already.

Unofficial acronyms

BANANA Build Absolutely Nothing Anywhere Near Anybody
NIMBY Not In My Backyard
NIMPOO Not in My Period of Office
CUCO Cough Up and Clear Off. A reference to the discredited policy of financial compensation 'for profits foregone'.

References

Adams, W.M. (1986). *Nature's Place. Conservation sites and countryside change.* Allen & Unwin, London.

Adams, W.M. (1996). *Future Nature: A vision for conservation.* Earthscan Publications, London.

Adams, W.M. (2001). Joined-up conservation. *Ecos*, 22(1), 22–27.

Asher, J., Warren, M., Fox, R., Harding, P., Jeffcoate, G. & Jeffcoate, S. (2001). *The Millennium Atlas of Butterflies in Britain and Ireland.* Oxford University Press, Oxford.

Avery, M. (1993). Arctic tern. In: Gibbons, D.W. (Ed.) *The New Atlas of Breeding Birds in Britain and Ireland: 1988–1991*, pp. 220–21. Poyser, London.

Baker, P., Newman, T. & Harris, S. (2001). Bristol's Foxes – 40 years of change. *British Wildlife* 12 (6), 411–417.

Banks, B., Foster, J., Langton, T. & Morgan, K. (2000). British Bullfrogs? *British Wildlife*, 11(5), 327–30.

Beardsall, J. (2001). 'Dam fine fellows.' *Daily Telegraph*, 12 May 2001.

Berry, R.J. (2000). *Orkney Nature.* Poyser Natural History, London.

Bignell, E. (1999). Agenda 2000. The Common Agricultural Policy reform proposals. *British Wildlife*, 10(3), 172–176.

Blunden, J. & Turner, G. (1985). *Critical Countryside.* BBC Books, London.

Boatman, N. & Stoate, C. (1999). Arable farming and wildlife – can they co-exist? *British Wildlife*, 10(4), 260–267.

Body, R. (1984). *Farming in the Clouds.* Temple Smith, London.

Bowen, H. (2001). *The Flora of Dorset.* Pisces Publications, Newbury.

Box, J. & Bramwell, H. (1998). Long-term changes in grazing in Sutton Park National Nature Reserve. *British Wildlife* 10(2), 69–75.

Boyd, J.M. (1999). *The Song of the Sandpiper. Memoirs of a Scottish naturalist.* Colin Baxter, Grantown-on-Spey.

Branson, A. (2000). The Wetland Centre, London. *British Wildlife* 11(4), 274–77.

Burton, N. (2001). Displaced redshank finding refuge. *BTO News*, 234, 10–11.

Colebourn, P. & Gibbons, B. (1990). *Britain's Countryside Heritage. A guide to the landscape.* Blandford and National Trust, London.

Conroy, J.W.H., Kitchener, A.C. & Gibson, J.A. (1998). The history of the Beaver in Scotland and its future reintroduction. In: Lambert, R.A. (Ed). *Species History in Scotland.* Scottish Cultural Press, Edinburgh.

Cosgrove, P., Hastie, L. & Young, M. (2000). Freshwater Pearl Mussels in peril. *British Wildlife* 11(5), 340–47.

Coulthard, N. & Scott, M. (2001). *Flowers of the forest. Managing Scottish woodlands for wild plant biodiversity.* Plantlife, London.

Coward, T.A. (1920). *Birds of the British Isles and Their Eggs.* Frederick Warne, London.

Cranbrook, Earl of (2001). Fifty years of statutory nature conservation in Great Britain. *In*: Hawksworth, D.L. (Ed.) *The*

Changing Wildlife of Great Britain and Ireland, 1–22. Taylor & Francis, London.

Dalyell, T. (1989). Thistle Diary. *New Scientist,* 2 September 1989.

Deere-Jones, T. (2001) 'Pair trawling kills dolphins'. *BBC Wildlife,* October 2001, p.38.

DETR (2000). *Countryside Survey 2000. Accounting for nature: assessing habitats in the UK countryside.* DETR, London.

Easterbrook, G. (1995). *A Moment on the Earth: The coming age of environmental optimism.* Penguin Books, London.

English Nature (2000). *Annual report 1999–2000.* English Nature, Peterborough.

Evans, D. (1997). *A History of Nature Conservation in Britain.* 2nd ed. Routledge, London.

Fairbrother, N. (1970). *New Lives, New Landscapes.* The Architectural Press.

Festing, S. (1996). The third battle of Newbury – war in the trees. *Ecos* 17(2), 41–49.

Firbank, L.G. *et al.* (2000). *Causes of change in British vegetation. Ecofact Volume 3.* Centre for Ecology and Hydrology, NERC.

French, C.N., Murphy, R.J. & Atkinson, M. (1999). *Flora of Cornwall.* Wheal Seton Press, Camborne.

Friday, L. (Ed.). (1997). *Wicken Fen. The making of a wetland nature reserve.* Harley Books, Colchester.

Gault, C. (1997). *A Moving Story. Species and community translocation in the UK: a review of policy, principle, planning and practice.* WWF-UK, Godalming.

George, M. (1990). 'A nightmare called nitrates'. *NCC News,* 6, April 1990. P.2.

Gibbons, D.W., Reid, J.B. & Chapman, R.A. (1993). *The New Atlas of Breeding Birds in Britain and Ireland*: 1988–1991. Poyser, London.

Gilbert, O.L. & Bevan, D. (1997). The effect of urbanisation on ancient woods. *British Wildlife,* 9(4), 213–218.

Gilbert, O.L. & Anderson, P. (1998). *Habitat creation and repair.* Oxford University Press, Oxford.

Harris, S. (2001). 'TB trial taken to task.' *BBC Wildlife,* June 2001, p.37.

Harris, S., Morris, P., Wray, S. & Yalden, D. (1995). *A review of British mammals: population estimates and conservation status of British mammals other than cetaceans.* JNCC, Peterborough.

Harvey, G. (1997). *The Killing of the Countryside.* Jonathan Cape, London.

Haskins, L. (2000). Heathlands in an urban setting – effects of urban development on heathlands of south-east Dorset. *British Wildlife,* 11(4), 229–237.

Henderson, S.A. (2001). The vegetation associated with *Spiranthes romanzoffiana* Cham., Irish Lady's-tresses, on the Isle of Coll, Inner Hebrides. *Watsonia,* 23(4), 493–503.

Hunter, J. (1995). *On the Other Side of Sorrow: Nature and people in the Scottish Highlands.* Mainstream, Edinburgh.

Johnston, J.L. (2000). *Scotland's Nature in Trust. The National Trust for Scotland and its wildlife and crofting management.* Poyser Natural History, London.

Jones, A. (2001). Comment: we plough the fields, but what do we scatter? A look at the science and practice of grassland restoration. *British Wildlife,* 12(4), 229–235.

Kirby, P. (1992). *Habitat Management for Invertebrates: a practical handbook.* JNCC, Peterborough.

Lack, P. (1986). *The Atlas of Wintering Birds in Britain and Ireland.* Poyser, London.

Latham, D. (1994). Translocation of amphibians. Letter to editor. *British Wildlife,* 5(4), p. 272.

Lawson, T. (1998). Badgers prove to be a marketing issue, not a conservation case. *Ecos,* 19(2), 71–76.

Lawson, T. (2001). Trial and error. *BBC Wildlife.* February 2001, 42–43.

Love, J. (1994). White-tailed Eagle. In: *The New Atlas of Breeding Birds.* BTO & Poyser, London. 100–101.

Lowe, P., Cox, G., MacEwen, M., O'Riordan, T. & Winter, M. (1986). *Countryside Conflicts. The Politics of farming, forestry and conservation.* Gower, Aldershot.

Mabey, R. (1980). *The Common Ground. A place for nature in Britain's future.* Hutchinson in association with NCC, London.

MacEwan, A. & MacEwan, M. (1982). An unprincipled Act? *The Planner,* May/June, 69–71.

Mackay, D. (1995). *Scotland's Rural Land Use Agencies. The history and effectiveness in Scotland of the Forestry Commission, Nature Conservancy Council and Countryside Commission for Scotland.* Scottish Cultural Press, Aberdeen.

McKibben, B. (1990). *The End of Nature.* Penguin Books, London.

Mantle, G., Power, J. & Jones, A. (1999). The conservation of grasslands – HLF's effects in Wiltshire. *Ecos* 20(3/4), 29–35.

Marren, P.R. (1979). *Muir of Dinnet. Portrait of a National Nature Reserve.* NCC, Edinburgh.

Marren, P.R. (1994). *England's National Nature Reserves.* Poyser Natural History, London.

Marren, P.R. (1995). *The New Naturalists.* HarperCollins, London.

Mather, A.S. (1987). The Greening of Scotland? *Scottish Geographical Magazine.*

Mead, C. (2000). *The State of the Nation's Birds.* Whittet Books, Stowmarket.

Mellanby, K. (1981). *Farming and Wildlife.* Collins New Naturalist, London.

Mitchell, I. (1999). *Isles of the West. A Hebridean Voyage.* Canongate, Edinburgh.

Moore, N.W. (1987). *The Bird of Time. The science and politics of nature conservation.* Cambridge University Press, Cambridge.

Morris, P.A. (1993). *A Red Data Book for British Mammals.* The Mammal Society, London.

Moss, B. (2001). *The Broads. The people's wetland.* HarperCollins, London, pp. 338–345.

Munro, L. & Munro, C. (2000). The East Tennants reef seafan study. *Marine Conservation,* 5(1), p.15.

Nature Conservancy (1968). *Progress 1964–1968.* Nature Conservancy, London.

Nature Conservancy Council (1978). Fourth Report. HMSO, London. p. 3.

NCC (1979). *Nature conservation in the marine environment.* NCC, London.

NCC (1984). *Nature conservation in Great Britain.* NCC, Peterborough.

NCC (1986). *Nature conservation and afforestation in Britain.* NCC, Peterborough.

NCC (1987). *Birds, bogs and forestry: The peatlands of Caithness and Sutherland.* NCC, Peterborough.

NCC (1987). Nature conservation and employment. In: *Topical Issues,* 3(1), 9–10. NCC, Peterborough.

NCC (1988). *The Flow Country. The peatlands of Caithness and*

Sutherland. NCC, Peterborough.

Nellist, K. & Crane, K. (2001). Sea Eagles – reclaiming the west. *British Wildlife,* 12(4), 264–270.

Nelson, E.C. (2000). *Sea beans and nickar nuts.* BSBI Handbooks, No. 10. BSBI, London.

Newby, H. (Ed.). (1995). *The National Trust.* National Trust, London. It contains a chapter on 'The first hundred years' by social historian David Cannadine, pp. 11–31.

Nicholson, E.M. (1957). *Britain's Nature Reserves.* Country Life, London.

Nicholson, E.M. (1970). *The Environmental Revolution.* Hodder & Stoughton, London.

Nicholson, E.M. (1987). *The New Environmental Age.* Cambridge University Press.

Oliver, G. (2001). The freshwater pearl mussel in Wales – a terminal decline? *Natur Cymru,* 1(1), 18–22.

Pennington, M. (1996). *Conservation and the Countryside: by Quango or Market?* IEA Studies on the Environment, 6. Institute of Economic Affairs, London.

Peterken, G.F. (1996). *Natural Woodland. Ecology and conservation in northern temperate regions.* Cambridge University Press, Cambridge.

Phillips, M. (1998). Preface in: Marren, P.R. *'Greater Protection for Wildlife'? Wildlife sites under threat in ministers' constituencies.* Friends of the Earth, London.

Phillips, M. & Huggett, D. (2001). From passive to positive – the Countryside Act 2000 and British wildlife. *British Wildlife,* 12(4), 237–243.

Plantlife (2000). *At War with Aliens. Changes needed to protect native plants from invasive species.* Plantlife, London.

Purseglove, J. (1988). *Taming the Flood: A history and natural history of rivers and wetlands.* Oxford University Press, Oxford.

Rackham, O. (1986). *The History of the Countryside.* Dent, London, p.28.

Rackham, O. (1990). *Trees and Woodland in the British Landscape.* Revised edition. Dent, London.

Ratcliffe, D.A. (Ed.). (1977). *A Nature Conservation Review. The selection of biological sites of national importance to nature conservation in Britain.* Cambridge University Press, Cambridge.

Ratcliffe, D.A. (2000). *In Search of Nature.* Peregrine Books, Leeds.

Ratcliffe, J. & Sandison, W. (2001). Puffin Island – will removing rats bring back the puffins? *Natur Cymru,* 1(1), 23–26.

Read, H.J. (Ed.). (1991). *Pollard and veteran tree management.* Corporation of London, London.

Ridley, M. (1996). *Down to Earth II. Combating Environmental Myths.* IEA Studies on the Environment, 7. Institute of Economic Affairs, London.

Rose, C. & Secrett, C. (1982). *Cash or crisis: the imminent failure of the Wildlife and Countryside Act.* Friends of the Earth and BANC, London.

Rothschild, M. & Marren, P. (1997). *Rothschild's Reserves. Time and fragile nature.* Balaban Publishers/ Harley Books, Colchester.

Rowell, T.A. (1993). *SSSIs: A Health Check. A review of the statutory protection afforded to Sites of Special Scientific Interest in Great Britain.* Wildlife Link, London.

Samstag, T. (1989). *For the Love of Birds: The story of the RSPB.* RSPB, Sandy.

Scott, M. (1992). NCC and after: the changing faces of British nature conservation. *British Wildlife*, 3(4), 214–221.

Scott, M. (1996). The Country Councils five years on – watchdogs or lapdogs? *British Wildlife*, 8(2), 81–86.

Scott, S. (2001). Marine finfish farming in Scotland. *British Wildlife*, 12(6), 394–401.

Scottish Executive (2001). *The Nature of Scotland. A policy statement.* Scottish Executive, Edinburgh.

Scottish Natural Heritage (2000). *Facts and Figures 1999/2000.* SNH, Edinburgh.

Scottish Natural Heritage (2001). *An Audit of Alien Species in Scotland.* SNH, Edinburgh.

Shaw, P. (1994). Orchid woods and floating islands – the ecology of fly ash. *British Wildlife*, 5(3), 149–157.

Sheail, J. (1976). *Nature in Trust. The history of nature conservation in Britain.* Blackie, Glasgow.

Sheail, J. (1998). *Nature Conservation in Britain. The formative years.* Stationery Office, London.

Shepherd, P. (1998). *The Plants of Nottingham. A city flora.* Wildtrack Publishing, Sheffield.

Shoard, M. (1980). *The Theft of the Countryside.* Temple Smith, London.

Shoard, M. (1999). *A Right to Roam. Should we open up Britain's countryside?* Oxford University Press.

Simpson, D. (2001). Whatever happened to the Ravenglass gullery? *British Wildlife* 12(3), 153–62.

Smith, K., Welch, G., Tyler, G., Gilbert, G., Hawkins, I. & Hirons, G. (2000). Management of RSPB Minsmere Reserve reedbeds and its impact on breeding Bitterns. *British Wildlife* 12(1), 16–21.

Smith, P. & Briggs, J. (1999). Zander – the hidden invader. *British Wildlife*, 11(1), 2–8.

Smout, T.C. (2000). *Nature Contested. Environmental history in Scotland and Northern England since 1600.* Edinburgh University Press, Edinburgh.

Smout, T.C. & Lambert, R.A. (1999). *Rothiemurchus. Nature and people on a Highland estate 1500–2000.* Scottish Cultural Press, Edinburgh.

Stamp, Sir Dudley (1969). *Nature Conservation in Britain.* Collins, London.

Steele, R.C. & Schofield, J.M. (1973). Conservation and management. In: Steele, R.C. & Welch, R.C. (Eds). *Monks Wood. A Nature Reserve Record*, 296–335. NERC, Huntingdon.

Stephenson, G. (1997). Is there life after subsidies? The New Zealand experience. *Ecos*, 18 (3/4), 22–26.

Steven, H.M. & Carlisle, A. (1959). *The Native Pinewoods of Scotland.* Oliver & Boyd, Edinburgh.

Stewart, A., Pearman, D.A. & Preston, C.D. (1994). *Scarce Plants in Britain.* JNCC, Peterborough.

Stubbs, A.E. (2001). Flies. In: Hawksworth, D.L. (Ed.). *The Changing Wildlife of Great Britain and Ireland*, 239–261.

Sullivan, F. (2001). In my view. *BBC Wildlife*, September 2001, 46–47.

Tansley, A.G. (1945). *Our Heritage of Wild Nature.* Cambridge University Press.

Taylor, M.B. (1996). *Wildlife Crime. A guide to wildlife law enforcement in the United Kingdom.* Stationery Office, London.

Teacher, J. & Raine, P. (Eds). (1998). *Wild Kent: its nature and landscape.* Kent Wildlife Trust, Maidstone.

Teagle, W.G. (1978). *The Endless Village.* NCC, London.

Thomas, K. (1983). *Man and the Natural World. Changing attitudes in England 1500–1800.* Allen Lane, London.

Tobin, B. & Taylor, B. (1996). Golf and wildlife. *British Wildlife,* 7(3), 137–146.

Tompkins, S. (1989). *Forestry in Crisis: the Battle for the Hills.* Christopher Helm, London.

Tubbs, C. (1999). *The ecology, conservation and history of The Solent.* Packard Publishing, Chichester.

UK Climate Impacts Programme (2001). *Climate Change and Nature Conservation in Britain and Ireland.* Summary Report. UKCIP, London.

Vittery, A. (1982). Wildlife Bill: inner workings. *Natural Selection,* 6/7, I–x.

Wheater, C.P. (1999). *Urban habitats.* Habitat guides series, Routledge, London.

Williams, G. & Bowers, J.K. (1987). Land drainage and birds in England and Wales. In: *RSPB Conservation Review,* pp.25–30. RSPB, Sandy.

Wilson, E.O. (1992). *The Diversity of Life.* Penguin Books, Harmonds-worth (first published by Harvard University Press 1992).

Wright, J.F. & Armitage, P.D. (2001). Freshwater invertebrates. *In:* Hawksworth, D.L. (Ed.) *The Changing Wildlife of Great Britain and Ireland.* 188–209, Taylor & Francis, London.

WWF-UK (1997). *A Muzzled Watchdog? Is English Nature protecting wildlife?* WWF-UK, Godalming.

Yalden, D. (1999). *The History of British Mammals.* Poyser Natural History, London.

Index